Maximum Entropy and Bayesian Methods
in Applied Statistics

Maximum Entropy and Bayesian Methods in Applied Statistics

Proceedings of the Fourth Maximum Entropy Workshop
University of Calgary, 1984

Edited by

JAMES H. JUSTICE

Chair in Exploration Geophysics, University of Calgary

CAMBRIDGE UNIVERSITY PRESS

Cambridge

London New York New Rochelle

Melbourne Sydney

CAMBRIDGE UNIVERSITY PRESS
Cambridge, New York, Melbourne, Madrid, Cape Town, Singapore, São Paulo, Delhi

Cambridge University Press
The Edinburgh Building, Cambridge CB2 8RU, UK

Published in the United States of America by Cambridge University Press, New York

www.cambridge.org
Information on this title: www.cambridge.org/9780521323802

First published 1986
This digitally printed version 2008

A catalogue record for this publication is available from the British Library

ISBN 978-0-521-32380-2 hardback
ISBN 978-0-521-09603-4 paperback

CONTENTS

PREFACE

The Fourth Workshop on Maximum Entropy and Bayesian Methods in Applied Statistics was held in Calgary, Alberta, at the University of Calgary, August 5-8, 1984. The workshop continued a three-year tradition of workshops begun at the University of Wyoming, in Laramie, attended by a small number of researchers who welcomed the opportunity to meet and to exchange ideas and opinions on these topics. From small beginnings, the workshop has continued to grow in spite of any real official organization or basis for funding and there always seems to be great interest in "doing it again next year."

This volume represents the proceedings of the fourth workshop and includes one additional invited paper which was not presented at the workshop but which we are pleased to include in this volume (Ellis, Gohberg, Lay). The fourth workshop also made a point of scheduling several exceptional tutorial lectures by some of our noted colleagues, Ed Jaynes, John Burg, John Shore, and John Skilling. These tutorial lectures were not all written up for publication and we especially regret that the outstanding lectures by John Burg and John Shore must go unrecorded.

The depth and scope of the papers included in this volume attest, I believe, to the growing awareness of the importance of maximum entropy and Bayesian methods in the pure and applied sciences and perhaps serve to indicate that much remains to be done and many avenues are yet to be explored. At the same time, it should be clear that significant inroads are being made and the results give credence to both the importance and the practicality of these methods.

In organizing this volume, I have tried to place papers of a general or tutorial nature first, followed by papers related to theoretical considerations, and finally papers dealing with applications in the applied sciences, more or less grouped by application, and a reasonable order for logical development of content.

I would like to express my sincere thanks to a number of people whose devotion and efforts made the workshop a success and the preparation of this volume possible. First among these was surely Mrs. Esther Cockburn who handled all arrangements, mailing, and organization for the conference, and much of the work related to the preparation of this volume. Her efforts assured our success. Pat Foster handled all of our accounting and insured our solvency. Special thanks are also due to Bonnie Sloan who expertly handled all of the typing for this volume and who met all deadlines with work of the highest quality.

Finally, but certainly not least, go our deepest expressions of gratitude to those donors in Calgary and Ottawa who answered our call in a time of need to generously provide all the funding required to insure the success of our efforts. The following section is set aside to call attention to these organizations and to express our thanks to them.
It is my hope that this volume will be of use not only to researchers in many scientific disciplines but also to those whose aspirations are to join their ranks and to carry on this tradition.

James H. Justice
Calgary, Alberta
April 24, 1986

DONORS

Our special thanks to the following organizations who made this work possible through financial support.

The Natural Sciences and Engineering Research Council of Canada (NSERC), Ottawa, Ontario

The Calgary, Alberta, Offices (in alphabetical order) of:

Anadarko Petroleum of Canada Ltd.
Asamera Inc.
BP Canada
Bralorne Resources Ltd.
Canada Northwest Energy Limited
Canadian Hunter Exploration Ltd.
Canadian Superior Oil
Dome Petroleum Limited
Energenics Exploration Ltd.
Gale Resources Ltd.
Golden Eagle Oil and Gas Limited
Grant Geophysical
Gulf Canada Resources Inc.
Harvard International Resources
Norcana Resource Services Ltd.
Pembina Resources Ltd.
Petty Ray Geophysical
Placer Cego Petroleum
Spitzee Resources Ltd.
Teledyne Exploration
Trans-Canada Resources Ltd.
Tricentrol Oils Limited
Union Oil Company of Canada Limited

BAYESIAN METHODS: GENERAL BACKGROUND
An Introductory Tutorial

E.T. Jaynes
St. John's College and Cavendish Laboratory
Cambridge CB2 1TP
England

We note the main points of history, as a framework on which to hang many background remarks concerning the nature and motivation of Bayesian/Maximum Entropy methods. Experience has shown that these are needed in order to understand recent work and problems. A more complete account of the history, with many more details and references, is given in Jaynes (1978).

The following discussion is essentially nontechnical; the aim is only to convey a little introductory "feel" for our outlook, purpose, and terminology, and to alert newcomers to common pitfalls of misunderstanding.

INDEX

HERODOTUS

The necessity of reasoning as best we can in situations where our information is incomplete is faced by all of us, every waking hour of our lives. We must decide what to do next, even though we cannot be certain what the consequences will be. Should I wear a raincoat today, eat that egg, cross that street, talk to that stranger, tote that bale, buy that book?

Long before studying mathematics we have all learned, necessarily, how to deal with such problems intuitively, by a kind of plausible reasoning where we lack the information needed to do the formal deductive reasoning of the logic textbooks. In the real world, some kind of extension of formal logic is needed.

And, at least at the intuitive level, we have become rather good at this extended logic, and rather systematic. Before deciding what to do, our intuition organizes the preliminary reasoning into stages: (a) try to foresee all the possibilities that might arise; (b) judge how likely each is, based on everything you can see and all your past experience; (c) in the light of this, judge what the probable consequences of various actions would be; (d) now make your decision.

From the earliest times this process of plausible reasoning preceding decisions has been recognized. Herodotus, in about 500 BC, discusses the policy decisions of the Persian kings. He notes that a decision was wise, even though it led to disastrous consequences, if the evidence at hand indicated it as the best one to make; and that a decision was foolish, even though it led to the happiest possible consequences, if it was unreasonable to expect those consequences.

So this kind of reasoning has been around for a long time, and has been well understood for a long time. Furthermore, it is so well organized in our minds in qualitative form that it seems obvious that: (a) the above stages of reasoning can be reproduced in a quantitative form by a mathematical model; (b) such an extended logic would be very useful in such areas as science, engineering, and economics, where we are also obliged constantly to reason as best we can in spite of incomplete information, but the number of possibilities and amount of data are far too great for intuition to keep track of.

BERNOULLI

A serious, and to this day still useful, attempt at a mathematical representation was made by James Bernoulli (1713), who called his work "Ars Conjectandi", or "The Art of Conjecture", a name that might well be revived today because it expresses so honestly and accurately what we are doing. But even though it is only conjecture, there are still wise and foolish ways, consistent and inconsistent ways, of doing it. Our extended logic should be, in a sense that was made precise only much later, an optimal or "educated" system of conjecture.

First one must invent some mathematical way of expressing a state of incomplete knowledge, or information. Bernoulli did this by enumerating a set of basic "equally possible" cases which we may denote by $(x_1, x_2 \ldots x_N)$, and which we may call, loosely, either events or propositions. This defines our "field of discourse" or "hypothesis space" $H\emptyset$. If we are concerned with two tosses of a die, $N = 6^2 = 36$.

Then one introduces some proposition of interest A, defined as being true on some specified subset H(A) of M points of $H\emptyset$, false on the others. M, the "number of ways" in which A could be true, is called the multiplicity of A, and the probability of A is defined as the proportion p(A) = M/N.

The rules of reasoning consist of finding the probabilities p(A), p(B), etc., of different propositions by counting the number of ways they can be true. For example, the probability that both A and B are true is the proportion of HØ on which both are true. More interesting, if we learn that A is true, our hypothesis space contracts to H(A) and the probability of B is changed to the proportion of H(A) on which B is true. If we then learn that B is false, our hypothesis space may contract further, changing the probability of some other proposition C, and so on.

Such elementary rules have an obvious correspondence with common sense, and they are powerful enough to be applied usefully, not only in the game of "Twenty Questions", but in some quite substantial problems of reasoning, requiring nontrivial combinatorial calculations. But as Bernoulli recognized, they do not seem applicable to all problems; for while we may feel that we know the appropriate HØ for dice tossing, in other problems we often fail to see how to define any set HØ of elementary "equally possible" cases. As Bernoulli put it, "What mortal will ever determine the number of diseases?" How then could we ever calculate the probability of a disease?

Let us deliver a short Sermon on this. Faced with this problem, there are two different attitudes one can take. The conventional one, for many years, has been to give up instantly and abandon the entire theory. Over and over again modern writers on statistics have noted that no general rule for determining the correct HØ is given, ergo the theory is inapplicable and into the waste-basket it goes.

But that seems to us a self-defeating attitude that loses nearly all the value of probability theory by missing the point of the game. After all, our goal is not omniscience, but only to reason as best we can with whatever incomplete information we have. To demand more than this is to demand the impossible; neither Bernoulli's procedure nor any other that might be put in its place can get something for nothing.

The reason for setting up HØ is not to describe the Ultimate Realities of the Universe; that is unknown and remains so. By definition, the function of HØ is to represent what we know; it cannot be unknown. So a second attitude recommends itself; define your HØ as best you can -- all the diseases you know -- and get on with the calculations.

Usually this suggestion evokes howls of protest from those with conventional training; such calculations have no valid basis at all, and can lead to grotesquely wrong predictions. To trust the results could lead to calamity.

But such protests also miss the point of the game; they are like the reasoning of a chess player who thinks ahead only one move and refuses to play at all unless the next move has a guaranteed win. If we think ahead two moves, we can see the true function and value of probability theory in inference.

When we first define HØ, because of our incomplete information we cannot be sure that it really expresses all the possibilities in the real world. Nor can we be sure that there is no unknown symmetry- breaking influence at work making it harder to realize some possibilities than others. If we knew of such an influence, then we would not consider the x_i equally likely.

To put it in somewhat anthropomorphic terms, we cannot be sure that our hypothesis space HØ is the same as Nature's hypothesis space HN. The conventional attitude holds that our calculation is invalid unless we know the "true" HN; but that is something that we shall never know. So, ignoring all protests, we choose to go ahead with that shaky calculation from HØ, which is the best we can actually do. What are the possible results?

Suppose our predictions turn out to be right; i.e. out of a certain set of propositions A_1, A_2, ... A_m the one A_k that we thought highly likely to be true (because it is true on a much larger subset of HØ than any other) is indeed confirmed by observation. That does not prove that our HØ represented correctly all those, and only those, possibilities that exist in Nature, or that no symmetry-breaking influences exist. But it does show that our HØ is not sufficiently different from Nature's HN to affect this prediction. Result: the theory has served a useful predictive purpose, and we have more confidence in our HØ. If this success continues with many different sets of propositions, we shall end up with very great confidence in HØ. Whether it is "true" or not, it has predictive value.

But suppose our prediction does indeed turn out to be grotesquely wrong; Nature persists in generating an entirely different A_j than the one we favoured. Then we know that Nature's HN is importantly different from our HØ, and the nature of the error gives us a clue as to how they differ. As a result, we are in a position to define a better hypothesis space H1, repeat the calculations to see what predictions it makes, compare them with observation, define a still better H2, ... and so on.

Far from being a calamity, this is the essence of the scientific method. HØ is only our unavoidable starting point.

As soon as we look at the nature of inference at this many-moves-ahead level of perception, our attitude towards probability theory and the proper way to use it in science becomes almost diametrically opposite to that expounded in most current textbooks. We need have no fear of making shaky calculations on inadequate knowledge; for if our predictions are indeed wrong, then we shall have an opportunity to improve that knowledge, an opportunity that would have been lost had we been too timid to make the calculations.

Instead of fearing wrong predictions, we look eagerly for them; it is only when predictions based on our present knowledge fail that probability theory leads us to fundamental new knowledge.

Bernoulli implemented the point just made, and in a more sophisticated way than we supposed in that little Sermon. Perceiving as noted that except in certain gambling devices like dice we almost never know Nature's set HN of possibilities, he conceived a way of probing HN, in the case that one can make repeated independent observations of some event A; for example, administering a medicine to many sick patients and noting how many are cured.

We all feel intuitively that, under these conditions, events of higher probability M/N should occur more often. Stated more carefully, events that have higher probability on HØ should be predicted to occur more often; events with higher probability on HN should be observed to occur more often. But we would like to see this intuition supported by a theorem.

Bernoulli proved the first mathematical connection between probability and frequency, today known as the weak law of large numbers. If we make n independent observations and find A true m times, the observed frequency $f(A) = m/n$ is to be compared with the probability $p(A) = M/N$. He showed that in the limit of large n, it becomes practically certain that f(A) is close to p(A). Laplace showed later that as n tends to infinity the probability remains more than 1/2 that f(A) is in the shrinking interval $p(A) \pm q$, where $q^2 = p(1-p)/n$.

There are some important technical qualifications to this, centering on what we mean by "independent"; but for present purposes we note only that often an observed frequency f(A) is in some sense a reasonable estimate of the ratio M/N in Nature's hypothesis space HN. Thus we have, in many cases, a simple way to test and improve our HØ in a semiquantitative way. This was an important start; but Bernoulli died before carrying the argument further.

BAYES

Thomas Bayes was a British clergyman and amateur mathematician (a very good one - it appears that he was the first to understand the nature of asymptotic expansions), who died in 1761. Among his papers was found a curious unpublished manuscript. We do not know what he intended to do with it, or how much editing it then received at the hands of others; but it was published in 1763 and gave rise to the name "Bayesian Statistics". For a photographic reproduction of the work as published, with some penetrating historical comments, see Molina (1963). It gives, by lengthy arguments that are almost incomprehensible today, a completely different kind of solution to Bernoulli's unfinished problem.

Where Bernoulli had calculated the probability, given N, n, and M, that we would observe A true m times (what is called today the "sampling distribution"), Bayes turned it around and gave in effect a formula for the probability, given N, n, and m, that M has various values. The method was long called "inverse probability". But Bayes' work had little if any direct influence on the later development of probability theory.

LAPLACE

In almost his first published work (1774), Laplace rediscovered Bayes' principle in greater clarity and generality, and then for the next 40 years proceeded to apply it to problems of astronomy, geodesy, meteorology, population statistics, and even jurisprudence. The basic theorem appears today as almost trivially simple; yet it is by far the most important principle underlying scientific inference.

Denoting various propositions by A, B, C, etc., let AB stand for the proposition "both A and B are true", $\bar{A}$ = "A is false", and let the symbol p(A:B) stand for "the probability that A is true, given that B is true". Then the basic product and sum rules of probability theory, dating back in essence to before Bernoulli, are

$$p(AB|C) = p(A|BC)p(B|C) \tag{1}$$

$$p(A|B) + p(\bar{A}|B) = 1 \tag{2}$$

But AB and BA are the same proposition, so consistency requires that we may interchange A and B in the right-hand side of (1). If $p(B|C) > 0$, we thus have what is always called "Bayes' Theorem" today, although Bayes never wrote it:

$$p(A|BC) = p(A|C)\ p(B|AC)/p(B|C) \tag{3}$$

But this is nothing more than the statement that the product rule is consistent; why is such a seeming triviality important?

In (3) we have a mathematical representation of the process of learning; exactly what we need for our extended logic. $p(A|C)$ is our "prior probability" of A, when we know only C. $p(A|BC)$ is its "posterior probability", updated as a result of acquiring new information B. Typically, A represents some hypothesis, or theory, whose truth we wish to ascertain, B represents new data from some observation, and the "prior information" C represents the totality of what we knew about A before getting the data B.

For example -- a famous example that Laplace actually did solve -- proposition A might be the statement that the unknown mass M_S of Saturn lies in a specified interval, B the data from observatories about the mutual perturbations of Jupiter and Saturn, C the common sense observation that M_S cannot be so small that Saturn would lose its rings; or so large that Saturn would disrupt the solar system. Laplace reported that, from the data available up to the end of the 18th

Century, Bayes' theorem estimates M_S to be (1/3512) of the solar mass, and gives a probability of .99991, or odds of 11,000:1, that M_S lies within 1 per cent of that value. Another 150 years' accumulation of data has raised the estimate 0.63 per cent.

The more we study it, the more we appreciate how nicely Bayes' theorem corresponds to -- and improves on -- our common sense. In the first place, it is clear that the prior probability $p(A|C)$ is necessarily present in all inference; to ask "What do you know about A after seeing the data B?" cannot have any definite answer -- because it is not a well-posed question -- if we fail to take into account, "What did you know about A before seeing B?".

Even this platitude has not always been perceived by those who do not use Bayes' theorem and go under the banner: "Let the data speak for themselves!" They cannot, and never have. If we want to decide between various possible theories but refuse to supplement the data with prior information about them, probability theory will lead us inexorably to favour the "Sure Thing" theory ST, according to which every minute detail of the data was inevitable; nothing else could possibly have happened. For the data always have a much higher probability on ST than on any other theory; ST is always the maximum likelihood solution over the class of all theories. Only our extremely low prior probability for ST can justify rejecting it.

Secondly, we can apply Bayes' theorem repeatedly as new pieces of information B_1, B_2, ... are received from the observatories, the posterior probability from each application becoming the prior probability for the next. It is easy to verify that (3) has the chain consistency that common sense would demand; at any stage the probability that Bayes' theorem assigns to A depends only on the total evidence $B_{tot} = B_1 \ldots B_k$ then at hand, not on the order in which the different updatings happened. We could reach the same conclusion by a single application of Bayes' theorem using B_{tot}.

But Bayes' theorem tells us far more than intuition can. Intuition is rather good at judging what pieces of information are relevant to a question, but very unreliable in judging the relative cogency of different pieces of information. Bayes' theorem tells us quantitatively just how cogent every piece of information is.

Bayes' theorem is such a powerful tool in this extended logic that, after 35 years of using it almost daily, I still feel a sense of excitement whenever I start on a new, nontrivial problem; because I know that before the calculation has reached a dozen lines it will give me some important new insight into the problem, that nobody's intuition has seen before. But then that surprising result always seems intuitively obvious after a little meditation; if our raw intuition was powerful enough we would not need extended logic to help us.

Two examples of the fun I have had doing this, with full technical details, are in my papers "Bayesian Spectrum and Chirp Analysis" given at the August 1983 Laramie Workshop, and "Highly Informative Priors" in the Proceedings Volume for the September 1983 International Meeting on Bayesian Statistics, Valencia, Spain (Jaynes, 1985). In both cases, completely unexpected new insight from Bayes' theorem led to quite different new methods of data analysis and more accurate results, in two problems (spectrum analysis and seasonal adjustment) that had been treated for decades by non-Bayesian methods. The Bayesian analysis took into account some previously neglected prior information.

Laplace, equally aware of the power of Bayes' theorem, used it to help him decide which astronomical problems to work on. That is, in which problems is the discrepancy between prediction and observation large enough to give a high probability that there is something new to be found? Because he did not waste time on unpromising research, he was able in one lifetime to make more of the important discoveries in celestial mechanics than anyone else.

Laplace also published (1812) a remarkable two-volume treatise on probability theory in which the analytical techniques for Bayesian calculations were developed to a level that is seldom surpassed today. The first volume contains, in his methods for solving finite difference equations, almost all of the mathematics that we find today in the theory of digital filters. An English translation of this work, by Professor and Mrs. A.F.M. Smith of Nottingham University, is in preparation.

Yet all of Laplace's impressive accomplishments were not enough to establish Bayesian analysis in the permanent place that it deserved in science. For more than a Century after Laplace, we were deprived of this needed tool by what must be the most disastrous error of judgement ever made in science.

In the end, all of Laplace's beautiful analytical work and important results went for naught because he did not explain some difficult conceptual points clearly enough. Those who came after him got hung up on inability to comprehend his rationale and rejected everything he did, even as his masses of successful results were staring them in the face.

JEFFREYS

Early in this Century, Sir Harold Jeffreys rediscovered Laplace's rationale and, in the 1930's, explained it much more clearly than Laplace did. But still it was not comprehended; and for thirty more years Jeffreys' work was under attack from the very persons who had the most to gain by understanding it (some of whom were living and eating with him daily here in St. John's College, and had the best possible opportunity to learn from him). But since about 1960 comprehension of what Laplace and Jeffreys were trying to say has been growing, at first slowly and today quite rapidly.

This strange history is only one of the reasons why, today, we Bayesians need to take the greatest pains to explain our rationale, as I am trying to do here. It is not that it is technically complicated; it is the way we have all been thinking intuitively from childhood. It is just so different from what we were all taught in formal courses on "orthodox" probability theory, which paralyze the mind into an inability to see the distinction between probability and frequency. Students who come to us free of that impediment have no difficulty in understanding our rationale, and are incredulous that anyone could fail to comprehend it.

My Sermons are an attempt to spread the message to those who labour under this handicap, in a way that takes advantage of my own experience at a difficult communication problem. Summarizing Bernoulli's work was the excuse for delivering the first Sermon establishing, so to speak, our Constitutional Right to use HØ even if it may not be the same as HN.

Now Laplace and Jeffreys inspire our second Sermon, on how to choose HØ given our prior knowledge; a matter on which they made the essential start. To guide us in this choice there is a rather fundamental "Desideratum of Consistency": in two problems where we have the same state of knowledge, we should assign the same probabilities.

As an application of this desideratum, if the hypothesis space HØ has been chosen so that we have no information about the x_i beyond their enumeration, then as an elementary matter of symmetry the only consistent thing we can do is to assign equal probability to all of them; if we did anything else, then by a mere permutation of the labels we could exhibit a second problem in which our state of knowledge is the same, but in which we are assigning different probabilities.

This rationale is the first example of the general group invariance principle for assigning prior probabilities to represent "ignorance". Although Laplace used it repeatedly and demonstrated its successful consequences, he failed to explain that it is not arbitrary, but required by logical consistency to represent a state of knowledge. Today, 170 years later, this is still a logical pitfall that causes conceptual hangups and inhibits applications of probability theory.

Let us emphasize that we are using the word "probability" in its original -- therefore by the usual scholarly standards correct -- meaning, as referring to incomplete human information. It has, fundamentally, nothing to do with such notions as "random variables" or "frequencies in random experiments"; even the notion of "repetition" is not necessarily in our hypothesis space.

In cases where frequencies happen to be relevant to our problem, whatever connections they may have with probabilities appear automatically, as mathematically derived consequences of our extended logic (Bernoulli's limit theorem being the first example). But, as shown in a discussion of fluctuations in time series (Jaynes, 1978), those connections are often of a very different nature than is supposed in conventional pedagogy; the predicted mean-square fluctuation is not the same as the variance of the first-order probability distribution.

So to assign equal probabilities to two events is not in any way an assertion that they must occur equally often in any "random experiment"; as Jeffreys emphasized, it is only a formal way of saying "I don't know". Events are not necessarily capable of repetition; the event that the mass of Saturn is less than (1/3512) had, in the light of Laplace's information, the same probability as the event that it is greater than (1/3512), but there is no "random experiment" in which we expect those events to occur equally often.

Of course, if our hypothesis space is large enough to accommodate the repetitions, we can calculate the *probability* that two events occur equally often.

To belabour the point, because experience shows that it is necessary: In our scholastically correct terminology, a *probability* p is an abstract concept, a quantity that we *assign* theoretically, for the purpose of representing a state of knowledge, or that we *calculate* from previously assigned probabilities using the rules (1) - (3) of probability theory. A *frequency* f is, in situations where it makes sense to speak of repetitions, a factual property of the real world, that we *measure* or *estimate*. So instead of committing the error of saying that the probability *is* the frequency, we ought to calculate the probability p(f)df that the frequency lies in various intervals df -- just as Bernoulli did.

In some cases our information, although incomplete, still leads to a very sharply peaked probability distribution p(f); and then we can indeed make very confident predictions of frequencies. in these cases, if we are not making use of any information other than frequencies, our conclusions will agree with those of "random variable" probability theory as usually taught today. Our results do not conflict with frequentist results whenever the latter are justified. From a pragmatic standpoint (i.e., ignoring philosophical stances and looking only at the actual results), "random variable" probability theory is contained in the Laplace-Jeffreys theory as a special case.

But the approach being expounded here applies also to many important real problems -- such as the "pure generalized inverse" problems of concern to us at this Workshop -- in which there is not only no "random experiment" involved, but we have highly cogent information that must be taken into account in our probabilities, but does not consist of frequencies.

A theory of probability that fails to distinguish between the notions of probability and frequency is helpless to deal with such problems. This is the reason for the present rapid growth of Bayesian/ Maximum Entropy methods -- which can deal with them, and with demonstrated success. And of course, we can deal equally well with the compound case where we have both random error and cogent non-frequency information.

COX

One reason for these past problems is that neither Laplace nor Jeffreys gave absolutely compelling arguments -- that would convince a person who did not want to believe it -- proving that the rules (1) - (3) of probability theory are uniquely favoured, by any clearly stated criterion of optimality, as the "right" rules for conducting inference. To many they appeared arbitrary, no better than a hundred other rules one could invent.

But those rules were -- obviously and trivially -- valid rules for combining frequencies, so in the 19th Century the view arose that a probability is not respectable unless it is also a frequency.

In the 1930's the appearance of Jeffreys' work launched acrimonious debates on this issue. The frequentists took not the slightest note of the masses of evidence given by Laplace and Jeffreys, demonstrating the pragmatic success of those rules when applied without the sanction of any frequency connection; they had their greatest success in just the circumstances where the frequentists held them to be invalid.

Into this situation there came, in 1946, a modest little paper by R.T. Cox, which finally looked at the problem in just the way everybody else should have. He issued no imperatives declaring that rules (1) - (3) were or were not valid for conducting inference. Instead he observed that, whether or not Laplace gave us the right "calculus of inductive reasoning", we can at least raise the question whether such a calculus could be created today.

Supposing that degrees of plausibility are to be represented by real numbers, he found the conditions that such a calculus be consistent (in the sense that if two different methods of calculation are permitted by the rules, then they should yield the same result). These consistency conditions took the form of two functional equations, whose general solution could be found. That solution uniquely determined the rules (1) and (2), to within a change of variables that can alter their form but not their content.

So, thanks to Cox, it was now a theorem that any set of rules for conducting inference, in which we represent degrees of plausibility by real numbers, is necessarily either equivalent to the Laplace-Jeffreys rules, or inconsistent. The reason for their pragmatic success is then pretty clear. Those who continued to oppose Bayesian methods after 1946 have been obliged to ignore not only the pragmatic success, but also the theorem.

SHANNON

Two years later, Claude Shannon (1948) used Cox's method again. He sought a measure of the "amount of uncertainty" in a probability distribution. Again the conditions of consistency took the form of functional equations, whose general solution he found. The resulting measure proved to be $-\sum p_i \log p_i$, just what physicists had long since related to the entropy of thermodynamics.

Gibbs (1875) had given a variational principle in which maximization of the phenomenological "Clausius entropy" led to all the useful predictions of equilibrium thermodynamics. But the phenomenological Clausius entropy still had to be determined by calorimetric measurements. Boltzmann (1877), Gibbs (1902) and von Neumann (1928) gave three variational principles in which the maximization of the "Shannon entropy" led, in both classical and quantum theory, to a theoretical prediction of the Clausius entropy; and thus again (if one was a good enough calculator) to all the useful results of equilibrium thermodynamics -- without any need for calorimetric measurements.

But again we had been in a puzzling situation just like that of Bayesian inference. Also here we had (from Gibbs) a simple, mathematically elegant, formalism that led (in the quantum theory version) to enormous pragmatic success; but had no clearly demonstrated theoretical justification. Also here, acrimonious debates over its rationale -- among others, does it or does it not require "ergodic" properties of the equations of motion? -- had been underway for decades. But again, Shannon's consistency theorem finally showed in what sense entropy maximization generated optimal inferences. Whether or not the system is ergodic, the formalism still yields the best predictions that could have been made on the information we had.

This was really a repetition of history from Bernoulli. Shannon's theorem only established again, in a new area, our Constitutional Right to use HØ based on whatever information we have, whether or not it is the same as HN. Our previous remarks about many-moves-ahead perception apply equally well here; if our HØ differs from HN, how else can we discover that fact but by having the courage to go ahead with the calculations on HØ to see what predictions it makes?

Gibbs had that courage; his HØ of classical mechanics was "terrible" by conventional attitudes, for it predicted only equations of state correctly, and gave wrong specific heats, vapor pressures, and equilibrium constants. Those terribly wrong predictions were the first clues pointing to quantum theory; our many-moves-ahead scenario is not a fancy, but an historical fact. It is clear from Gibbs' work of 1875 that he understood that scenario; but like Laplace he did not explain it clearly, and it took a long time for it to be rediscovered.

There is one major difference in the two cases. The full scope and generality of Bayesian inference had been recognized already by

Jeffreys, and Cox's theorem only legitimized what he had been doing all along. But the new rationale from Shannon's theorem created an enormous expansion of the scope of maximum entropy.

It was suddenly clear that, instead of applying only to prediction of equilibrium thermodynamics, as physicists had supposed before Shannon, the variational principles of Gibbs and von Neumann extended as well to nonequilibrium thermodynamics, and to any number of new applications outside of thermodynamics. As attested by the existence of this Workshop, it can be used in spectrum analysis, image reconstruction, crystallographic structure determination, econometrics; and indeed any problem, whatever the subject-matter area, with the following logical structure:

We can define an hypothesis space HØ by enumerating some perceived possibilities ($x_1 \ldots x_N$); but we do not regard them as equally likely, because we have some additional evidence E. It is not usable as the "data" B in Bayes' theorem (3) because E is not an event and does not have a "sampling distribution" $p(E|C)$. But E leads us to impose some constraint on the probabilities $P_i = P(x_i)$ that we assign to the elements of HØ, which forces them to be nonuniform, but does not fully determine them (the number of constraints is less than N).

We interpret Shannon's theorem as indicating that, out of all distributions P_i that agree with the constraints, the one that maximizes the Shannon entropy represents the "most honest" description of our state of knowledge, in the following sense: it expresses the enumeration of the possibilities, the evidence E; and assumes nothing beyond that.

If we subsequently acquire more information B that can be interpreted as an event then we can update this distribution by Bayes' theorem. In other words, MAXENT has given us the means to escape from the "equally possible" domain of Bernoulli and Laplace, and construct nonuniform prior distributions.

Thus came the unification of these seemingly different fields. Boltzmann and Gibbs had been, unwittingly, solving the prior probability problem of Bernoulli and Laplace, in a very wide class of problems. The area of useful applications this opens up may require 100 years to explore and exploit.

But this has only scratched the surface of what can be done in inference, now that we have escaped from the errors of the past. We can see, but only vaguely, still more unified, more powerful, and more general theories of inference which will regard our present one as an approximate special case.

COMMUNICATION DIFFICULTIES

Our background remarks would be incomplete without taking note of a serious disease that has afflicted probability theory for 200 years. There is a long history of confusion and controversy, leading in some cases to a paralytic inability to communicate. This has been caused, not only by confusion over the notions of probability and frequency, but even more by repeated failure to distinguish between different problems that happen to lead to similar mathematics. We are concerned here with only one of these failures.

Starting with the debates of the 1930's between Jeffreys and Fisher in the British Statistical Journals, there has been a puzzling communication block that has prevented orthodoxians from comprehending Bayesian methods, and Bayesians from comprehending orthodox criticisms of our methods. On the topic of how probability theory should be used in inference, L.J. Savage (1954) remarked that "there has seldom been such complete disagreement and breakdown of communication since the Tower of Babel".

For example, the writer recalls reading, as a student in the late 1940's, the orthodox textbook by H. Cramér (1946). His criticisms of Jeffreys' approach seemed to me gibberish, quite unrelated to what Jeffreys had done. There was no way to reply to the criticisms, because they were just not addressed to the topic. The nature of this communication block has been realized only quite recently.

For decades Bayesians have been accused of "supposing that an unknown parameter is a random variable"; and we have denied hundreds of times, with increasing vehemence, that we are making any such assumption. We have been unable to comprehend why our denials have no effect, and that charge continues to be made.

Sometimes, in our perplexity, it has seemed to us that there are two basically different kinds of mentality in statistics; those who see the point of Bayesian inference at once, and need no explanation; and those who never see it, however much explanation is given.

But a Seminar talk by Professor George Barnard, given in Cambridge in February 1984, provided a clue to what has been causing this Tower of Babel situation. Instead of merely repeating the old accusation (that we could only deny still another time), he expressed the orthodox puzzlement over Bayesian methods in a different way, more clearly and specifically than we had ever heard it put before.

Barnard complained that Bayesian methods of parameter estimation, which present our conclusions in the form of a posterior distribution, are illogical; for "How could the distribution of a parameter possibly become known from data which were taken with only one value of the parameter actually present?"

This extremely revealing -- indeed, appalling -- comment finally gave some insight into what has been causing our communication problems. Bayesians have always known that orthodox terminology is not well adapted to expressing Bayesian ideas; but at least this writer had not realized how bad the situation was.

Orthodoxians try to understand Bayesian methods have been caught in a semantic trap by their habitual use of the phrase "distribution of the parameter" when one should have said "distribution of the probability". Bayesians had supposed this to be merely a figure of speech; i.e. that those who used it did so only out of force of habit, and really knew better. But now it seems that our critics have been taking that phraseology quite literally all the time.

Therefore, let us belabour still another time what we had previously thought too obvious to mention. In Bayesian parameter estimation, both the prior and posterior distribution represent, not any measurable property of the parameter, but only our own state of knowledge about it. The width of the distribution is not intended to indicate the range of variability of the true values of the parameter, as Barnard's terminology led him to suppose. It indicates the range of values that are consistent with our prior information and data, and which honesty therefore compels us to admit as possible values. What is "distributed" is not the parameter, but the probability.

Now it appears that, for all these years, those who have seemed immune to all Bayesian explanation have just misunderstood our purpose. All this time, we had thought it clear from our subject-matter context that we are trying to estimate the value that the parameter had when the data were taken. Put more generally, we are trying to draw inferences about what actually did happen in the experiment; not about what might have happened but did not.

But it seems that, all this time, our critics have been trying to interpret our calculations in a different way, imposed on them by their habits of terminology, as an attempt to solve an entirely different problem.

With this realization, our past communication difficulties become understandable: the problem our critics impute to us has -- as they correctly see -- no solution from the information at hand. The fact that we nevertheless get a solution then seems miraculous to them, and we are accused of trying to get something for nothing.

Re-reading Cramér and other old debates in the literature with this in mind, we can now see that this misunderstanding of our purpose was always there, but covertly. Our procedure is attacked on the overt grounds that our prior probabilities are not frequencies -- which seemed to us a mere philosophical stance, and one that we rejected. But to our critics this was far more than a philosophical difference; they took it to mean that, lacking the information to solve the problem, we are contriving a false solution.

What was never openly brought out in these frustrating debates was that our critics had in mind a completely different problem than the one we were solving. So to an orthodoxian our denials seemed dishonest; while to a Bayesian, the orthodox criticisms seemed so utterly irrelevant that we could see no way to answer them. Our minds were just operating in different worlds.

Perhaps, realizing this, we can now see a ray of light that might, with reasonable good will on both sides, lead to a resolution of our differences. In the future, it will be essential for clear communication that all parties see clearly the distinction between these two problems:

(i) In the Bayesian scenario we are estimating, from our prior information and data, the unknown constant value that the parameter had when the data were taken.

(ii) In Barnard's we are deducing, from prior knowledge of the frequency distribution of the parameter over some large class C of repetitions of the whole experiment, the frequency distribution that it has in the subclass C(D) of cases that yield the same data D.

The problems are so different that one might expect them to be solved by different procedures. Indeed, if they had led to completely different algorithms, the problems would never have been confused and we might have been spared all these years of controversy. But it turns out that both problems lead, via totally different lines of reasoning, to the same actual algorithm; application of Bayes' theorem.

However, we do not know of any case in which Bayes' theorem has actually been used for problem (ii); nor do we expect to hear of such a case, for three reasons: (a) in real problems, the parameter of interest is almost always an unknown constant, not a "random variable"; (b) even if it is a random variable, what is of interest is almost always the value it had during the real experiment that was actually done; not its frequency distribution in an imaginary subclass of experiments that were not done; (c) even if that imaginary frequency distribution were the thing of interest, we never know the prior frequency distribution that would be needed.

That is, even if the orthodoxian wanted to solve problem (ii), he would not have, any more than we do, the information needed to use Bayes' theorem for it, and would be obliged to seek some other method.

This unfortunate mathematical accident reinforced the terminological confusion; when orthodoxians saw us using the Bayesian algorithm, they naturally supposed that we were trying to solve problem (ii). Still more reinforcement came from the fact that, since orthodoxy sees no meaning in a probability which is not also a frequency, it is unable to see Bayes' theorem as the proper procedure for solving problem (i). Indeed for that problem it is obliged to seek, not just other methods than Bayes' theorem; but other tools than probability theory.

For the Bayesian, who does see meaning in a probability that is not a frequency, all the theoretical principles needed to solve problem (i) are contained in the product and sum rules (1), (2) of probability theory. He views them, not merely as rules for calculating frequencies (which they are, but trivially); but also rules for conducting inference -- a nontrivial property requiring mathematical demonstration.

But orthodoxians do not read the Bayesian literature, in which it is a long since demonstrated fact, not only in the work of Cox but in quite different approaches by de Finetti, Wald, Savage, Lindley, and others, that Bayesian methods yield the optimal solution to problem (i), by some very basic, and it seems to us inescapable, criteria of optimality. Pragmatically, our numerical results confirm this; whenever the Bayesian and orthodoxian arrive at different results for problem (i), closer examination has always shown the Bayesian result to be superior (Jaynes, 1976).

In view of this, we are not surprised to find that, while orthodox criticisms of Bayesian and/or Maximum Entropy methods deplore our philosophy and procedure, they almost always stop short of examining our actual numerical results in real problems and comparing them with orthodox results (when the latter exist).

The few who take the trouble to do this quickly become Bayesians themselves. Today, it is our pragmatic results, far more than the philosophy or even the optimality theorems, that is making Bayesianity a rapidly growing concern, taking over one after another of the various areas of scientific inference.

So, rising above past criticisms which now appear to have been only misunderstandings of our purpose, Bayesians are in a position to help orthodox statistics in its current serious difficulties. In the real problems of scientific inference, it is almost invariably problem (i) that is in need of solution. But aside from a few special (and to the Bayesian, trivial) cases, orthodoxy has no satisfactory way to deal with problem (i).

In our view, then, the orthodoxian has a great deal to gain in useful results (and as far as we can see, nothing to lose but his ideological chains) by joining the Bayesian camp and finally taking advantage of this powerful tool. Failing to do this, he faces obsolescence as new applications of the kind we are discussing at this Workshop pass far beyond his domain.

There have been even more astonishing misunderstandings and misrepresentations of the nature and purpose of the Maximum Entropy principle. These troubles just show that statistics is different from other fields. Our problems are not only between parties trying to communicate with each other; not all readers are even trying to understand your message. It's a jungle out there, full of predators tensed like coiled springs, ready and eager to pounce upon every opportunity to misrepresent your meaning, your purpose, and even your results. I am by no means the first to observe this; both H. Jeffreys and R.A. Fisher complained about it in the 1930's.

These are the main points of the historical background of this field; but the important background is not all historical. The general nature of Bayesian/Maximum Entropy methods still needs some background clarification, because their logical structure is different not only from that of orthodox statistics, but also from that of conventional mathematical theories in general.

IS OUR LOGIC OPEN OR CLOSED?

Let us dwell a moment on a circumstance that comes up constantly in this field. It cannot be an accident that almost all of the literature on Bayesian methods of inference is by authors concerned with specific real problems of the kind that arise in science, engineering, and economics; while most mathematicians concerned with probability theory take no note of the existence of these methods. Mathematicians and those who have learned probability theory from them seem uncomfortable with Bayesianity, while physicists take to it naturally. This might be less troublesome if we understood the reason for it.

Perhaps the answer is to be found, at least in part, in the following circumstances. A mathematical theory starts at ground level with the basic axioms, a set of propositions which are not to be questioned within the context of the theory. Then one deduces various conclusions from them; axioms 1 and 2 together imply conclusion A, axioms 2 and 3 imply conclusion B, conclusion A and axiom 4 imply conclusion C, - - -, and so on. From the ground level axioms one builds up a long chain of conclusions, and there seems to be no end to how far the chain can be extended. In other words, mathematical theories have a logical structure that is open at the top, closed at the bottom.

Scientists and others concerned with the real world live in a different logical environment. Nature does not reveal her "axioms" to us, and so we are obliged to start in the middle, at the level of our direct sense perceptions. From direct observations and their generalizations we can proceed upward, drawing arbitrarily long chains of conclusions, by reasoning which is perforce not as rigorous as that of Gauss and Cauchy, but with the compensation that our conclusions can be tested by observation. We do not proceed upward very far before the subject is called "engineering".

But from direct observations we also proceed downward, analyzing things into fundamentals, searching for deeper propositions from which all those above could have been deduced. This is called "pure science" and this chain of interlocking inferences can, as far as we know, also be extended indefinitely. So the game of science is, unlike that of mathematics, played on a logical field that is open at both the top and the bottom.

Consider now the two systems or probability theory created by mathematicians and scientists. The Kolmogorov system is a conventional mathematical theory, starting with ground axioms and building upward from them. If we start with probabilities of elementary events, such as "heads at each toss of a coin", we can proceed to deduce probabilities of more and more complicated events such as "not more than 137 or less than 93 occurrences of the sequence HHTTTHH in 981 tosses". In content, it resembles parts of the Bernoulli system, restated in set and measure-theory language.

But nothing in the Kolmogorov system tells us what probability should, in fact, be assigned to any real event. Probabilities of elementary events are simply given to us in the statement of a problem, as things determined elsewhere and not to be questioned. Probabilities of more complicated events follow from them by logical deduction following the postulated rules. Only one kind of probability (additive measure) exists, and the logical structure is closed at the bottom, open at the top. In this system, all is safe and certain and probabilities are absolute (indeed, so absolute that the very notion of conditional probability is awkward, unwanted, and avoided as long as possible).

In contrast, the Bayesian system of Laplace and Jeffreys starts with rules applied to some set of propositions of immediate interest to us. if by some means we assign probabilities to them, then we can build upwards, introducing more propositions whose probabilities can be found as in the Kolmogorov system. The system is, as before, open at the top.

But it is not closed at the bottom, because in our system it is a platitude that all probabilities referring to the real world are, of necessity, conditional on our state of knowledge about the world; they cannot be merely postulated arbitrarily at the beginning of a problem. Converting prior information into prior probability assignments is an open-ended problem; you can always analyze further by going into deeper and deeper hypothesis spaces. So our system of probability is open at both the top and the bottom.

The downward analysis, that a scientist is obliged to carry out, represents for him fully half of probability theory, that is not present at all in the Kolmogorov system. Just for that reason, this neglected half is not as fully developed, and when we venture into it with a new application, we may find ourselves exploring new territory.

I think that most mathematicians are uncomfortable when the ground opens up and that safe, solid closed bottom is lost. But their security is bought only at the price of giving up contact with the real world, a price that scientists cannot pay. We are, of necessity, creatures of the bog.

Mathematicians sometimes dismiss our arguments as nonrigorous; but the reasoning of a physicist, engineer, biochemist, geologist, economist -- or Sherlock Holmes -- cannot be logical deduction because the necessary information is lacking. A mathematician's reasoning is no more rigorous than ours when he comes over and tries to play on our field. Indeed, it has been demonstrated many times than an experienced scientist could reason confidently to correct conclusions, where a mathematician was helpless because his tools did not fit the problem. Our reasoning has always been an intuitive version of Bayes' theorem.

DOWNWARD ANALYSIS IN STATISTICAL MECHANICS

To illustrate this open-ended descent into deeper and deeper hypothesis spaces, a physicist or chemist considering a common object, say a sugar cube, might analyze it into a succession of deeper and deeper hypothesis spaces on which probabilities might be defined. Our direct sense perceptions reveal only a white cube, with frosty rather than shiny sides, and no very definite hypothesis space suggests itself to us. But examination with a low power lens is sufficient to show that it is composed of small individual crystals. So our first hypothesis space H1 might consist of enumerating all possible sizes and orientations of those crystals and assigning probabilities to them. Some of the properties of the sugar cube, such as its porosity, could be discussed successfully at that level.

On further analysis one finds that crystals are in turn composed of molecules regularly arranged. So a deeper hypothesis space H2 is formed by enumerating all possible molecular arrangements. Before an x-ray structure determination is accomplished, our state of knowledge would be represented by a very broad probability distribution on H2, with many arrangements "equally likely" and almost no useful predictions. After a successful structure analysis the "nominal" arrangement is known and it is assigned a much higher probability than any other; then in effect the revised probabilities on H2 enumerate the possible departures from perfection, the lattice defects. At that level, one would be able to say something about other properties of sugar, such as cleavage and heat conductivity.

Then chemical analysis reveals that a sucrose molecule consists of 12 carbon, 22 hydrogen and 11 oxygen atoms; a deeper hypothesis space H3 might then enumerate their positions and velocities (the "phase space" of Maxwell and Gibbs). At this level, many properties of the sugar cube, such as its heat capacity, could be discussed with semiquantitative success; but full quantitative success would not be achieved with any probability distribution on that space.

Learning that atoms are in turn composed of electrons and nuclei suggests a still deeper space H4 which enumerates all their possible positions and velocities. But as Arnold Sommerfeld found, H4 leads us to worse inferences than H3; fabulously wrong specific heats for metals. In this way Nature warns us that we are going in the wrong direction.

Still further analysis shows that the properties of atoms are only approximately, and those of electrons are not even approximately, describable in terms of positions and velocities. Rather, our next deeper hypothesis space H5 is qualitatively different, consisting of the enumeration of their quantum states. This meets with such great success that we are still exploring it. In principle (i.e. ignoring computational difficulties) it appears that all thermodynamic and chemical properties of sugar could be inferred quantitatively at that level.

Our present statistical mechanics stops at the level H5 of enumerating the "global" quantum states of a macroscopic system. At that deepest level yet reached, simple counting of those states (multiplicity factors) is sufficient to predict all equilibrium macrostates; they are the ones with greatest multiplicity W (thus greatest entropy log W) compatible with our macroscopic data. Thus while "equally likely" on H2 had almost no predictive value, and "equally likely" on H3 was only partially successful, "equally likely" on H5 leads to what is probably the greatest predictive success yet achieved by any probabilistic theory.

Presumably, simple counting of quantum states will also suffice to predict all reproducible aspects of irreversible processes; but the computations are so huge that this area is still largely unexplored. We cannot, therefore, rule out the possibility that new surprises, and resulting further analysis, may reveal still deeper hypothesis spaces H6, H7, and so on (hidden variables?). Indeed, the hope that this might happen has motivated much of the writer's work in this field.

But the fact that we do have great success with H5 shows that still deeper spaces cannot have much influence on the predictions we are now making. As Henri Poincaré put it, rules which succeed "-- will not cease to do so on the day when they become better understood". Even if we knew all about H6, as long as our interest remained on the current predictions, we would have little to gain in pragmatic results, and probably much to lose in computational cost, by going to H6. So, although in principle the downward analysis is open ended, in practice there is an art in knowing when to stop.

CURRENT PROBLEMS

In newer problems (image reconstruction, spectrum analysis, geophysical inverse problems, etc.), the analysis of deeper hypothesis spaces H1, H2, --- , is still underway, and we don't know how far it will go. It would be nice if we could go down to a space Hx deep enough so that on it some kind of "symmetry" points to what seems "equally possible" cases, with predictive value. Then prior probabilities on the space M of the observable macroscopic things would be just multiplicities on the deeper space Hx, and inference would reduce to maximizing entropy on Hx, subject to constraints specifying regions of M.

The program thus envisaged would be in very close analogy with thermodynamics. Some regard our efforts to cling to this analogy as quaint; in defense we note that statistical mechanics is the only example we have thus far of that deeper analysis actually carried through to a satisfactory stopping point, and it required over 100 years of effort to accomplish this. So we think that we had better learn as much as we can from this example.

But in the new problems we have not yet found any Liouville theorem to guide us to the appropriate hypothesis space, as Gibbs had to guide him to H3 and was then generealized mathematically to H5. For Gibbs, invariance of phase volume under the equations of motion and under canonical transformations -- which he took great care to demonstrate and discuss at some length before entering into his thermodynamic applications -- meant that assigning uniform prior probability, or weight, to equal phase volumes had the same meaning at all times and in all canonical coordinate systems. This was really applying the principle of group invariance, in just the way advocated much later by the writer (Jaynes, 1968).

Specifying our deepest hypothesis space, on which we assign uniform weight before adding any constraints to get the nonuniform MAXENT prior, is the means by which we define our starting point of complete ignorance but for enumeration of the possibilities, sometimes called "pre-prior" analysis. Long ago, Laplace noted this problem and stated that the exact appreciation of "equally possible" is "one of the most delicate points in probability theory". How right he was! Two hundred years later, we are still hung up on this "exact appreciation" in every new application.

For a time, writers thought they had evaded this delicate point by redefining a probability as a frequency; but in fact they had only restricted the range of applications of probability theory. For the general problems of inference now being attacked, the need to define what we mean by "complete ignorance" -- complete, that is, but for enumeration of the possibilities to be considered in our problem -- cannot be evaded, any more than the notion of zero could be evaded in arithmetic.

Today this is not just a puzzle for philosophers. It is crucially important that we learn how to build more prior information into our prior probabilities by developing that neglected half of probability theory. All inverse problems need this, and the possibility of any major progress in pattern recognition or artificial intelligence depends on it.

But in each area, pending a satisfactory analysis to a satisfactory stopping point, we can take some comfort in Tukey pragmatism (don't confuse the procedure with the reason):

> "A procedure does not have hypotheses. Rather, there are circumstances where it does better, and others where it does worse". (John W. Tukey, 1980)

Our present maximum entropy procedure is supported by many different rationales, including:

(1)	Combinatorial	Boltzmann, Darwin, Fowler
(2)	Information Theory	Shannon, Jaynes
(3)	Utility	Good, Skilling
(4)	Logical Consistency	Shore, Johnson, Gull
(5)	Coding Theory	Rissanen
(6)	Pragmatic Success	Gibbs, Papanicolaou, Mead

Of these, (1) is easy to explain to everybody, while (2) is more general, but hard to explain to those with orthodox statistical training, (3) and (4) are currently popular, and (5) shows great long-range theoretical promise, but is not yet well explored.

Most of the writer's recent discussions have concentrated on (1) rather than (2) in the belief that, after one has become comfortable with using an algorithm in cases where it has a justification so clear and simple that everybody can understand it, he will be more disposed to see a broader rationale for what he is doing.

It might be thought that, if many rationales all point to the same procedure, it is idle to argue about their relative merits. Indeed, if we were to stay forever on the current problems, different rationales would be just different personal tastes without real consequences.

But different rationales generalize differently. In the current problems all these rationales happen to come together and point to the same procedure; but in other problems they would go their separate ways and point to different procedures. Therefore we think it is important in each application to understand the rationale and the circumstances as well as the procedure.

Of course, in case of doubt one can always fall back on (6). Doubtless, those who write the specific computer programs have done this a great deal, sometimes just trying out everything one can think of and seeing what works. We agree with Tukey that the theoretical justification of a procedure is often a mere tidying-up that takes place after the successful procedure has been found by intuitive trial and error.

But too much of that basically healthy Tukey pragmatism can lead one to take a negative view of theoretical efforts in general. The excessive disparagement of all theory, characteristic of that school, has been very costly to the field of data analysis; for Bayesian theory has a demonstrated ability to discover -- in a few lines -- powerful and useful procedures that decades of intuitive ad hockery did not find.

We have already noted the writer's "Bayesian Spectrum and Chirp Analysis" given at the 1983 Laramie meeting on Maximum Entropy, where the Schuster periodogram acquires a new significance, leading to a very different way of using it in data analysis. Basically the same thing was noted by Steve Gull, who perceived the real Bayesian significance of the "dirty map" of radio astronomy (a two-dimensional analog of the periodogram), and therefore the proper way of using it in data analysis.

Other examples are Litterman's (1985) Bayesian economic forecasting method and the writer's Bayesian seasonal adjustment method noted above, both of which process the data in a way that takes into account previously neglected prior information.

G.E.P. Box (1982) also observes: "--- recent history has shown that it is the omission in sampling theory, rather than the inclusion in Bayesian analysis, of an appropriate prior distribution, that leads to trouble."

In our next talk, "Monkeys, Kangaroos, and N," we want to continue this line of thought, with more specific details about hypothesis spaces and rationales, for the particular case of image reconstruction. We want to make a start on the question whether some of that deeper analysis might have helped us. Our hypothesis spaces are still at the "Boltzmann level"; if we can understand exactly what is happening there, it might become evident that we need to go down at least to the "Gibbs level" and possibly beyond it, before finding a satisfactory stopping point for current problems.

REFERENCES

1 Box, G.E.P. (1982). An Apology for Ecumenism in Statistics. NCR Technical Report 2408, Mathematics Research Center, University of Wisconsin, Madison.

2 Cox, R.T. (1946). Probability, Frequency, and Reasonable Expectation. Am. Jour. Phys. 17, 1-13. Expanded in The Algebra of Probable Inference, John Hopkins University Press, Baltimore, 1961. Reviewed by E.T. Jaynes, Am. Jour. Phys. 31, 66, 1963.

3 Cramér, H. (1946). Mathematical Methods of Statistics. Princeton University Press.

4 Gibbs, J. Willard (1875). On the Equilibrium of Heterogeneous Substances. Reprinted in The Scientific Papers of J. Willard Gibbs, Vol. I, Longmans, Green & Co., 1906 and by Dover Publications, Inc., 1961.

5 Gibbs, J. Willard (1902). Elementary Principles in Statistical Mechanics. Yale University Press, New Haven, Connecticut. Reprinted in The Collected Works of J. Willard Gibbs, Vol. 2, by Longmans, Green & Co., 1928, and by Dover Publications, Inc., New York, 1960.

6 Jaynes, E.T. (1968). Prior Probabilities. IEEE Trans. Systems Science and Cybernetics SSC-4, 227-241. Reprinted in V.M. Rao Tummala and R.C. Henshaw, eds., Concepts and Applications of Modern Decision Models. Michigan State University Business Studies Series, 1976; and in Jaynes, 1983.

7 Jaynes, E.T. (1976). Confidence Intervals vs Bayesian Intervals. In W.L. Harper & C.A. Hooker, eds., Foundations of Probability Theory, Statistical Inference, and Statistical Theories of Science, Vol. II, Reidel Publishing Co., Dordrecht-Holland, pp. 175-257. Reprinted in Jaynes, 1983.

8 Jaynes, E.T. (1978). Where do we Stand on Maximum Entropy? In The Maximum Entropy Formalism, R.D. Levine and M. Tribus, eds., M.I.T. Press, Cambridge, Mass., pp. 15-118. Reprinted in Jaynes, 1983.

9 Jaynes, E.T. (1982). On the Rationale of Maximum-Entropy Methods. Proc. IEEE 70, pp. 939-982.

10 Jaynes, E.T. (1983). Papers on Probability, Statistics, and Statistical Physics. R.D. Rosenkrantz, ed., D. Reidel Publishing Co., Dordrecht-Holland.

11 Jaynes, E.T. (1985). In Bayesian Statistics 2. J.M. Bernardo et al., eds., Elseviev Science Publishers, Amsterdam, pp. 329-360.

12 Jeffreys, H. (1939). Theory of Probability. Oxford University Press. Later editions, 1948, 1961, 1983.

13 Laplace, P.S. (1812). Theorie analytique des probabilites. 2 Vols. Reprints of this work are available from Editions Culture et Civilisation, 115 Ave. Gabriel Lebron, 1160 Brussels, Belgium.

14 Litterman, R.B. (1985). Vector Autoregression for Macroeconomic Forecasting. In Bayesian Inference and Decision Techniques, P. Zellner and P. Goel, eds. North-Holland Publishers, Amsterdam.

15 Molina, E.C. (1963). Two Papers by Bayes with Commentaries. Hafner Publishing Co., New York.

16 Savage, L.J. (1954). Foundations of Statistics. J. Wiley & Sons. Second Revised Edition, 1972, by Dover Publications, Inc., New York.

17 Shannon, C.E. (1948). A Mathematical Theory of Communication. Bell Systems Technical Jour., 27, 379, 623. Reprinted in C.E. Shannon and W. Weaver, The Mathematical Theory of Communication. University of Illinois Press, Urbana, 1949.

18 Tukey, J.W. (1980). In The Practice of Spectrum Analysis. University Associates, Princeton, New Jersey.

MONKEYS, KANGAROOS, AND N

E.T. Jaynes*
St. John's College and Cavendish Laboratory
Cambridge CB2 1TP
England

*Visiting Fellow, 1983-1984. Permanent Address:
Dept. of Physics, Washington University
St. Louis, MO 63130

ABSTRACT

We examine some points of the rationale underlying the choice of priors for MAXENT image reconstruction. The original combinatorial (monkey) and exchangeability (kangaroo) approaches each contains important truth. Yet each also represents in a sense an extreme position which ignores the truth in the other. The models of W.E. Johnson, I.J. Good, and S. Zabell provide a continuous interpolation between them, in which the monkeys' entropy factor is always present in the prior, but becomes increasingly levelled out and disappears in the limit.

However, it appears that the class of interpolated priors is still too narrow. A fully satisfactory prior for image reconstruction, which expresses all our prior information, needs to be able to express the common-sense judgment that correlations vary with the distance between pixels. To do this, we must go outside the class of exchangeable priors, perhaps into an altogether deeper hypothesis space.

INDEX

INTRODUCTION

Image reconstruction is an excellent ground for illustrating the generalities in our Tutorial Introduction. Pedagogically, it is an instructive and nontrivial example of the open-ended problem of determining priors which represent real states of knowledge in the real world. In addition, better understanding of this truly deep problem should lead in the future to better reconstructions and perhaps improvements in results for other problems of interest at this Workshop.

In discussing the various priors that might be used for image reconstruction, it should be emphasized that we are not dealing with an ideological problem, but a technical one. We should not think that any choice of prior hypothesis space and measure on that space is in itself either right or wrong. Presumably, any choice will be "right" in some circumstances, "wrong" in others. It is failure to relate the choices to the circumstances that gives the appearance of arbitrariness.

In a new problem, it is inevitable that different people have in the back of their minds different underlying hypothesis spaces, for several reasons:

(1) Different prior knowledge of the phenomenon.
(2) Different amounts or kinds of practical experience.
(3) Some have thought more deeply than others.
(4) Past training sets their minds in different channels.
(5) Psychological quirks that can't be accounted for.

Therefore, rather than taking a partisan stand for one choice against another, we want to make a start on better relating the choices to the circumstances.

This means that we must learn to define the problem much more carefully than in the past. If you examine the literature with this in mind, I think you will find that 90% of the past confusions and controversies in statistics have been caused, not by mathematical errors or even ideological differences; but by the technical difficulty that the two parties had different problems in mind, and failed to realize this. Thinking along different lines, each failed to perceive at all what the other considered too obvious to mention. If you fail to specify your sample space, sampling distribution, prior hypothesis space, and prior information, you may expect to be misunderstood -- as I have learned the hard way.

We are still caught up to some degree in the bad habits of orthodox statistics, taught almost exclusively for decades. For example, denoting the unknown true scene by $[p(i), 1 \le i \le n]$, we specify the mock data

$$M_k = \sum_{i=1}^{n} A(k,i)p(i), \qquad 1 \le k \le m \tag{1}$$

confidently, as if the point-spread function $A(k,i)$ were known exactly, and pretend it is a known, "objectively real" fact that the measurement errors were independent gaussian with known standard deviation.

But we say nothing about the prior information we propose to use -- not even the underlying hypothesis space on which the prior probabilities are to exist. Then in applying Bayes' theorem (I = prior information):

$$p(\text{Scene}|\text{Data},I) = p(\text{Scene}|I)\ P(\text{Data}|\text{Scene},I)/p(\text{Data}|I) \quad (2)$$

the likelihood of a scene,

$$p(\text{Data}|\text{Scene},I) = \exp[-\sum_{k=1}^{m}(d_k - M_k)^2/2\sigma^2] = \exp[-\ \text{Chi}^2/2] \quad (3)$$

has been fully specified in the statement of the problem; while its prior probability p(Scene | I) is left unspecified by failure to complete the statement of the problem.

In effect, we are claiming more knowledge than we really have for the likelihood, and less than we really have for the prior; just the error that orthodox statistics has always made. This makes it easy to say, "The data come first" and dismiss p(Scene | I) by declaring it to be completely uninformative. Yet in generalized inverse problems we usually have prior information that is fully as cogent as the data.

We need a more balanced treatment. A major point of Bayesian analysis is that it combines the evidence of the data with the evidence of the prior information. Unless we use an informative prior probability, Bayes' theorem can add nothing to the evidence of the data, and its advantage over sampling theory methods lies only in its ability to deal with technical problems like nuisance parameters.

To repeat the platitudes: in image reconstruction the data alone, whether noisy or not, cannot point to any particular scene because the domain R of maximum likelihood, where $\text{Chi}^2 = 0$, is not a point but a manifold of high dimensionality, every point of which is in the "feasible set" R' (which we may think of as R enlarged by adding all points at which Chi^2 is less than some specified value). An uninformative prior leaves us, inevitably, no wiser. So if entropy is denied a role in the prior probability, it must then be invoked in the end as a value judgment in addition to Bayes' theorem, to pick out one point in R'.

This does not necessarily lead to a difference in the actual algorithm, for it is well known that in decision theory the optimal decision depends only on the product of the prior probability and the utility function, not on the functions separately. But it does leave the question of rationale rather up in the air.

We want, then, to re-examine the problem to see whether some of that deeper analysis might have helped us; however, the following could hardly be called an analysis in depth. For lack of time and space we can indicate only how big the problem is, and note a few places where more theoretical work is needed.

This is, in turn, only one facet of the general program to develop that neglected half of probability theory. We are not about to run out of jobs needing to be done.

MONKEYS

In the pioneering work of Gull and Daniell (1978) the prior probability of a scene (map of the sky) with n pixels of equal area and N_i units of intensity in the i'th pixel, was taken proportional to its multiplicity:

$$p(\text{Scene} \mid I_o) \propto W = \frac{N!}{N_1! \ldots N_n!} \tag{4}$$

One could visualize this by imagining the proverbial team of monkeys making test maps by strewing white dots at random, N_i being the number that happen to land in the i'th pixel.

If the resulting map disagrees with the data it is rejected and the monkeys try again. Whenever they succeed in making a map that agrees with the data, it is saved. Clearly, the map most likely to result is the one that has maximum multiplicity W, or equally well maximum entropy per dot, $H = (\log W)/N$, while agreeing with the data.

If the N_i are large, then as we have all noted countless times, H goes asymptotically into the "Shannon entropy":

$$H \to - \Sigma \, (N_i/N) \log (N_i/N) \tag{5}$$

and by the entropy concentration theorem (Jaynes, 1982) we expect that virtually all the feasible scenes generated by the monkeys will be close to the one of maximum entropy.

Mathematically, this is just the combinatorial argument by which Boltzmann (1877) found his most probable distribution of molecules in a force field. But in Boltzmann's problem, $N = \Sigma N_i$ was the total number of molecules in the system, a determinate quantity.

In the image reconstruction problem, definition of the monkey hypothesis space stopped short of specifying enough about the strewing process to determine N. As long as the data were considered noiseless this did no harm, for then the boundary of the feasible set, or class C of logically possible scenes, was sharply defined (the likelihood was rectangular, so $C = R = R'$) and Bayes' theorem merely set the posterior probability of every scene outside C equal to zero, leaving the entire decision within C to the entropy factor. The value of N did not matter for the actual reconstruction.

But if we try to take into account the fact that real data are contaminated with noise, while using the same "monkey hypothesis space" H1 with n^N elements, the most probable scene is not the one that maximizes H subject to hard constraints from the data; it maximizes the sum (NH + log L) where L(Scene) = p(Data | Scene, I), the likelihood that allows for noise, is no longer rectangular but might, for example, be given by (3). Then N matters, for it determines the relative weighting of the prior probability and the noise factors.

If L is nonzero for all scenes and we allow N to become arbitrarily large, the entropy factor exp(NH) will overwhelm the likelihood L and force the reconstruction to the uniform grey scene that ignores the data. So if we are to retain the hypothesis space H1, we must either introduce some cutoff in L that places an upper limit on the possible noise magnitude; or assign some definite finite value of N.

Present practice -- or some of it -- chooses the former alternative by placing an upper limit on the allowable value of Chi^2. Although this leads, as we all know, to very impressive results, it is clearly an ad hoc device, not a true Bayesian solution. Therefore we ought to be able to do still better -- how much better, we do not know.

Of course, having found a solution by this cutoff procedure, one can always find a value of N for which Bayes' theorem would have given the same solution without the cutoff. It would be interesting, for diagnostic purposes, to know what these after-the-fact N values are, particularly the ratios N/n; but we do not have this information.

In different problems of image reconstruction (optics, radio astronomy, tomography, crystallography) the true scene may be generated by Nature in quite different ways, about which we know something in advance. In some circumstances, this prior information might make the whole monkey rationale and space H1 inappropriate from the start; in others it would be clearly "right".

In a large class of intermediate cases, H1 is at least a usable starting point from which we can build up to a realistic prior. In these cases, multiplicity factors are always cogent, in the sense that they always appear as a factor in the prior probability of a scene. Further considerations may "modulate" them by additional factors.

What are we trying to express by a choice of N? There can be various answers to this. One possible interpretation is that we are specifying something about the fineness of texture that we are asking for in the reconstruction. On this view, our choice of N would express our prior information about how much fineness the data are capable of giving. We discuss only this view here, and hope to consider some others elsewhere.

A preliminary attempt to analyze the monkey picture more deeply with this in mind was made by the writer at the 1981 Laramie Workshop. If the measurement errors σ are generated in the variability of the scene itself:

$$N_i \pm \sqrt{N_i}$$

there is a seemingly natural choice of N that makes $N\sigma^2$ = const. Varying N and σ then varies only the sharpness of the peak in the posterior probability space, not its location; with more accurate measurements giving smaller σ and larger N, we do not change our reconstruction but only become more confident of its accuracy, and so display it with a finer texture.

However, it appears that in the current applications we are closer to the truth if we suppose that the errors are generated independently in the measurement apparatus. Then there is a seemingly natural choice of N that expresses our prior information about the quality of the data by making the typical smallest increments in the mock data M_k due to changes in the scene, of the same order of magnitude as the smallest increment σ that the real data could detect.

If $p_i = N_i/N$, this increment is $dM_k \simeq A/N$, where A is a typical large element of A; and $N\sigma \simeq A$. Smaller values of N will yield an unnecessarily coarse reconstruction, lacking all the density gradations that the data give evidence for; while larger values in effect ask for finer gradations than the data can justify. The reconstruction depends on σ for the intuitive reason that if, for given data, we learned that the noise level is smaller than previously thought, then some details in the data that were below the noise level and ignored, now emerge above the noise and so are believed, and appear in the reconstruction.

At present, we have no actual reconstructions based on this idea, and so do not know whether there are unrecognized difficulties with it. In one highly oversimplified case, where the data give evidence only for p_1, John Skilling concludes that the $N\sigma \simeq A$ choice leads to absurd conclusions about (p_2-p_3). Yet there are at least conceivable, and clearly definable, circumstances in which they are not absurd. If the true scene is composed of N_i quanta of intensity in the i'th pixel (whether placed there by monkeys or not) then p_1 cannot be measured more accurately -- because it is not even defined more accurately -- than $dM_k/A = N^{-1}$. It is not possible to measure p_1 to one part in 100 unless N_1 is at least 100.

Then if we specify that p_1 is measured more and more accurately without limit, we are not considering a single problem with fixed N and a sequence of smaller and smaller values of a parameter σ. We are considering a sequence of problems with different N, in which we are drawing larger and larger samples, of size $N_1 = Np_1$. From this one expects to estimate other quantities to a relative accuracy improving like $N^{-1/2}$.

This is not to say that we are "measuring" (p_2-p_3) more and more accurately; we are not measuring it at all. In a sequence of different states of knowledge we are inferring it more and more confidently, because the statement of the problem -- that p_1 was measured very accurately -- implies that N must have been large.

Doubtless, there are other conceivable circumstances (i.e., other states of knowledge about how Nature has generated the scene) in which our conclusion about (p_2-p_3) would indeed be absurd. Any new information which could make our old estimate seem absurd would be, to put it mildly, highly cogent; and it would seem important that we state explicitly what this information is so we can take full advantage of it. But at present, not having seen this information specified, we do not know how to use it to correct our estimate of (p_2-p_3); no alternative estimate was proposed.

This situation of unspecified information -- intuition feels it but does not define it -- is not anomalous, but the usual situation in exploring this neglected part of probability theory. It is not an occasion for dispute, but for harder thinking on a technical problem that is qualitatively different from the ones scientists are used to thinking about. One more step of that harder thinking, in a case very similar to this, appears in our discussion of the kangaroo problem below.

In any event, as was stressed at the 1981 Laramie Workshop and needs to be stressed again, the question of the choice of N cannot be separated from the choices of m and n, the number of pixels into which we resolve the blurred image and the reconstruction, and u, v, the quantizing increments that we use to represent the data d(k) and the reconstruction p(i) for calculational purposes.

In most problems the real and blurred scenes are continuous, and the binning and digitization are done by us. Presumably, our choices of (N, m, n, u, v) all express something about the fineness of texture that the data are capable of supporting; and also some compromises with computation cost. Although computer programmers must necessarily have made decisions on this, we are not aware of any discussion of the problem in the literature, and the writer's thinking about it thus far has been very informal and sketchy. More work on these questions seems much needed.

In this connection, we found it amusing to contemplate going to the "Fermi statistics" limit where n is very large and we decree that each pixel can hold only one dot or none, as in the halftone method for printing photographs.

Also one may wonder whether there would be advantages in working in a different space, expanding the scene in an orthogonal basis and estimating the expansion coefficients instead of the pixel intensities. A particular orthogonal basis recommends itself; that generated by the singular-value decomposition of the smearing matrix A_{ki}. Our data comprise an (m x 1) vector: $d = Ap + e$, where e is the vector of "random errors". Supposing the (m x n) matrix A to be of rank m, it can be factored:

$$A = V\,D\,U^T \tag{7}$$

where U and V are (n x n) and (m x m) orthogonal matrices that diagonalize A^TA and AA^T, and $D^2 = V^T AA^T V$ is the positive definite (m x m) diagonalized matrix. $D = V^T A U$ is its square root, padded with (n-m) extra columns of zeroes. Label its rows and columns so that $D_{11}^2 \geq D_{22}^2 \geq \ldots$ Then if we use the columns of U as our basis:

$$p_i = \sum_{j=1}^{n} U_{ij} a_j, \qquad 1 \leq i \leq n \tag{8}$$

our data equation $d = Ap + e$ collapses to

$$d_k - e_k = \sum_{j=1}^{m} V_{kj} D_{jj} a_j \qquad 1 \leq k \leq m \tag{9}$$

Only the first m expansion coefficients ($a_1 \ldots a_m$) appear; in this coordinate system the relevance of the data is, so to speak, not spread all over the scene, but cleanly separated off into a known m-dimensional subregion. The likelihood (3) of a scene becomes

$$L(\text{Scene}) = L(a_1 \ldots a_m) = \exp[- \sum_{j=1}^{m} D_{jj}^{\;2}(a_j - b_j)^2/2\sigma^2] \tag{10}$$

where $b = dVD^{-1}$ is the data vector in the new coordinates. The expansion coefficients a_j belonging to large eigenvalues of AA^T are determined quite accurately by the data (to $\pm\, \sigma/D_{jj}$). But the data give no evidence at all about the last (n-m) coordinates ($a_{m+1} \ldots a_n$).

There might be advantages in a computational scheme that, by working in these coordinates, is able to deal differently with those a_j that are well determined by the data, and those that are undetermined. Perhaps we might decree that for the former "the data come first". But for the latter, the data never come at all.

In any event, whatever our philosophy of image reconstruction, the coordinates ($a_{m+1} \ldots a_n$) must be chosen solely on grounds of prior information. If ($a_1 \ldots a_m$) are specified first, the problem reverts to a pure generalized inverse problem (i.e., one with hard constraints). The scene which has maximum entropy subject to prescribed ($a_1 \ldots a_m$)

is determined without any reference to N. Computational algorithms for carrying out the decomposition (7) are of course readily available (Chambers, 1977).

As we see from this list of unfinished projects, there is room for much more theoretical effort, which might be quite pretty analytically and worthy of a Ph.D. thesis or two; even the specialized monkey approach is open-ended.

KANGAROOS

A different rationale for maximizing entropy was illustrated by Steve Gull, on the occasion of a talk in Australia in 1983, by supposing it established by observation that 3/4 of the kangaroos are left-handed, and 3/4 drink Foster's; from which we are to infer what fraction of them are both right-handed and Foster's drinkers, etc.; that is, to reconstruct the (2 x 2) table of proportions p(i,j)

$$\begin{array}{cccc} & L & R & \\ F & p(11) & p(12) & 3/4 \\ \text{no } F & p(21) & p(22) & 1/4 \\ & 3/4 & 1/4 & \end{array} \qquad (11)$$

from the specified marginal row and column totals given to the right and below the table.

It is interesting to compare the solutions of this problem given by various algorithms that have been proposed. Gull and Skilling (1984), applying the work of Shore and Johnson, find the remarkable result that if the solution is to be found by maximizing some quantity, entropy is uniquely determined as the only choice that will not introduce spurious correlations in the matrix (11), for which there is no evidence in the data. The maximum entropy solution is then advocated on grounds of logical consistency rather than multiplicity.

I want to give an analysis of the kangaroo problem, with an apology in advance to Steve Gull for taking his little scenario far more literally and seriously than he ever expected or wanted anybody to do. My only excuse is that it is a conceivable problem, so it provides a specific example of constructing priors for real problems, exemplifying some of our Tutorial remarks about deeper hypothesis spaces and measures. And, of course, the principles are relevant to more serious real problems -- else the kangaroo problem would never have been invented.

What bits of prior information do we all have about kangaroos, that are relevant to Gull's question? Our intuition does not tell us this immediately, but a little pump priming analysis will make us aware of it.

In the first place, it is clear from (11) that the solution must be of the form:

$$\begin{bmatrix} (.50 + q) & (.25 - q) \\ (.25 - q) & q \end{bmatrix} \quad 0 \le q \le .25 \tag{12}$$

But, kangaroos being indivisible, it is required also that the entries have the form p(i,j) = N(i,j)/N with N(i,j) integers, where N is the number of kangaroos. So for any finite N there are a finite number of integer solutions N(i,j). Any particular solution will have a multiplicity

$$W = \frac{N!}{N(11)!N(12)!N(21)!N(22)!} \tag{13}$$

This seems rather different from the image reconstruction problem; for there it was at least arguable whether N makes any sense at all. The maximum entropy scene was undeniably the one the monkeys would make; but the monkeys were themselves only figments of our imagination.

Now, it is given to us in the statement of the problem that we are counting and estimating attributes of kangaroos, which are not figments of our imagination; their number N is a determinate quantity. Therefore the multiplicities W are now quite real, concrete things; they are exactly equal to the number of possibilities in the real world, compatible with the data. It appears that, far from abandoning monkeys, if there is any place where the monkey (combinatorial) rationale seems clearly called for, it is in the kangaroo problem!

Let us see some exact numerical solutions. Suppose N = 4; then there are only two solutions:

$$N(i,j) = \begin{bmatrix} 2 & 1 \\ 1 & 0 \end{bmatrix}, \begin{bmatrix} 3 & 0 \\ 0 & 1 \end{bmatrix} \tag{14}$$

with multiplicities W = 12, 4 respectively. The solution with greater entropy comprises 75% of the feasible set of possibilities consistent with the data.

If N = 16, there are five integer solutions:

$$N(i,j) = \begin{bmatrix} 8 & 4 \\ 4 & 0 \end{bmatrix}, \begin{bmatrix} 9 & 3 \\ 3 & 1 \end{bmatrix}, \begin{bmatrix} 10 & 2 \\ 2 & 2 \end{bmatrix}, \begin{bmatrix} 11 & 1 \\ 1 & 3 \end{bmatrix}, \begin{bmatrix} 12 & 0 \\ 0 & 4 \end{bmatrix} \tag{15}$$

W = 900900, 1601600, 720720, 87360, 1820
36%, 64%, 29%, 3.5%, .07%

The single maximum entropy solution comprises nearly two-thirds of the feasible set.

But there are many kangaroos; when $N \gg 1$ the multiplicities go asymptotically into $W \sim \exp(NH)$ where from (12), the entropy is

$$H = -(.5+q)\log(.5+q) - 2(.25-q)\log(.25-q) - q\log q \tag{16}$$

This reaches its peak at $q = 1/16$, corresponding as noted to no correlations between the attributes of kangaroos. For $q < 1/16$ we have negative correlations (drinkers tend to be right handed, etc.); while the solutions with $q > 1/16$ give positive correlations. Near the peak, a power series expansion yields the asymptotic formula

$$W \sim \exp\left[-(128N/9)(q - 1/16)^2\right] \tag{17}$$

which would lead us to the (mean $\pm$ standard deviation) estimate of q:

$$q(\text{est}) = (1/16)(1 \pm 3/\sqrt{N}) \tag{18}$$

Thus if there are $N = 900$ kangaroos the last factor in (18) is (1 ± 0.1); if $N = 90{,}000$ it is (1 ± 0.01); and if there are $N = 9{,}000{,}000$ kangaroos it becomes (1 ± 0.001). These are the predictions made by uniform weighting on our first (monkey) hypothesis space H1.

Here we can start to discover our own hidden prior information by introspection; at what value of N do you begin feeling unhappy at this result? Most of us are probably willing to believe that the data reported by Steve Gull could justify an estimate of q for which we could reasonably claim 10% accuracy; but we may be reluctant to believe that they could determine it to one part in 1,000, however many kangaroos there are.

Eq. (18) is essentially the same kind of result discussed above, that John Skilling called "absurd"; but he could dismiss it before on the grounds that N was only an imagined quantity. Now that argument is unavailable; for N is a real, determinate quantity. So what has gone wrong this time? I feel another Sermon coming on.

SERMON ON THE MULTIPLICITY

However large N, it is a combinatorial theorem that most of the possibilities allowed by the data are within that shrinking interval (18). But at some point someone says: "This conclusion is absurd; I don't believe it!" What is he really saying?

It is well established by many different arguments that Bayesian inference yields the unique consistent conclusions that follow from the model, the data, and the prior information <u>that was actually used</u> in the calculation. Therefore, if anyone accepts the model and the data but rejects the estimate (18), there are two possibilities: either he is reasoning inconsistently and his intuition needs educating; or else he has extra prior information.

We have met nobody who claims the first distinction for himself, although we all have it to some degree. Many times, the writer has been disconcerted by a Bayesian result on first finding it, but realized on deeper thought that it was correct after all; his intuition thereby became a little more educated.

The same policy -- entertain the possibility that your intuition may need educating, and think hard before rejecting a Bayesian result -- is recommended most earnestly to others. As noted in our Tutorial, intuition is good at perceiving the relevance of information, but bad at judging the relative cogency of different pieces of information. If our intuition was always trustworthy, we would have no need for probability theory.

Over the past 15 years many psychological tests have shown that in various problems of plausible inference with two different pieces of evidence to consider, intuition can err -- sometimes violently and in opposite directions -- depending on how the information is received. Some examples are noted in Appendix A.

This unreliability of intuition is particularly to be stressed in our present case, for it is not limited to the untrained subjects of psychological tests. Throughout the history of probability theory, the intuition of those familiar with the mathematics has remained notoriously bad at perceiving the cogency of multiplicity factors. Some expositions of probability theory start by pointing to the fact that observed frequencies tend to remain within the $\pm n^{-1/2}$ "random error" bounds. This observed property of frequencies, to become increasingly stable with increasing number of observations, is seen as a kind of Miracle of Nature -- the empirical fact underlying probability theory -- showing that probabilities are physically real things.

Yet as Laplace noted, those frequencies are only staying within the interval of high multiplicity; far from being a Miracle of Nature, the great majority of all things that could have happened correspond to frequencies remaining in that interval. If one fails to recognize the cogency of multiplicity factors, then virtually every "random experiment" does indeed appear to be a Miracle of Nature, even more miraculous than (18).

In most of the useful applications of direct probability calculations -- the standard queueing, random walk, and stochastic relaxation problems -- the real function of probability theory is to correct our faulty intuition about multiplicities, and restore them to their proper strength in our predictions. In particular, the Central Limit Theorem expresses how multiplicities tend to pile up into a Gaussian under repeated convolution.

Present orthodox statistics takes multiplicity into account correctly in sampling distributions, but takes no note of multiplicity on parameter spaces. This can lead to very bad estimates of a parameter whose multiplicity varies greatly within the region of high likelihood. It behooves us to be sure that we are not committing a similar error here.

Bear in mind, therefore, that in this problem the entire population of kangaroos is being sampled; as N increases, so does the amount of data that is generating that estimate (18). Estimates which improve as the square root of the number of observations are ubiquitous in all statistical theory.

But if, taking note of all this, you still cannot reconcile (18) to your intuition, then realize the implications. Anyone who adamantly refuses to accept (18) is really saing: "I have extra prior information about kangaroos that was not taken into account in the calculation leading to (18)."

More generally, having done any Bayesian calculation, if you can look at the result and know it is "wrong"; i.e., the conclusion does not follow reasonably from your information, then you must have extra information that was not used in the calculation. You should have used it.

Indeed, unless you can put your finger on the specific piece of information that was left out of the calculation, and show that the revised calculation corrects the difficulty, how can you be sure that the fault is in the calculation and not in your intuition?

HIDDEN PRIOR INFORMATION

The moral of the Sermon was that, if we react to (18) by casting out the whole monkey picture and calculation, and starting over from the beginning without asking what that extra information is, we are losing the whole value and point of the calculation. The monkey calculation on H1 has only primed the mental pump; at this point, the deep thought leading us down to H2 is just ready to begin:

> What do we know about kangaroos, that our common sense suddenly warns us was relevant, but we didn't think to use at first?

There are various possibilities; again, intuition feels them but does not define them. Consider first an extreme but conceivable state of prior knowledge:

(H2a): If we knew that the left-handed gene and the Foster's gene were linked together on the same chromosome, we would know in advance that these attributes are perfectly correlated and the data are redundant:

$q = 1/4$. In the presence of this kind of prior information the "logical consistency" argument pointing to $q = 1/16$ would be inapplicable.

Indeed, any prior information that establishes a logical link between these two attributes of kangaroos will make that argument inapplicable in our problem. Had our data or prior information been different, in almost any way, they would have given evidence for correlations and MAXENT would exhibit it.

The "no correlations" phemonenon emphasized by the kangaroo rationale is a good illustration of the "honesty" of MAXENT (i.e., it does not draw conclusions for which there is no evidence in the data) in one particular case. But it seems to us a useful result -- a reward for virtue -- rather than a basic desideratum for all MAXENT.

Of course, if we agree in advance that our probabilities are always to be found by maximizing the same quantity whatever the data, then a single compelling case like this is sufficient to determine that quantity, and the kangaroo argument does pick out entropy in preference to any proposed alternative. This seems to have been Steve Gull's purpose, and it served that purpose well.

The H2a case is rather unrealistic, but as we shall see it is nevertheless a kind of caricature of the image reconstruction problem; it has, in grossly exaggerated form, a feature that was missing from the pure monkey picture.

(H2b): More realistically, although there are several species of kangaroos with size varying from man to mouse, we assume that Gull intended his problem to refer to the man-sized species (who else could stand up at a bar and drink Foster's?). The species has a common genetic pool and environment; one is much like another. But we did not have any prior information about left/right-handedness or drinking habits.

In this state of prior knowledge, learning that one kangaroo is left-handed makes it more likely that the next one is also left-handed. This positive correlation (not between attributes, but between kangaroos) was left out of the monkey picture.

The same problem arises in survey sampling. Given that, in a sample of only 1,000 kangaroos, 750 were left-handed, we would probably infer at once that about 3/4 of the millions of unsampled kangaroos are also left-handed. But as we demonstrate below, this would not follow from Bayes' theorem with the monkey prior (13), proportional only to multiplicities. In that state of prior knowledge (call it I_0), every kangaroo is a separate, independent thing; whatever we learn about one specified individual is irrelevant to inference about any other.

Statisticians involved in survey sampling theory noticed this long ago and reacted in the usual way: if your first Bayesian calculation contradicts intuition, do not think more deeply about what prior information your intuition was using but your calculation was not; just throw out all Bayesian notions. Thus was progress limited to the bits of Bayesian analysis (stratification) that intuition could perceive without any theory, and could be expressed in non-Bayesian terms by putting it into the model instead of the prior probability.

Following Harold Jeffreys instead, we elect to think more deeply. Our state of knowledge anticipates some positive correlation between kangaroos, but for purpose of defining H2, suppose that we have no information distinguishing one kangaroo from another. Then whatever prior we assign over the 4^N possibilities, it will be invariant under permutations of kangaroos.

This reduces the problem at once; our neglected prior information about kangaroos must be all contained in a single function $g(x_1,x_2,x_3)$ of three variables (the number of attributes minus one) rather than N (the number of kangaroos). For it is a well-known theorem that a discrete distribution over exchangeable kangaroos (or exchangeable anything else) is a de Finetti mixture of multinomial distributions, and the problem reduces to finding the weighting function of that mixture.

For easier notation and generality, let us now label the four mutually exclusive attributes of kangaroos by (1, 2, 3, 4) instead of (11, 12, 21, 22), and consider instead of just 4 of them, any number n of mutually exclusive attributes, one of which kangaroos must have. Then de Finetti's famous theorem (Kyburg and Smokler, 1981) says that there exists a generating function $G(x_1 \ldots x_n)$ such that the probability that N_1 of them have attribute 1, and so on, is

$$p(N_1 \ldots N_n|I) = W(N) \int x_1^{N_1} \ldots x_n^{N_n} G(x_1 \ldots x_n) dx_1 \ldots dx_n \quad (19)$$

where W(N) is the monkey multiplicity factor (4). Normalization for all N requires that G contain a delta-function:

$$G(x_1 \ldots x_n) = \delta\,(\Sigma x_i - 1) g(x_1 \ldots x_n) \quad (20)$$

Since g need be defined only when $\Sigma x_i = 1$, it really depends only on (n-1) variables, but it is better for formal reasons to preserve symmetry by writing it as in (20).

As it stands, (19) expresses simply a mathematical fact, which holds independently of whatever meaning you or I choose to attach to it. But it can be given a natural Bayesian interpretation if we think of $(x_1 \ldots x_N)$ as a set of "real" parameters which define a class of hypotheses about what is generating our data. Then the factor

$$p(N_1 \ldots N_n|x_1 \ldots x_n) = W(N)\ x_1^{N_1} \ldots\ x_n^{N_n} \quad (21)$$

is the multinomial sampling distribution conditional on those parameters; the hypothesis indexed by $(x_1 \ldots x_n)$ assigns a probability numerically equal to x_1 that any specified kangaroo has attribute 1, and so on.

This suggests that we interpret the generating function as

$$G(x_1 \ldots x_n) = p(x_1 \ldots x_n \mid I), \tag{22}$$

the prior probability density of those parameters, following from some prior information I. Note, to head off a common misconception, that this is in no way to introduce a "probability of a probability". It is simply convenient to index our hypotheses by parameters x_i chosen to be numerically equal to the probabilities assigned by those hypotheses; this avoids a doubling of our notation. We could easily restate everything so that the misconception could not arise; it would only be rather clumsy notationally and tedious verbally.

However, this is a slightly dangerous step for a different reason; the interpretation (21), (22) has a mass of inevitable consequences that we might or might not like. So before taking this road, let us note that we are here choosing, voluntarily, one particular interpretation of the theorem (19). But the choice we are making is not forced on us, and after seeing its consequences we are free to return to this point and make a different choice.

That this choice is a serious one conceptually is clear when we note that (22) implies that we had some prior knowledge about the x_i. But if the x_i are merely auxiliary mathematical quantities defined from $p(N_1 \ldots N_n \mid I)$ through (19), then they are, so to speak, not real at all, only figments of our imagination. They are, moreover, not necessary to solve the problem, but created on the spot for mathematical convenience; it would not make sense to speak of having prior knowledge about them. They would be rather like normalization constants or MAXENT Lagrange multipliers, which are also created on the spot only for mathematical convenience, so one would not think of assigning prior probabilities to them.

But if we do consider the x_i as "real" enough to have some independent existence justifying a prior probability assignment, (19) becomes a standard relation of probability theory:

$$p(N_1 \ldots N_n|I) = \int d^n x \; p(N_1 \ldots N_n|x_1 \ldots x_n) \; p(x_1 \ldots x_n|I) \tag{23}$$

in which the left-hand side has now become the joint <u>predictive</u> prior probability that exactly N_i kangaroos have attribute i, $1 \le i \le n$.

This choice is also serious functionally, because it opens up a long avenue of mathematical development. We can now invoke the Bayesian apparatus to calculate the joint posterior probability distribution for the parameters and the posterior predictive distribution for $(N_1 \ldots N_n)$ given some data D. Without the choice (22) of interpretation it would hardly make sense to do this, and we would not see how (19) could lead us to any such notion as a posterior predictive distribution. Any modification of (19) to take account of new data would have to be done in some other way.

But let us see the Bayesian solution. Suppose our data consist of sampling M kangaroos, $M<N$, and finding that M_1 have attribute 1, and so on. Then its sampling distribution is

$$p(D|x_1 \ldots x_n) = W(M)\, x_1^{M_1} \ldots x_n^{M_n} \tag{24}$$

where W(M) is the multiplicity factor (4) with N's replaced by M's everywhere. The posterior distribution is

$$\begin{aligned} p(x_1 \ldots x_n|DI) &= p(x_1 \ldots x_n|I)\,p(D|x_1 \ldots x_n)/p(D|I) \\ &= A\, G(x_1 \ldots x_n)\, x_1^{M_1} \ldots x_n^{M_n} \end{aligned} \tag{25}$$

where A is a normalizing constant, independent of the x_i, and by G we always mean g with the delta function as in (20). This leads to a predictive posterior distribution for future observations; if we sample K more kangaroos, the probability that we shall find exactly K_1 with attribute 1, and so on, is

$$p(K_1 \ldots K_n|DI) = A\, W(K) \int G(x_1 \ldots x_n)\, x_1^{M_1+K_1} \ldots x_n^{M_n+K_n}\, d^n x \tag{26}$$

These generalities will hold in any exchangeable situation where it makes sense to think of G as a prior probability.

Now, our aim being to relate the choices to the circumstances, we need to think about specific choices of g to represent various kinds of prior information. Some suggestions are before us; a generating function of the form

$$g = A\, x_1^{k-1}\, x_2^{k-1} \ldots x_n^{k-1} \tag{27}$$

is often called a "Dirichlet prior", although I do not know what Dirichlet had to do with it. For the case k=1 it was given by Laplace (1778) and for general k by Hardy (1889). However, they gave only the choices, not the circumstances; intuitively, just what prior information is being expressed by (27)?

A circumstance was given by the Cambridge philosopher W.E. Johnson (1924); he showed, generalizing an argument that was in the original work of Bayes, that if in (19) all choices of $(N_1 \ldots N_n)$ satisfying $N_i \geq 0$, $\Sigma N_i = N$ are considered equally likely for all N, this uniquely determines Laplace's prior. In a posthumously published work (Johnson, 1932) he gave a much more cogent circumstance, which in effect asked just John Skilling's question: "Where would the next photon come from?".

Defining the variables: $y_k = i$ if the k'th kangaroo has attribute i, $(1 \leq k \leq N,\ 1 \leq i \leq n)$, Johnson's "sufficientness postulate" is that

$$p(y_{N+1}=i | y_1 \ldots y_n, I) = f(N, N_i) \tag{28}$$

Let us state what this means intuitively in several different ways: (a) The probability that the next kangaroo has attribute i should depend only on how many have been sampled thus far, and how many had attribute i; (b) If a sampled kangaroo did not have attribute i, then it is irrelevant what attribute it had; (c) A binary breakdown into (i)/(not i) captures everything in the data that is relevant to the question being asked; (d) Before analyzing the data, it is permissible to pool the data that did not yield (i).

Johnson showed that if (28) is to hold for all (N, N_i), this requires that the prior must have the Dirichlet-Hardy form (27) for some value of k. For recent discussions of this result, with extensions and more rigorous proofs, see Good (1965), Zabell (1982). In particular, an extension we need is that the function $f(N, N_i)$ need not be the same for all i; we may express prior information that is not symmetric over all attributes, without losing either Johnson's basic idea or the symmetry over kangaroos, by using n different functions $f_i(N, N_i)$, which leads to n different values $(k_1 \ldots k_n)$ of k in the factors of (27).

This intuitive insight of Johnson still does not reveal the meaning of the parameter k. Most discussions have favored small values, in $(0 \leq k \leq 1)$, on the grounds of being uninformative. Let us look at the specific details leading to the function $f(N, N_i)$ in (28). Analytically, everything follows from the generalized Beta function integral

$$\int_0^\infty dx_1 \ldots \int_0^\infty dx_n \; x_1^{k_1-1} \ldots x_n^{k_n-1} \; \delta(\Sigma x_i - a) = \frac{\Gamma(k_1) \ldots \Gamma(k_n)}{\Gamma(k_1 + \ldots + k_n)} a^{K-1} \tag{29}$$

where $K = \Sigma k_i$. Thus a properly normalized generating function is

$$g(x_1 \ldots x_n) = \frac{\Gamma(K)}{\Gamma(k_1) \ldots \Gamma(k_n)} x_1^{k_1-1} \ldots x_n^{k_n-1} \quad . \tag{30}$$

Denote by I_D the prior information leading to (30). Conditional on I_D, the probability of obtaining the data $(N_1 \dots N_n)$ in N observations is given by (19); using (29) and rearranging, we have

$$p(N_1 \dots N_n | I_D) = \frac{N!\Gamma(K)}{\Gamma(N+K)} \frac{\Gamma(N_1+k_1)}{N_1!\Gamma(k_1)} \dots \frac{\Gamma(N_n+k_n)}{N_n!\Gamma(k_n)} \tag{31}$$

Note that the monkey multiplicity factor W(N) is still contained in (31). For Laplace's prior (all $k_i = 1$) it reduces to

$$p(N_1 \dots N_n | I_D) = \frac{N!(n-1)!}{(N+n-1)!} \tag{32}$$

independent of the N_i, in accordance with Johnson's 1924 circumstance.

This is the reciprocal of the familiar Bose-Einstein multiplicity factor (number of linearly independent quantum states that can be made by putting N bosons into n single-particle states). Indeed, the number of different scenes that can be made by putting N dots into n pixels or N kangaroos into n categories, is combinatorially the same problem; one should not jump to the conclusion that we are invoking "quantum statistics" for photons. Note that the monkey multiplicity factor W(N) is the solution to a very different combinatorial problem, namely the number of ways in which a given scene can be made by putting N dots into n pixels.

In the "uninformative" limit where one or more of the $k_i \to 0$, the integral (29) becomes singular. However, the relevant quantity (31) is a ratio of such integrals, which does not become singular. In the limit it remains a proper (i.e., normalized) distribution, for a typical factor of (31) behaves as follows: as $k \to 0$,

$$\frac{\Gamma(N+k)}{N!\Gamma(k)} \to \begin{cases} k/N, & N > 0 \\ 1, & N = 0 \end{cases} \tag{33}$$

Therefore, for example, as $k_1 \to 0$, (31) goes into

$$p(N_1 \dots N_n | I_D) \to \begin{cases} 0, & N_1 > 0 \\ p(N_2 \dots N_n | I_D), & N_1 = 0 \end{cases} \tag{34}$$

The probability is concentrated on the subclass of cases where $N_1 = 0$. In effect, attribute #1 is removed from the menu available to the kangaroos (or pixel #1 is removed from the set of possible scenes). Then if any other $k_i \to 0$, attribute i is removed, and so on.

But if all k_i tend to zero simultaneously in a fixed proportion; for example, if we set

$$k_i = k\, a_i, \qquad a_i > 0, \qquad 1 \le i \le n \tag{35}$$

and let $k \to 0+$, (31) goes into

$$p(N_1 \dots N_n | I_D) \to \begin{cases} a_i/\Sigma a_i & \text{if } N_i = N, \text{ all other } N_j = 0 \\ 0, & \text{otherwise} \end{cases} \tag{36}$$

and the probability is concentrated entirely on those cases where all kangaroos have the same attribute (or those scenes with all the intensity in a single pixel); i.e., the extreme points of the sample space which have the minimum possible multiplicity $W = 1$.

But these results seem even more disconcerting to intuition than the one (18) which led us to question the pure monkey rationale. There we felt intuitively that the parameter q should not be determined by the data to an accuracy of 1 part in 1000. Does it seem reasonable that merely admitting the possibility of a positive correlation between kangaroos, should totally wipe out multiplicity ratios of $10^{100}:1$, as it appears to be doing in (32), and even more strongly in (36)?

In the inference called for, relative multiplicities are cogent factors. We expect them to be moderated somewhat by the knowledge that kangaroos are a homogeneous species; but surely multiplicities must still retain much of their cogency. Common sense tells us that there should be a smooth, continuous change in our results starting from the pure monkey case to a more realistic one as we allow the possibility of stronger and stronger correlations. Instead, (32) represents a discontinuous jump to the opposite extreme, which denies entropy any role at all in the prior probability. Eq. (36) goes even further and violently reverses the entropy judgments, placing all the prior probability on the situations of zero entropy.

In what sense, then, can we consider small values of k to be "uninformative"? In view of (34), (36) they are certainly not uninformative about the N_i.

A major thing to be learned in developing this neglected half of probability theory is that the mere unqualified epithet "uninformative" is meaningless. A distribution which is uninformative about variables in one space need not be in any sense uninformative about related variables in some other space. As we learn in quantum theory, the more sharply peaked one function, the broader is its Fourier transform; yet both are held to be probability amplitudes for related variables.

Our present problem exhibits a similar "uncertainty relation". The monkey multiplicity prior is completely uninformative on the sample space S of n^N possibilities. But on the parameter space X of the x_i

it corresponds to an infinitely sharply peaked generating function G, a product of delta functions $\delta(x_i - n^{-1})$. Conversely, small values of k are uninformative about the x_i but highly informative about the different points in S, in the limit (36) tying the sample numbers N_i rigidly together.

It is for us to say which, if either, of these limits represents our state of knowledge. This depends, among other things, on the meaning we attach to the variables. In the present problem the x_j are only provisionally "real" quantities, introduced for mathematical convenience, the integral representation (19) being easy to calculate with. But we have avoided saying anything about what they really mean.

We now see one of those inevitable consequences of assigning priors to the x_i, that the reader was warned he might or might not like.

In the kangaroo problem it is the N_i that are the truly, unquestionably real things about which we are drawing inferences. Prior to de Finetti, nobody's intuition had perceived that exchangeability alone, without knowledge of the x_i, is such a strong condition that a broad generating function can force such correlations between all the N_i.

If our prior information was that the x_j are themselves the "real physical quantities" of interest and the N_i only auxiliary quantities representing the exigencies of real data, then a prior that is uninformative about the x_i might be just what we need. This observation opens up another interpretive question about the meaning of a de Finetti mixture, that we hope to consider elsewhere.

Now let us examine the opposite limit of (31). As $k \to \infty$, the LHS of (33) goes into $k^N/N!$. Thus as $k_1 \to \infty$, (31) goes into

$$p(N_1 \ldots N_n | I_D) \to \begin{cases} 1, & N_1 = N, \text{ all other } N_i = 0 \\ 0, & \text{otherwise} \end{cases} \tag{37}$$

All categories except the first are removed from the menu. But if all k_i increase in a fixed proportion by letting $k \to \infty$ in (35), the limiting form of (31) is

$$p(N_1 \ldots N_n | I_D) \to W(N)\ (k_1/K)^{N_1} \ldots (k_n/K)^{N_n} \tag{38}$$

just the multinomial distribution with selection probabilities k_i/K. If the k_i are all equal, this reverts to a constant times the pure monkey multiplicity from whence we started. So it is the region of large k, not small, that provides the smooth, continuous transition from the "too good" prediction (18).

One way to define an intuitive meaning for the parameters k_i is to calculate Johnson's predictive function $f(N,N_i)$ in (28) or its generalization $f_i(N,N_i)$. With any initial generating function G, (26) shows that, having observed M kangaroos and finding sample numbers $(M_1 \ldots M_n)$, the probability that the next kangaroo sampled will be found to have attribute i is proportional to

$$\int G\, x_1^{M_1} \ldots x_i^{M_i+1} \ldots x_n^{M_n}\, d^n x \tag{39}$$

but for the particular generating function (30) the result is given, with the correct normalization factor, by the RHS of (31) after the appropriate changes of notation:

$$k_i \rightarrow k_i + M_i; \qquad N_i = 1, \text{ all other } N_j = 0$$

We find

$$f_i(M, M_i) = \frac{\Gamma(M+K)}{\Gamma(M+K+1)} \; \frac{\Gamma(M_i+k_i+1)}{\Gamma(M_i+k_i)} = \frac{M_i + k_i}{M + K} \tag{40}$$

a generalized form of Laplace's famous Rule of Succession; it has a strange history.

THE RULE OF SUCCESSION

Given by Laplace in the 18th Century, this rule came under scathing attack in the 19th Century from the philosopher John Venn (here in Cambridge, where his portrait can be seen in the Caius College Hall). Although the incident happened a long time ago, some comments about it are still needed because the thinking of Venn persists in much of the recent statistical literature.

With today's hindsight we can see that Venn suffered from a massive confusion over "What is the Problem?" Laplace derived the mathematical result as the solution of one problem. Venn (1866), not a mathematician, ignored his derivation -- which might have provided a clue as to what the problem is -- and tried to interpret the result as the solution to a variety of very different problems. Of course, he chose his problems so that Laplace's solution was indeed an absurd answer to every one of them.

Apparently, it never occurred to Venn that he himself might have misunderstood the circumstances in which the solution applies. R.A. Fisher (1956), pointed this out and expressed doubt as to whether Venn was even aware that Laplace's Rule had a mathematical basis and like other mathematical theorems had "stipulations specific for its validity".

Fisher's testimony is particularly cogent here, for he was an undergraduate in Caius College when Venn was still alive (Venn eventually became the President of Caius College), and they must have known each other. Furthermore, Fisher was himself an opponent of Laplace's methods; yet he is here driven to defending Laplace against Venn.

Indeed, it apparently never occurred to Venn that no single result -- Laplace's or anybody else's -- could possibly have provided the solution to all of the great variety of problems where he tried to use it. Yet we still find Venn's arguments repeated uncritically in some recent "orthodox" textbooks; so let the reader beware.

Now in the 1910's and 1920's Laplace's result became better understood by many: C.D. Broad, H. Jeffreys, D. Wrinch, and W.E. Johnson (all here in Cambridge also). Their work being ignored, it was rediscovered again in the 1930's by de Finetti, who added the important observation that the results apply to all exchangeable sequences. de Finetti's work being in turn ignored, it was partly rediscovered still another time by Carnap and Kemeny, whose work was in turn ignored by almost everybody in statistics, still under the influence of Venn.

It was only through the evangelistic efforts of I.J. Good and L.J. Savage in the 1950's and 1960's and D.V. Lindley in the 1960's and 1970's, that this exchangeability analysis finally became recognized as a respectable and necessary working part of statistics. Today, exchangeability is a large and active area of research in probability theory, much as Markov chains were thirty years ago.

We think, however, that the autoregressive models, in a sense intermediate between exchangeable and markoffian ones, that were introduced in the 1920's by G. Udny Yule (also here in Cambridge, and living in the same room that John Skilling now occupies), offer even greater promise for future applications.

In the 1980's, more than 200 years after Laplace started it, great mathematical generalizations are known but we are still far from understanding the useful range of application of exchangeability theory, because the problem of relating the choices to the circumstances is only now being taken seriously and studied as a technical problem of statistics, rather than a debating point for philosophers. Indeed, our present problem calls for better technical understanding than we really have at the moment. But at least the mathematics flows on easily for some distance more.

Thinking of the x_i as "real" parameters, we have a simple intuitive meaning of the hyperparameters $(k_1 \ldots k_n)$ if we denote the observed proportion of attribute i in the sampled population by $p_i = M_i/M$, and define a fictious prior proportion by $g_i = k_i/K$. Then (40) can be written

$$f_i(M,M_i) = \frac{Mp_i + Kg_i}{M + K} \qquad (41)$$

a weighted average of the observed proportion and an initial estimate of it. Thus we may regard $K = \Sigma k_i$ as the "weight" we attach to our prior information, measured in equivalent number of observations; i.e., the prior information I_D that leads to (30) has the same cogency as would K observations yielding the proportions $g_i = k_i/K$, starting from a state of complete ignorance about the x_i.

We may interpret the k's also in terms of the survey sampling problem. Starting from the prior information I_D and considering the data $(M_1 \ldots M_n)$ to be the result of a survey of $M<N$ kangaroos as in (24)-(26), what estimate should we now make of the proportion of kangaroos with attribute i? What accuracy are we entitled to claim for this estimate?

The answer is given by (26) with $L = N-M$, $L_i = N_i-M_i$. Substituting (30) into (26), sum out $(L_2 \ldots L_n)$ before doing the integrations using (29). The probability that exactly L_1 unsampled kangaroos have attribute 1 is found to be a mixture of binomial distributions:

$$p(L_1|DI) = \int_0^1 p(L_1|x)\ g(x)\ dx \tag{42}$$

where

$$p(L_1|x) = \frac{L!}{L_1!(L-L_1)!}\ x^{L_1}\ (1-x)^{L-L_1} \tag{43}$$

and a generating function

$$g(x) = \frac{\Gamma(b)}{\Gamma(a)\Gamma(b-a)}\ x^{a-1}\ (1-x)^{b-a-1} \tag{44}$$

Where $a = (M_1 + k_1)$, $b = (M+K)$. The first two factorial moments of (42) are then

$$\langle L_1\rangle = L \int_0^1 x\ g(x)\ dx = L\ a/b$$
$$\langle L_1(L_1-1)\rangle = L(L-1) \int_0^1 x^2\ g(x)\ dx \tag{45}$$

$$= L(L-1)\ \frac{a(a+1)}{b(b+1)} \tag{46}$$

from which the (mean) $\pm$ (standard deviation) estimate of the number L_1 of unsampled kangaroos with attribute 1 is

$$(L_1)_{est} = L\left[p \pm \sqrt{\frac{p(1-p)}{M+K+1}\ \frac{(M+K+L)}{N^2}}\right] \tag{47}$$

where $p = a/b = (M_1 + k_1)/(M+K)$. Comparing with (40) we see that the Rule of Succession has two different meanings; this estimated fraction p is numerically equal to the probability that the next kangaroo sampled will have attribute 1. As we have stressed repeatedly, such connections between probability and frequency always appear automatically, as a consequence of Bayesian theory, whenever they are justified.

Generally, the results of survey samplings are reported as estimated fractions of the total population N, rather than of the unsampled part $L = N-M$. Since $(N_1)_{est} = (L_1)est + M_1$, we find from (17), after a little algebra, the estimated fraction of all kangaroos with attribute 1:

$$(N_1/N)_{est} = p + p' \pm \sqrt{\frac{p(1-p)}{M+K+1} \frac{(N+K)}{N} \frac{(N-M)}{N}} \qquad (48)$$

where $p' = (Kp-k_1)/N$.

Examining the dependence of (48) on each of its factors, we see what Bayes' theorem tells us about the interplay of the size of the population, the prior information, and the amount of data.

Suppose we have sampled only a small fraction of the population, $M<<N$. If we also have relatively little prior information about the x_i, $K<<N$, the accuracy of the estimate depends basically on $(M+K+1)$, the number of actual observations plus the effective number implied by the weight of prior information; and depends little on N. Thus the "too good" estimates implied by (18) as $N \to \infty$ are now corrected.

But if $K>>N$ (the monkey multiplicity factor limit), the accuracy goes into the limiting form $p(1-p)/N$ and a result like (18) is recovered. The changeover point from one regime to the other is at about $K=N$. Note, however, that (48) is not directly comparable to (18) because in (18) we used Steve Gull's data on kangaroos to restrict the sample space before introducing probabilities.

Now suppose we have sampled an appreciable fraction of the entire population. Our estimates must perforce become more accurate, and the $(N-M)/N = 1 - (M/N)$ factor so indicates. When we have sampled the entire population, $M=N$, then we know the exact N_1, so the error vanishes, the prior information becomes irrelevant, and the RHS of (48) reduces to $M_1/M \pm 0$, as it should.

Thus if we admit the x_i as real quantities, so that it makes sense to apply Bayes' theorem in the way we have been doing, then Bayes' theorem tells us in quantitative detail -- just as it always does -- what our common sense might have perceived if our intuition was powerful enough.

THE NEW KANGAROO SOLUTION

We started considering Steve Gull's kangaroo problem on the original monkey hypothesis space H1, were somewhat unhappy at the result (18), and have now seen some of the general consequences of going down into H2. How does this affect the answer to the original kangaroo problem, particularly in the region of large N where we were unhappy before?

When the N_i and k_i are large enough to use the Stirling approximation for all terms, a typical term in the exchangeable prior (31) goes into the form

$$L = \log\,[\Gamma(N+k)/N!\,\Gamma(k)] \sim \log\,[(N+h)^{N+h}/N^N h^h] + \text{const.} \quad (49)$$

where $h = k - (1/2)$. Thus, when N and k are quite different we have for all practical purposes

$$L \sim \begin{bmatrix} N\log(ke/N), & N << k \\ k\log(Ne/k), & k << N \end{bmatrix} \quad (50)$$

So if $N_i << k_i$, call it prior information I_2, the prior (31) is given by

$$\log p(N_1 \ldots N_n | I_2) \sim - \Sigma N_i \log(N_i/k_i) + \text{const.} \quad (51)$$

and the most probable sample numbers $(N_1 \ldots N_n)$ subject to any data D that imposes a "hard" constraint on them, are the ones that maximize the entropy relative to the "prior prejudice" (k_i/K). With no prior prejudice, $k_i = k$, this will just lead us back to the original solution (18) from the pure monkey multiplicity factors, confirming again that the region of large k is the one that connects smoothly to the previous solution.

When N>>k, call it prior information I_2', instead of (51) we have the limiting form

$$\log p(N_1 \ldots N_n | I_2') \sim \Sigma k_i \log N_i + \text{const.} \quad (52)$$

and the solution will be the one that maximizes this expression, which resembles the "Burg entropy" of spectrum analysis.

So applying Bayes' theorem with n = 4, the exchangeable prior (52) and Steve Gull's hard constraint data

$$D:\ N_1 + N_2 = N_1 + N_3 = 3N/4$$

the posterior probability of the parameter $q = N_4/N$ can be read off from (12):

$$p(q|DI_2') \propto (0.5 + q)^{k_1} (0.25 - q)^{k_2+k_3} q^{k_4} \tag{53}$$

When all $k_i = k$, this is proportional to

$$p(q|DI_2') \propto (q - 6q^2 + 32q^4)^k \tag{54}$$

This reaches its peak at $q = 0.0915$, and yields the (mean) $\pm$ (standard deviation) estimate

$$q(est) = 0.0915(1 \pm 0.77/\sqrt{k}). \tag{55}$$

The "too good" estimate (18) where we had the accuracy factor $(1 \pm 3/\sqrt{N})$, is indeed corrected by this prior information on H2; however large N, the accuracy cannot exceed that corresponding to an effective value

$$N_{eff} = (3/.77)^2\ k = 15.2\ k = 3.8\ K\ . \tag{56}$$

These comparisons have been quite educational; we had from the start the theorem that maximizing any quantity other than entropy will introduce correlations in the 2x2 table (12), for which there is no evidence in the data D. That is, starting from the pure monkey solution with $q = 1/16$, learning that one kangaroo is left handed makes no difference; the odds on his being a Foster's drinker remains 3:1.

But now, admitting the possibility of a positive correlation between kangaroos must, from the theorem, induce some correlation between their attributes. With the new solution, q is increased to about 1/11; so learning that a kangaroo is left-handed increases the odds on his being a drinker to 3.73:1; while learning that he is right-handed reduces them to only 1.73:1.

At this point our intuition can again pass judgment; we might or might not be happy to see such correlations. Our first analysis of the monkey rationale on H1 was a mental pump-priming that made us aware of relevant information (correlations between kangaroos) that the monkey rationale did not recognize, and led us down into H2. Now the analysis on H2 has become a second mental pump-priming that suddenly makes us aware of still further pertinent prior information that we had not thought to use, and leads us down into H3.

When we see the consequences just noted, we may feel that we have over-corrected by ignoring a nearness effect; it is relevant that correlations between kangaroos living close together must be stronger than between those at opposite ends of the Austral continent. In the U.S.A. there are very marked differences in the songs and other behavior of birds of the same species, living in New Hampshire and Iowa. But an exchangeable model insists on placing the same correlations between all individuals.

In image reconstruction, we feel intuitively that this nearness effect must be more important than it is for kangaroos; in most cases we surely know in advance that correlations are to be expected between nearby pixels, but not between pixels far apart. But in this survey we have only managed to convey some idea of the size of the problem. To find the explicit hypothesis space H3 on which we can express this prior information, add the features that the data are noisy and N is unknown; and work out the quantitative consequences, are tasks for the future.

CONCLUSION: RESEMBLANCE TO THE TRUTH

However far we may go into deeper spaces, we can never escape entirely from the original monkey multiplicity factors, because counting the possibilities is always relevant to the problem, whatever other considerations may also be relevant. Therefore, however you go at it, when you finally arrive a satisfactory prior, you are going to find that monkey multiplicity factor sitting there, waiting for you. This is more than a mere philosophical observation, for the following reason.

In image reconstruction or spectrum analysis, if entropy were not a factor at all in the prior probability of a scene, then we would expect that MAXENT reconstructions from sparse data, although they might be "preferred" on other grounds, would seldom resemble the true scene or the true spectrum.

This would not be an argument against MAXENT in favor of any alternative method, for it is a theorem that no alternative using the same information could have done better. Resemblance to the truth is only a reward for having used good and sufficient information, whether it comes from the data or the prior. If the requisite information is lacking, neither MAXENT nor any other method can give something for nothing.

But if the MAXENT reconstruction seldom resembled the truth, neither would we have a very good argument for MAXENT in preference to alternatives; there would be small comfort in the admittedly correct value judgment that MAXENT was the only consistent thing we could have done.

More important, the moral of our Sermons on this in the Tutorial was that if such a discrepancy should occur, far from being a calamity, it might enable us to repeat the Gibbs scenario and find a better hypothesis space. In many cases, empirical evidence on this resemblance to the truth or lack of it for image reconstruction can be obtained.

It might be thought that there is no way to do this with astronomical sources, since there is no other independent evidence. For an object of a previously uncharted kind, this is of course true, but we already know pretty well what galaxies look like. If Roy Frieden's MAXENT reconstruction of a galaxy was no more likely to be true than any other, then would we not expect it to display any one of a variety of weird structures different from spiral arms?

We need hardly ask whether MAXENT reconstructions of blurred auto license plates do or do not resemble the true plates, or whether MAXENT tomographic or crystal structure reconstructions do or do not resemble the true objects. If they did not, nobody would have any interest in them.

The clear message is this: if we hold that entropy has no role in the prior probability of a scene, but find that nevertheless the MAXENT reconstructions consistently resemble the true scene, does it not follow that MAXENT was unnecessary? Put differently, if any of the feasible scenes is as likely to be true as the MAXENT one, then we should expect any feasible scene to resemble the truth as much as does the MAXENT one; resemblance to the truth would not be ascribable to the use of MAXENT at all.

It seems to us that there is only one way this could happen. As the amount of data increases, the feasible set contracts about the true scene, and we might conjecture (by analogy with John Parker Burg's shrinking circles for reflection coefficients in spectrum analysis) that eventually all the feasible scenes would resemble the true one very closely, making MAXENT indeed superfluous; any old inversion algorithm, such as the canonical generalized inverse matrix $R = A^T(AA^T)^{-1}$ for eq. (1), would do as well. If so, how much data would we need to approach this condition?

In March 1984 the writer found, in a computer study of a one-dimensional image reconstruction problem, that when the number of constraints was half the number of pixels the feasible set had not contracted very much; it still contained a variety of wildly different scenes, having almost no resemblance to the true one. The canonical inverse (which picks out the feasible scene of minimum Σf_i^2) was about the wildest of all, grossly underestimating every pixel intensity that was not forced to be large by the data, and having no aversion to negative estimates had the program allowed them.

So this amount of data still seems "sparse" and in need of MAXENT; any old algorithm would have given any old result, seldom resembling the truth. Perhaps the conjecture is wrong; more ambitious computer studies and analytical work will be needed to understand this.

To say: "The MAXENT reconstruction is <u>no more likely</u> to be true than any other" can be misleading to many, including this writer, because it invites us to interpret "likely" in the colloquial sense of the word. After months of puzzlement over this statement, I finally learned what John Skilling meant by it, through some close interrogation just before leaving Cambridge. Indeed, it requires only a slight rephrasing to convert it into a technically correct statement: "The MAXENT reconstruction has <u>no more likelihood</u> than any other with equal or smaller Chisquared." Then it finally made sense.

The point is that "likelihood" is a well-defined technical term of statistics. What is being said can be rendered, colloquially, as "The MAXENT reconstruction is not indicated by the data alone any more strongly than any other with equal or smaller Chisquared." But that is just the statement that we are concerned with a generalized inverse problem, from whence we started.

In any such problem, a specific choice within the feasible set must be made on other considerations than the data; prior information or value judgments. Procedurally, it is possible to put the entropy factor in either. The difference is that is we consider entropy only a value judgment, it is still "preferred" on logical consistency grounds, but we have less reason to expect that our reconstruction resembles the true scene because we have invoked only our wishes, not any actual information, beyond the data.

In my view, the MAXENT reconstruction is far more "likely" (in the colloquial sense of that word) to be true than any other consistent with the data, precisely because it does take into account some highly cogent prior information in addition to the data. MAXENT images and spectrum estimates should become still better in the future, as we learn how to take into account other prior information not now being used.

Indeed, John Skilling's noting that bare MAXENT is surprised to find isolated stars, but astronomers are not; and choosing "prior prejudice" weighting factors accordingly, has already demonstrated this improvement.

Pragmatically, all views about the role of entropy seem to lead to the same actual class of algorithms for the current problems; different views have different implications for the future. For diagnostic purposes in judging future possibilities it would be a useful research project to explore the full feasible set very carefully to see just how great a variety of different scenes it holds, how it contracts with increasing data, and whether it ever contracts enough to make MAXENT unnecessary as far as resemblance to the truth is concerned. We conjecture that it will not, because as long as $m<n$ it has not contracted in all directions; i.e., the coordinates $(a_{m+1} \cdots a_n)$ of Eq. (8) remain undetermined by the data, and confined only by nonnegativity.

In the meantime, we think there is still some merit in monkeys, and no one needs to be apologetic for invoking them. If they are not the whole story, they are still relevant and useful, providing a natural starting point from which to construct a realistic prior. For very fundamental reasons they will continue to be so.

APPENDIX A: PSYCHOLOGICAL TESTS

Kahneman and Tversky (1973) report on tests in which subjects were given the prior information: I = "In a certain city, 85% of the taxicabs are blue, 15% green"; and then the data: D = "A witness to a crash who is 80% reliable (i.e., who in the lighting conditions prevailing can distinguish correctly green and blue 80% of the time) reports that the cab involved was green." The subjects are then asked to judge the probability that the cab was actually blue.

From Bayes' theorem, the correct answer is

$$p(B|DI) = .85 \times .2/(.85 \times .2 + .15 \times .8) = 17/29 = .59$$

This is easiest to reason out in one's head in terms of odds; since the statement of the problem told us that the witness was equally likely to err in either direction (G -> B or B -> G), Bayes' theorem reduces to simple multiplication of odds. The prior odds in favor of blue are 85:15, or nearly 6:1; but the odds on the witness being right are only 80:20 = 4:1, so the posterior odds on blue are 85:60 = 17:12.

Yet the subjects in the test tended to guess $p(B|DI)$ as about .2, corresponding to odds of 4:1 in favor of green, thus ignoring the prior information. For them, "the data come first" with a vengeance, even though the prior information implies many more observations than the single datum.

The opposite error -- clinging irrationally to prior opinions in the face of massive contrary evidence -- is equally familiar to us; that is the stuff of which fundamentalist religious/political stances are made. The field is reviewed by Donmell and Du Charme (1975). It is perhaps not surprising that the intuitive force of prior opinions depends on how long we have held them.

Persons untrained in inference are observed to commit wild irrationalities of judgment in other respects. Slovic et al (1977) report experiments in which subjects, given certain personality profile information, judged the probability that a person is a Republican lawyer to be greater than the probability that he is a lawyer.

Hacking (1984) surveys the history of the judicial problem and notes that the Bayesian probability models of jury behavior given by Laplace and long ignored, account very well for the performance of modern English juries. L.J. Cohen (1984) reports on controversy in the medical profession over whether one should, in defiance of Bayesian principles, test first for rare diseases before common ones.

Such findings not only confirm our worst fears about the soundness of jury decisions, but engender new ones about medical decisions. These studies have led to proposals -- doubtless 100 years overdue -- to modify current jury systems. The services of some trained Bayesians are much needed wherever important decisions are being made.

REFERENCES

1 Boltzmann, L. (1877). Wiener Berichte, 76, p. 373.

2 Chambers, J.M. (1977). Computational Methods for Data Analysis. J. Wiley & Sons, Inc., New York.

3 Cohen, L.J. (1984). Epistemologia, VII, Special Issue on Probability, Statistics, and Inductive Logic, pp. 68-71 and 213-222.

4 Donmell, M.L. and W.M. du Charme (1975). The Effect of Bayesian Feedback on Learning in an Odds Estimation Task. Organizational Behavior and Human Performance, 14, 305-313.

5 Einhorn, H.J. and R.M. Hogarth (1981), "Behavioral Decision Theory: Processes of Judgment and Choice". Annual Review of Psychology, 32, pp. 53-88.

6 Fisher, R.A. (1956). Statistical Methods and Scientific Inference. Hafner Publishing Co., New York.

7 Good, I.J. (1965). The Estimation of Probabilities. Research Monographs #30, MIT Press, Cambridge, Mass.

8 Gull, S.F. and G.J. Daniell (1978). Image Reconstruction with Incomplete and Noisy Data. Nature, 272, 686.

9 Gull, S.F. and J. Skilling (1984). The Maximum Entropy Method. In Indirect Imaging, J.A. Roberts, ed., Cambridge University Press, U.K.

10 Hacking, Ian (1984). Historical Models for Justice. Epistemologia, VII, Special Issue on Probability, Statistics, and Inductive Logic, pp. 191-212.

11 Hardy, G.F. (1889). Letter in Insurance Record, p. 457. Reprinted in Trans. Fac. Actuaries, 8, 180 (1920).

12 Jaynes, E.T. (1982). On the Rationale of Maximum-Entropy Methods. Proc. IEEE, 70, pp. 939-952.

13 Johnson, W.E. (1924). Logic, Part III: The Logical Foundations of Science. Cambridge University Press. Reprinted by Dover Publishing Co., 1964.

14 Johnson, W.E. (1932). Probability, the Deduction and Induction Problems. Mind, 44, pp. 409-413.

15 Kahneman, D. and A. Tversky (1973). On the Psychology of Prediction. Psychological Review 80, pp. 237-251.

16 Kyburg, H.E. and H.I. Smokler (1981). Studies in Subjective Probability. 2nd Edition, J. Wiley & Sons, Inc., New York.

17 Laplace, P.S. (1778). Mem. Acad. Sci. Paris, 227-332; Oeuvres Completes, 9, 383-485.

18 Slovic, P., B. Fischhoff and S. Lichtenstein (1977). Behavioral Decision Theory. Annual Review of Psychology, 28, pp. 1-39.

19 Venn, John (1866). The Logic of Chance. MacMillan & Co., London.

20 Zabell, S. (1982). W.E. Johnson's 'Sufficientness Postulate'. Annals of Statistics, 10, 1091-1099.

THE THEORY AND PRACTICE OF THE MAXIMUM ENTROPY FORMALISM

R.D. Levine
The Fritz Haber Research Center for Molecular Dynamics
The Hebrew University
Jerusalem 91904, Israel

1 INTRODUCTION AND OVERVIEW

We consider three aspects of the maximum entropy formalism [1-3]. Our purpose is to dispel the three more common objections raised against the rationale and results of the approach. To do so we restrict the scope of the formalism: We consider only such experiments that can be repeated N, (N not necessarily large), times.

(a) Consistent inference: The probabilities determined using the maximum entropy formalism are shown to have the interpretation of the mean frequency. Their value is independent of the number, N, of repetitions of the experiment. What very much does depend on N is the variance of the frequency. The larger N is, the smaller is the variance and the less likely are the actual, observed, frequencies to deviate from the mean. Here (following [4]), we shall show that the maximum entropy formalism does have the stated consistency property. Elsewhere [5,6] we have shown that it is the only algorithm with that property. In Ref. 6 there are additional arguments which are also based on the need for consistency of predictions in reproducible experiments.

The maximum entropy approach dates at least as far back as Boltzmann [7]. He showed that, in the $N \to \infty$ limit, the maximum entropy formalism determines the most probable frequencies. Ever since, the approach has been plagued by the criticism that it is only valid in the $N \to \infty$ limit. The present [4-6] results should put an end to such arguments. What we show here is the following. Consider an experiment with n mutually exclusive and collectively exhaustive outcomes. Let there be many experimentalists. Each one repeats the experiment N times and thereby observes the number, N_i, $\Sigma N_i = N$, of times that the specific outcome i $i=1,\ldots,n$, occurs. For any finite N, the frequencies $f_i = N_i/N$ reported by the different experimentalists can be somewhat different. Our best bet is then to average the results of the different experimentalists. The rationale here is the same as always - the average is the best estimate (in the sense of least square error). What is provided by the maximum entropy formalism is shown (in Section 2) to be precisely that average frequency $\langle f_i \rangle$. This conclusion is independent of the number N of repetitions of the experiment by any individual experimentalist. What will depend on N is the extent to which the results obtained by any particular experimentalist deviate from the mean (over all experimentalists). Such deviances are shown to diminish as N increases. We need a large N not for the validity of the predicted probabilities but only for an agreement between the measured frequency and the theoretical probability. Moreover, for any finite N, we can compute the variance (and higher moments) of the predicted results.

Keeping a clear distinction between theoretical probabilities and measured frequencies allows one also to consider other sources (beside the finite value of N) of uncertainty in the input (the 'constraints'). Some remarks on that point are provided in Section 3 and elsewhere [6,8,9].

There are two additional criticisms, both related to the 'subjective' label which is sometimes attached to the formalism, which are refuted by the present point of view. The first objection concerns the use of entropy as a measure of 'uncertainty' or of 'missing information'. The question is why should the description of a system which can be measured in somebody's laboratory be dependent in any way on the uncertainty of the person doing the prediction. We have shown elsewhere [5,6] that one need not invoke any extremum property in deriving the maximum entropy formalism. One can, of course, incorporate the axioms used (e.g., in [10]) to define entropy as part of a set of axioms which are then used to characterize the formalism. But the approach used in the past (e.g., in [11]) did assume that some functional need be extremized. It does not therefore address the objection. The second objection in the same vein is more subtle. Constraints are central to the maximum entropy formalism. We are to maximize the entropy subject however to constraints. Nowhere however is the origin of the constraints spelled out. Different people may therefore legitimately impose entirely different constraints. The approach is subjective in that you impose your own constraints. That is not quite possible in the present approach. Constraints are (sample) average values as measured by any one of the experimentalists mentioned above. This point is further discussed in Ref. 6.

(b) The adjoint problem: One of the central applications of the maximum entropy formalism is to thermodynamics [12,13]. Yet in classical thermodynamics the constraints and their Lagrange multipliers appear on a much more 'symmetric' footing. One considers, for example [14], the fluctuation in the energy or the fluctuation in the corresponding Lagrange multiplier which in this case is known as the temperature. This complementarity is absent in the strict maximum entropy formalism. Given are the constraints and from them one is to compute the Lagrange multipliers which will have sharp values. That there is indeed a conjugate relationship between the constraints and their multipliers is shown in Section 3. The approach presented therein can also be used to advantage as a computational tool and to place error bars (due to experimental uncertainty in the values of the constraints) on the values of the Lagrange multipliers or vice versa.

(c) Novel applications: The claim is frequently made that in the natural sciences the maximum entropy point of view is a useful didactic approach in the teaching of statistical mechanics but that it has provided no new results. For over a dozen years we have provided many diverse applications to problems as far removed from equilibrium thermodynamics as one could possibly expect [15-18]. Both experimental results and computer simulation were used for comparison with the results of the maximum entropy formalism. A survey of such applications is given in Section 4.

The road ahead. There are (at least) two major problems to tackle before one could be satisfied that the formalism is ready to address most of the relevant questions which can be raised about it in the natural sciences. The two that we identify are:

(a) Chaos and the origin of the constraints. The discussion of Section 2 and elsewhere [4-6] shows that the 'entropy' in the maximum entropy formalism comes in not via the 'physics' of the problem but from consistency considerations for a reproducible experiment. The physics is introduced via the constraints. To understand 'why' the formalism works we have to understand the constraints rather than the entropy. What we have to ask is why is it that in many interesting and relevant situations very few constraints completely govern the distribution. To show that this is the valid question we have shown that with the proper choice of constraints, the maximum entropy formalism is strictly equivalent to solving the problem exactly using quantum (or classical) mechanics [17-20]. We hasten to caution the reader that the statement above is not trivial. The number of constraints required can be much smaller than the number required to uniquely determine the state of the system. What determines then the number of required constraints? Two aspects: The first is the initial state [19,20,21]. To see this consider the evolution in time of a (closed or open) system which is not in equilibrium. The subsequent evolution must surely depend on the initial state. A limiting case is a system at equilibrium. The very few constraints which specify the initial state will continue to suffice to specify the system at any future time. Now let's displace the system somewhat from its equilibrium distribution. Does it now follow that very many additional constraints are required to specify its evolution? For some concrete examples that this is not the case see Refs. 17-20.

The second aspect which governs the number of effective constraints is the nature of the interaction. Even fully deterministic (e.g., Hamiltonian) systems can have a chaotic time evolution. Speaking loosely, this means that details of the initial state get erased in the course of the time evolution [22]. Unless the initial state is specified with infinite precision then fewer details of the initial state remain relevant as time progresses. The stability of the initial constraints and their relevance to predictions at later times is a subject of our current research [23,24].

(b) Interacting systems. The technical discussion in this review is limited to such experiments where the individual repetitions are independent of one another. (So called, Bernouli trials.) Say our interest is in the distribution of states of a molecule in a non ideal gas. Then the molecule is not independent of its neighbours and the entire volume of gas in the system. The outcomes in any particular repetition refer therefore to the states of the entire gas rather than to the states of a single molecule. Yet on practical grounds what we really wish to know are the distribution of the latter rather than of the former. Proving that the states of the whole gas have a distribution of maximal entropy is therefore not quite enough. Assuming that the molecules are hardly interacting simply avoids the problem. What is missing is a theory for a subsystem of a system whose entropy is maximal.

2 REPRODUCIBLE EXPERIMENTS

The purpose of this section is twofold: To show that the maximum entropy probabilities are the mean frequencies in a reproducible experiment and to show that the inference is consistent [4]. That it is the only consistent inference is shown elsewhere [5,6].

Consider N sequential but independent repetitions of an experiment which has n mutually exclusive and collectively exhaustive alternatives. (We shall refer to these n alternatives as states.) The result is a sequence of N entries, each entry being the index i, i=1,...,n of the particular state that did occur. The sequence can be thought of as a word of length N in an alphabet of n letters.

Different experimentalists may observe different sequences. Designating a particular sequence by $\underset{\sim}{N}$, our problem is to determine its probability, $P(\underset{\sim}{N})$. Given that probability we can compute the mean number of times, $\langle N_i \rangle$,

$$\langle N_i \rangle = \sum_{\underset{\sim}{N}} N_i P(\underset{\sim}{N}) \tag{2.1}$$

that the i'th outcome occurred in N repetitions. Summation in (2.1) is over all possible sequences. Note that any possible sequence must satisfy

$$\sum_{i=1}^{n} N_i = N \tag{2.2}$$

and that there will be $g(\underset{\sim}{N})$,

$$g(\underset{\sim}{N}) = N!/\prod_{i=1}^{n} N_i! \tag{2.3}$$

distinct sequences which correspond to a given distribution $\{\underset{\sim}{N}\} \equiv (N_1,\ldots,N_n)$ of events. Of course,

$$\sum_{\underset{\sim}{N}} P(\underset{\sim}{N}) = 1 . \tag{2.4}$$

Let p_i be the probability that event i might occur. For a reproducible experiment we take it that

$$p_i = \langle N_i \rangle / N . \tag{2.5}$$

Those in favor of a frequency interpretation of probability [25], will regard (2.5) as essentially self evident. To see this recall that when the probabilities are given then [25]

$$P(\underset{\sim}{N}) = g(\underset{\sim}{N}) \prod_{i=1}^{n} p_i^{N_i} . \tag{2.6}$$

The equality in (2.5) follows then directly from the definitions (2.1) and (2.3) and the multinomial theorem (cf. (2.4))

$$\begin{aligned} \left(\sum_{i=1}^{n} p_i\right)^N &= \sum_{\{\underset{\sim}{N}\}} g(\underset{\sim}{N}) \prod_{i=1}^{n} p_i^{N_i} \\ &= \sum_{\underset{\sim}{N}} \prod_{i=1}^{n} p_i^{N_i} . \end{aligned} \tag{2.7}$$

It is important however to note that one can also understand (2.5) as the mean being the best estimate in the sense of least square error.

2.1 Equality of the mean and the most probable inference

The probabilities p_i are not however given. Hence one cannot use (2.6). What are given are m expectation values $\langle A_r \rangle$,

$$\langle A_r \rangle = \sum_{i=1}^{n} A_{ri} p_i, \qquad r=1,\ldots,m. \tag{2.8}$$

of the observables A_r which obtain the value A_{ri} for the state i. Since $m \leq n-1$ and typically $m \ll n$, (2.8) cannot be inverted so as to uniquely determine the probabilities.

In view of (2.5) we can write

$$N\langle A_r \rangle = \sum_{\underset{\sim}{N}} \left(\sum_{i=1}^{n} A_{ri} N_i\right) P(\underset{\sim}{N}) \tag{2.9}$$

In other words, $\langle A_r \rangle$ is, in a reproducible experiment, equal to the mean of the 'sample average' $\bar{A}_r$,

$$\bar{A}_r = \frac{1}{N}\sum_{i=1} N_i A_{ri} . \tag{2.10}$$

Here $\bar{A}_r$ is the average value of the observable A_r as measured by some particular experimentalist. (2.9) is the condition that $\langle A_r \rangle$ equals the value of $\bar{A}_r$ averaged over all experimentalists,

$$\langle A_r \rangle = \sum_{\underset{\sim}{N}} \bar{A}_r P(\underset{\sim}{N}), \quad r=1,\ldots,m. \tag{2.11}$$

We determine $P(\underset{\sim}{N})$ by the procedure of maximum entropy subject to normalization, (2.4), and the m constraints (2.11). The result, as usual, is

$$\begin{aligned} P(\underset{\sim}{N}) &= \exp(-\mu_0 - \sum_{r=1}^{m} \mu_r \bar{A}_r) \\ &= \exp\{-\mu_0 - \sum_{r=1}^{m}\mu_r[\sum_{i=1}^{n} A_{ri}(N_i/N)]\} . \end{aligned} \tag{2.12}$$

Here μ_0 is the Lagrange multiplier for the normalization constraint and is a function (determined by the normalization condition (2.4)) of the other m Lagrange multipliers. Explicitly

$$\begin{aligned} \exp(\mu_0) &= \sum_{\underset{\sim}{N}}\exp\{-\sum_{r=1}^{m}\mu_r[\sum_{i=1}^{n} A_{ri}(N_i/N)]\} \\ &= \sum_{\{\underset{\sim}{N}\}} g(\underset{\sim}{N})\exp\{-\sum_{r=1}^{m} \mu_r[\sum_{i=1}^{n} A_{ri}(N_i/N)]\} . \end{aligned} \tag{2.13}$$

Note that in the second line of (2.13) we have made explicit use of the observation that $P(\underset{\sim}{N})$ as given by (2.12) is independent of the <u>order</u> of the N repetitions. $P(\underset{\sim}{N})$ depends only on the number, N_i, of times that event i, i=1,...,n, occurred. All sequences which correspond to the same set of 'occupation numbers' $\{\underset{\sim}{N}\} \equiv (N_1,\ldots,N_n)$ have therefore the same probability. Rather than considering the probability $P(\underset{\sim}{N})$ of a sequence we can consider the probability $P(\{\underset{\sim}{N}\})$ of the distribution

$$\begin{aligned} P(\{\underset{\sim}{N}\}) &= g(\{\underset{\sim}{N}\})P(\underset{\sim}{N}) \\ &= g(\{\underset{\sim}{N}\})\exp\{-\mu_0 - \sum_{r=1}^{m} \mu_r[\sum_{i=1}^{n} A_{ri}(N_i/N)]\}. \end{aligned} \tag{2.14}$$

It is important to note that the dependence of $P(\underset{\sim}{N})$ only on the distribution and not on the order is a direct result of our assumption of independent repetitions of the experiment. Elsewhere [4] and in future publications we consider the more general case where the events are correlated.

To evaluate the 'partition function' exp (μ_0) we define

$$\gamma_i \equiv \frac{1}{N} \sum_{r=1}^{m} \mu_r A_{ri} \ . \tag{2.15}$$

Then (2.13) is just a multinomial expansion (cf. (2.7))

$$\exp(\mu_0) = \sum_{\{N\}} g(\underset{\sim}{N}) \prod_{i=1}^{n} [\exp(-\gamma_i)]^{N_i} \tag{2.16}$$

or

$$\exp(\lambda_0) = \sum_{i=1}^{n} \exp(-\gamma_i)$$

$$= \sum_{i=1}^{n} \exp(- \sum_{r=1}^{m} \lambda_r A_{ri}) \ . \tag{2.17}$$

Here

$$\lambda_r = N^{-1} \mu_r \qquad r = 0,1,\ldots,m. \tag{2.18}$$

so that

$$P(\underset{\sim}{N}) = \exp(-N\lambda_0 - \sum_{i=1}^{n} N_i \gamma_i) \tag{2.19}$$

with

$$\exp(N\lambda_0) = \sum_{\underset{\sim}{N}} \exp(- \sum_{i=1}^{n} N_i \gamma_i) \ . \tag{2.20}$$

It follows from (2.17), (2.19) and (2.20) that

$$\begin{aligned} \frac{\langle N_i \rangle}{N} &\equiv \frac{1}{N} \sum_{\underset{\sim}{N}} N_i \, P(\underset{\sim}{N}) \\ &= -\partial\lambda_0/\partial\gamma_i \\ &= -\exp(-\lambda_0)\,\partial\exp(\lambda_0)/\partial\gamma_i \\ &= \exp(-\lambda_0-\gamma_i) \ . \end{aligned} \tag{2.21}$$

The final result for the mean occupation number (2.21), is (using (2.15) and (2.18))

$$\langle N_i\rangle/N = \exp(-\lambda_0 - \sum_{r=1}^{m} \lambda_r A_{ri}) \ . \tag{2.22}$$

But (2.22) is precisely the result of directly determining p_i as the normalized probability distribution subject to the m constraints (2.8). Determining the probability of the different events by maximizing the entropy of the distribution $\{p_i\}$ subject to m constraints (given by 2.8)) yields the very same values as determining the p_i's as the mean frequency. That the values of the Lagrange multipliers are the same follows from p_i and $\langle f_i\rangle = \langle N_i\rangle/N$ being consistent with the same data. Since

$$\sum_i p_i \ln(p_i/\langle f_i\rangle) \geqq 0 \tag{2.23}$$

with equality iff $p_i \equiv \langle f_i\rangle$ it follows, using (2.22) in (2.23), that the p_i's obtained directly by the procedure of maximum entropy are not only equal to the $\langle f_i\rangle$'s but are the unique and only distribution which satisfies the m constraints (2.8) and which equals the $\langle f_i\rangle$'s. Q.E.D.

2.2 The variance

Given the distribution $P(\underset{\sim}{N})$ of the results obtained by different experimentalists one can compute not only $\langle N_i\rangle$ but other averages as well. Foremost amongst those is the variance

$$\sigma_i^2 = (\langle N_i^2\rangle - \langle N_i\rangle^2)/N^2 \ . \tag{2.24}$$

Proceeding as in the derivation of (2.21) one readily establishes that

$$\begin{aligned} N\sigma_i^2 &= -\partial^2\lambda_0/\partial\gamma_i^2 \\ &= p_i(1-p_i) \ . \end{aligned} \tag{2.25}$$

In contrast to the probabilities p_i whose value is independent of the value of N, the variance does very much depend on N. The dependence is in the expected direction, with N_i having a sharper distribution about $\langle N_i\rangle$ as N increases. For a finite value of N it is possible, for a particular experimentalist to observe a value of N_i which differs from $\langle N_i\rangle$. But his probability of doing so diminishes (by the Chebychev inequality, see for example [25]) as N increases. The law of large numbers is not required to conclude that $p_i = \langle N_i\rangle/N$. That is true for any value of N. Where the law is required is for the secondary conclusion that the sample frequency approaches the probability as N increases.

One can equally consider the variance in other observables. For example, from (2.11), (2.12) and (2.18),

$$\begin{aligned}\sigma_r &= \sum_{\underset{\sim}{N}} (\bar{A}_r - \langle\bar{A}_r\rangle)^2 P(\underset{\sim}{N}) \\ &= -\partial^2\mu_0/\partial\mu_r^2 \\ &= -N^{-1}\partial^2\lambda_0/\partial\lambda_r^2 \\ &= N^{-1}\sum_i (A_{ri} - \langle A_r\rangle)^2 p_i \ .\end{aligned} \tag{2.26}$$

Since the summation on the right hand side of (2.26) is independent of N and hence remains finite as $N \to \infty$, we have that $\sigma_r \to 0$ as 1/N. That is a key practical conclusion since it enables us in the large N limit to use the m sample averages $\bar{A}_r$ (defined in (2.10)) in place of the m averages $\langle A_r\rangle$ to compute the numerical values of the Lagrange multipliers. See also [26].

In practice we almost always have only sample averages. We must therefore be prepared for the possibility (further discussed in Sec. 3.4 below) that our input data (i.e., the values of the constraints) has some experimental uncertainty. What (2.26) guarantees is that as the sample size increases, using the observed sample averages is an increasingly better approximation.

A special case of our considerations is when the experimentalist actually measures the frequency of the different states. Then the number m of observables equals the number n of states $A_{ri} = \delta_{r,i}$. Then $\bar{A}_i$, the sample average (cf. (2.10)) is obviously just $f_i = N_i/N$, the sample frequency and $\langle A_i\rangle = \langle N_i\rangle/N$. It might appear then that there is no need for the maximum entropy formalism for there are enough equations in (2.8) to solve for a unique probability distribution p_i. On the one hand that is indeed so but on the other, it is not. By using the mean frequencies $\langle N_i\rangle/N$ as constraints on the distribution $P(\underset{\sim}{N})$ we get more than just the p_i's. We get, for example, the variance of the p_i's and hence of the mean frequency.

Finally, we turn to the entropy itself. Consider the entropy per repetition

$$S = -N^{-1}\sum_{\underset{\sim}{N}} P(\underset{\sim}{N})\,\ell n P(\underset{\sim}{N}) \tag{2.27}$$

For the distribution $P(\underset{\sim}{N})$ of maximum entropy we have, using (2.19) and

$$\begin{aligned}S &= \lambda_0 + \sum_{i=1}^{n} \gamma_i \langle N_i\rangle/N \\ &= \lambda_0 + \sum_{r=1}^{m} \lambda_r \sum_{i=1}^{n} A_{ri}\langle N_i\rangle/N \\ &= \lambda_0 + \sum_{r=1}^{m} \lambda_r \langle A_r\rangle .\end{aligned} \tag{2.28}$$

The result (2.28) is also the entropy of the distribution $p_i = \langle N_i \rangle / N$. In other words, the maximum entropy inference insures that the entropy per repetition of the experiment equals (for any N) the entropy of the distribution of states.

The result (2.28) for the maximal entropy enables us to compare it with the information provided by a particular experimentalist [27]. Consider a measurement (by N repetitions) of a set of occupation numbers $\{N_i\}$. The information provided (per repetition) is (using (2.12) and (2.18))

$$
\begin{aligned}
-N^{-1} \ell n P(\underset{\sim}{N}) &= -\sum_i (N_i/N) \ell n p_i \\
&= \sum_{r=0}^{m} \lambda_r \sum_{i=1}^{n} A_{ri} (N_i/N) \qquad (2.29) \\
&= \sum_{r=0}^{m} \lambda_r \overline{A}_r \quad .
\end{aligned}
$$

Here any particular experiment leads to an entropy which upon averaging over all experiments yields (2.28). The difference between (2.29) and (2.30) is $O(N^{-1})$.

2.3 Discussion

The derivation of Sec. 2.1 showed an equivalence between two routes for the determination of the probabilities, p_i, of n states given m average values $\langle A_r \rangle$ measured in a reproducible experiment which was repeated N times. In one route the p_i's were computed as the average frequency. Note that since N is finite there is, in a sense, a 'double' average. First we consider one experimentalist who has repeated the experiment N times. We count the number of times, N_i, that outcome i occurred and take N_i/N as the frequency of outcome i. Then we average this 'sample average' over all possible experimentalists to obtain $\langle N_i \rangle / N$ which we take to be p_i. To perform the latter average we need the distribution $P(\underset{\sim}{N})$ of the (ordered in a sequence) results as measured by different experimentalists. That distribution is determined by maximizing the entropy subject to the m average values $N\langle A_r \rangle$.

In the second route we took the m average values (per repetition) $\langle A_r \rangle$, and determined the p_i's directly, by the procedure of maximal entropy. The two routes gave the same result for the p_i's and the result is independent of the value of N.

Applying the maximum entropy formalism directly to the p_i's is usually described as 'determining the most probable distribution'. The reason is well known. As $N \to \infty$, $P(\underset{\sim}{N})$ has an increasingly (with N) sharp maximum and the values $\hat{N}_i$ of the occupation numbers at the maximum are readily computed (using the Stirling approximation) to be

$$\hat{N}_i = N p_i \quad . \qquad (2.30)$$

What we have shown is that the p_i computed via the reasoning leading to (2.30) is the very same p_i computed via (2.5). It is not necessary to have a large value of N to assign a significance to the p_i's computed via maximum entropy.

Whether we first use the maximum entropy formalism (to obtain the distribution P(N)) and then take an average over all possible experiments (to determine the mean frequency $\langle N_i \rangle / N$) or first take an average over all experiments (to obtain the m $\langle A_r \rangle$'s) and then use the maximum entropy formalism (to determine the state probabilities p_i) was shown to lead to the same results for the state probabilities. Elsewhere [5,6] we have argued that maximum entropy is the only method of inference that will satisfy this requirement. Other considerations, all of which pertain to the relation between the sample average A_r and the mean over all samples, $\langle A_r \rangle$ will also be found therein.

3 THE ADJOINT PROBLEM

So far in the development, the Lagrange multipliers and the constraints appear on unequal footing. The values of the Lagrange multipliers are to be determined in terms of the values of the constraints. Furthermore, while the variance in the values of the constraints can be readily computed (e.g., as in (2.26)), the Lagrange multipliers appear to have sharp given values. This asymmetry is however only skin deep. It is possible to formulate the problem so that the strictly equivalent role of the constraints and the Lagrange multipliers is explicitly apparent. This reformulation will be carried out in this section. The two sets of variables will be shown to be 'conjugate' to one another in the usual sense of that terminology in science. In particular, the Lagrange multipliers do not have sharp values (only their mean values are sharp, cf. [9]). Indeed the variance in the Lagrange multipliers and the variance in the constraints satisfy a ('Heisenberg type') uncertainty relation (i.e., one is the inverse of the other [8,9]).

3.1 The Lagrangian

Let $\{\underset{\sim}{A}\}$ be a set of m mean values, $\langle A_1 \rangle$ to $\langle A_m \rangle$, of the constraints. Furthermore, let {p} be the normalized distribution of maximal entropy subject to these constraints. (We assume throughout that $\{\underset{\sim}{A}\}$ is a feasible set of constraints and that the same is true for the set $\{\underset{\sim}{A}'\}$.) The entropy of p is then a function of $\{\underset{\sim}{A}\}$ (cf. (2.28)), denoted here by $S(\{\underset{\sim}{A}\})$. For a different distribution the mean values of the constraints may be different, say $\{\underset{\sim}{A}'\}$. Let p' be the normalized distribution whose entropy is maximal subject to $\{\underset{\sim}{A}'\}$. Rather than using the set of m mean values we shall now use the set of m values of the corresponding Lagrange multipliers, $\{\underset{\sim}{\lambda}'\}$,

$$p_i' = \exp[-\lambda_0'(\{\underset{\sim}{\lambda}'\}) - \sum_{r=1}^{m} \lambda'_r A_{ri}] \tag{3.1}$$

In (3.1) we have explicitly indicated that λ_0' is a function, determined by the normalization condition

$$\exp[\lambda_0'(\{\underset{\sim}{\lambda}'\})] = \sum_{i=1}^{n} \exp(-\sum_{r=1}^{m} \lambda_r' A_{ri}) , \tag{3.1}$$

of the m values, $\{\underset{\sim}{\lambda}'\}$, of the Lagrange multipliers. As usual (and as follows from (3.1) and (3.2))

$$-\partial\lambda_0(\{\underset{\sim}{\lambda}'\})/\partial\lambda_r' = \langle A_r\rangle' . \tag{3.3}$$

The Lagrangian of the problem (first introduced in [28]) is

$$L = \lambda_0'(\{\underset{\sim}{\lambda}'\}) - S(\{\underset{\sim}{A}\}) + \underset{\sim}{\lambda}'.\underset{\sim}{A} . \tag{3.4}$$

That $L \geq 0$ with equality if and only if $p_i = p_i'$ for all i follows from the equivalent form

$$L = \sum_i p_i \ln(p_i/p_i') . \tag{3.5}$$

The equivalence of (3.5) and (3.4) is shown directly by substituting (3.1) in (3.5). The thermodynamic interpretation of the Lagrangian (3.4) as the maximal available work is provided in [28,29].

3.2 Variational principle

Two (conjugate) variational principles follow from the inequality $L \geq 0$ which implies the bounds

$$S(\{\underset{\sim}{A}\}) \leq \lambda_0'(\{\underset{\sim}{\lambda}'\}) + \underset{\sim}{\lambda}'\cdot\underset{\sim}{A} \tag{3.6}$$

and

$$\lambda_0'(\{\underset{\sim}{\lambda}'\}) \geq S(\{\underset{\sim}{A}\}) - \underset{\sim}{\lambda}'\cdot\underset{\sim}{A} . \tag{3.7}$$

In (3.6) it is the m λ' 's (or the m mean values $\{\underset{\sim}{A}'\}$) which are arbitrary while in (3.7) it is the m mean values $\{\underset{\sim}{A}\}$ (or the conjugate m Lagrange multipliers) which are arbitrary. Note also the direction of the inequalities. The familiar maximum entropy variational principle is complementary to (3.6). If q is a normalized distribution which is consistent with the set of m mean values $\{\underset{\sim}{A}\}$, then $S(\{\underset{\sim}{A}\})$ is an upper bound

$$-\sum_i q_i \ln q_i \leq S(\{\underset{\sim}{A}\}) . \tag{3.8}$$

In (3.6), $S(\{\underset{\sim}{A}\})$ is the lower bound.

To find the set of m values of the Lagrange multipliers corresponding to the set of m mean values, $\{\underset{\sim}{A}\}$, of the constraints, we use (3.6). Regarding the m λ' 's as subject to arbitrary variation we have

$$S(\{\underset{\sim}{A}\}) = \min_{\underset{\sim}{\lambda}'}[\lambda_0'(\{\underset{\sim}{\lambda}'\})+\underset{\sim}{\lambda}'\cdot\underset{\sim}{A}]. \tag{3.9}$$

The variational principle is then that the set of m λ' 's are to be varied until $\lambda_0(\underset{\sim}{\lambda}')+\underset{\sim}{\lambda}'.\underset{\sim}{A}$ has its minimal value. In Section 3.3 it will be shown that there are no local minima. There is only one (global) minimum and it is the required solution.

It should be noted that: (a) The only input required in (3.9) about the 'unknown' distribution p are the m expectation values, $\underset{\sim}{A}$ over that distribution. It is not necessary to know the actual distribution. The most direct proof of this result is using (3.5) and (3.1). (3.6) is seen to be true for any distribution p' which is consistent with the constraints (i.e., has the set $\{\underset{\sim}{A}\}$ of expectation values). What distinguishes the particular distribution p which is both consistent with the constraints and of maximal entropy is that equality in (3.6) can be obtained; (b) Varying the Lagrange multipliers is varying the distribution p' (cf. (3.1)). What the minimum in (3.9) seeks to locate is that distribution p' which is identical to p.

To show that (3.9) does yield the desired solution consider the variation with respect to λ_r'. At the extremum, (using (3.3))

$$\begin{aligned} 0 &= \partial\lambda_0'(\{\underset{\sim}{\lambda}'\})/\partial\lambda_r'+\langle A_r\rangle \\ &= \langle A_r\rangle-\langle A_r\rangle' . \end{aligned} \tag{3.10}$$

At the extremum p' is of maximal entropy and has the same set of m expectation values as p. The Lagrange multipliers at the extremum have therefore the required values.

That the extremum (3.10) is indeed a minimum will be shown in Section 3.3.

The adjoint problem is that of determining the set of expectation values $\{\underset{\sim}{A}'\}$ corresponding to a given set of m values, $\{\underset{\sim}{\lambda}'\}$, for the Lagrange multipliers. Now it is the m expectation values $\{\underset{\sim}{A}\}$ (or, equivalently the distribution p) which is subject to variation. From (3.7) the optimal set is given by

$$\lambda_0'(\{\underset{\sim}{\lambda}'\})= \max_{\underset{\sim}{A}}[S(\{\underset{\sim}{A}\})-\underset{\sim}{\lambda}'.\underset{\sim}{A}] . \tag{3.11}$$

The right hand side of (3.11) is to be evaluated for different sets $\{\underset{\sim}{A}\}$ and the one that gives rise to the (one and only) minimum is the desired set.

The extremum of (3.11) is determined by

$$\begin{aligned} 0 &= \partial S(\{\underset{\sim}{A}\})/\partial\langle A_r\rangle-\lambda_r' \\ &= \lambda_r-\lambda_r' . \end{aligned} \tag{3.12}$$

That the extremum is a minimum follows from the proof in Sec. 3.3.

3.3 Computations

The variational principle of Sec. 3.2 (proposed in [28]) has been applied as a computational tool in [30]. The essential point is the proof that if there is a feasible solution then the Lagrangian is a function which is everywhere convex (for (3.9)) or concave (for (3.11)) as a function of the variational parameter. Hence there is one and only one extremum and steepest descent (or ascent) can be readily used to numerically determine the optimal set of values.

We present the proof for the case where the Lagrange multipliers are the variational parameters: That the true minimum is unique follows from the result that $L \geq 0$ with equality if and only if $p_i = p_i'$, for all i. Now at any set of $\{\underset{\sim}{\lambda}'\}$

$$\begin{aligned} &\partial^2[\lambda_0'(\{\underset{\sim}{\lambda}'\})+\underset{\sim}{\lambda}'.\underset{\sim}{A}]/\partial\lambda_r\partial\lambda_s \\ &= -\partial\langle A_r\rangle'/\partial\lambda_s' \\ &\equiv M_{rs}' \quad . \end{aligned} \tag{3.13}$$

Here the symmetric $\underset{\sim}{M}'$ matrix is defined by

$$\begin{aligned} M_{rs}' &\equiv -\partial\langle A_r\rangle'/\partial\lambda_s' \\ &\equiv -\partial\langle A_s\rangle'/\partial\lambda_r' \\ &= \sum_i p_i'(A_{ri}-\langle A_r\rangle')(A_{si}-\langle A_s\rangle') \ . \end{aligned} \tag{3.14}$$

Now $\lambda_0'+\underset{\sim}{\lambda}'.\underset{\sim}{A}$ is everywhere convex (as a function of the m values $\{\underset{\sim}{\lambda}'\}$) if $\underset{\sim}{M}'$ is a positive definite matrix, i.e., if for any non-null vector $\underset{\sim}{x}$ (of m components)

$$\underset{\sim}{x}^T\underset{\sim}{M}'\underset{\sim}{x} > 0. \tag{3.15}$$

Now

$$\begin{aligned} \underset{\sim}{x}^T\underset{\sim}{M}'\underset{\sim}{x} &= \sum_{rs} x_r M_{rs}' x_s \\ &= \sum_{i=1}^{n} p_i'[\sum_{r=1}^{m} x_r(A_{ri}-\langle A_{ri}\rangle)]^2 \ . \end{aligned} \tag{3.16}$$

$\underset{\sim}{M}'$ is thus positive definite (being the expectation value of a non negative quantity) provided that there is no non-null vector $\underset{\sim}{x}$ such that

$$\sum_{r=1}^{m} x_r (A_{ri} - \langle A_{ri} \rangle) = 0, \qquad m \leq n-1 \ . \tag{3.17}$$

But (3.17) is the condition that the constraints (for each i, i=1,...m, m>n) are linearly independent, which we take to be always the case. If in practice it is not, then the dependent constraints should be eliminated. 'Near' linear dependence is also unwelcome in practice (since the minimum then is very shallow [31]). Hence our computer algorithm ([30], actual program available upon request) begins by orthogonalizing the constraints.

3.4 Surprisal analysis

The practical situation is often somewhat different than that considered so far. The experimental distribution p_i is often directly measurable and is available as a table of numerical entries (vs. i). The real question is then what constraints govern that distribution, i.e, which observables A_r are such that the 'theoretical' distribution p_i' as given by (3.1) fits the data. If $m < n-1$ such constraints can be found then the observations can be accounted for using fewer constraints than data points via the procedure of maximal entropy. If $m \ll n$, the compaction that can be achieved is considerable. (Imagine the radio or tv announcer who has to read the population in each and every quantum state of the molecules in the air rather than to merely state the temperature.)

Of course, the compaction is only the first step. Since, as was argued in Section 2, the physics enters in via the specification of the constraints, the next step is to understand on physical grounds why it is that those m constraints govern the distribution (in the sense that they suffice to specify the p_i's.)

The Lagrangian (3.5) with the p_i' 's given by (3.1) provide again the required theoretical framework. Using (3.1)

$$L = -(\sum_i p_i \ln p_i) + \lambda_0' + \sum_{r=1}^{m} \lambda_r' (\sum_{i=1}^{n} A_{ri} p_i) \ . \tag{3.18}$$

The difference between (3.4) and (3.18) is that here there is no assumption about the distribution p_i. It is simply given to us in numerical form. Hence all terms in (3.18) can be computed, once the set of m observables has been specified. In practice, the first term in (3.18) is just a number which will not change as we vary the λ_r' 's (or change m) hence one might just as well work only with the other two terms.

One now selects a set of m observables A_r, an arbitrary set of m trial Lagrange multipliers λ_r''s and computes λ_0' via (3.2). As before, we vary the trial distribution p_i' by varying the m trial multipliers λ_r' (recomputing λ_0' after each variation). The extremum of L is at (cf. (3.3))

$$\partial L/\partial\lambda_r = 0 \qquad r=1,\ldots,m$$

$$= -\sum_{i=1}^{n} A_{ri}p_i' + \sum_{i=1}^{n} A_{ri}p_i, \tag{3.19}$$

i.e., at the point where p_i' and p_i have the same expectation values for all m constraints. The derivation of Sec. 3.3 insures that the extremum is a minimum.

Having reached the minimal value of L does not necessarily imply a perfect fit. At the minimum, the experimental (p_i) and theoretical (p_i') distributions yield the same expectation values for m constraints, (3.19), but $m < n-1$. Hence the two distributions can still yield different results for other observables (linearly independent on the m A_r's). Only if at the minimum $L = 0$ is the fit perfect.

Should one then increase m (i.e., add more constraints) until the fit is perfect? In practice, the answer is not necessarily. The experimental distribution p_i will invariably not be known with perfect numerical accuracy. There are experimental uncertainties ($\pm\delta p_i$) in the values of the p_i's. By increasing m there will always come a point when we are no longer fitting the real data but the noise [32].

The real world situation is then that one should compute not only the values of the Lagrange multipliers (at the minimal value of L) but also their uncertainty, $\pm\delta\lambda_r$'. Those observables for which $|\lambda_r'| \leq |\delta\lambda_r|$ should be assigned the numerical value zero for their Lagrange multiplier (i.e., they should not constrain the distribution at the stated level of experimental accuracy). There is only one problem: how to compute the $\delta\lambda_r$'s, since we do not know now to assign signs to the δp_i's. That is the topic of the next section.

3.5 Experimental and inherent uncertainties

Consider an observable A_r (which can but need not be the indicator function, $\delta_{r,i}$, for the state i). The experimental uncertainty in $\langle A_r\rangle$ is

$$\delta\langle A_r\rangle = \sum_{i=1}^{n} A_{ri}\delta p_i \,. \tag{3.20}$$

Since we do not know the absolute sign of the δp_i's, (3.20) cannot be used to compute $\delta\langle A_r\rangle$. Putting $\delta p_i = p_i\delta\ell n p_i$ and using the normalization condition ($\Sigma\delta p_i = 0$)

$$\delta\langle A_r\rangle = \sum_{i=1}^{n} (A_{ri}-\langle A_{ri}\rangle)\,\delta\ell n p_i p_i \tag{3.21}$$

The Cauchy-inequality can now be invoked [8] to yield

$$(\delta\langle A_r\rangle)^2 \leqq \left[\sum_{i=1}^{n} (A_{ri}-\langle A_{ri}\rangle)^2 p_i\right]\cdot\left[\sum_{i=1}^{n} p_i(\delta p_i/p_i)^2\right] . \tag{3.22}$$

Everything in (3.22) can now be computed! The first term on the right is the variance of A_r and the second is the mean squared error in the data.

The result (3.22) is a practical bound on $\delta\langle A_r\rangle$. A simple route (which turns out to be correct [8,9]) for estimating the $\delta\lambda_r$'s is, when m=1,

$$\begin{aligned}\delta\lambda_r &\approx (\partial\lambda_r/\partial\langle A_r\rangle)\,\delta\langle A_r\rangle \\ &= (-\partial^2\lambda_0/\partial\lambda_r^2)^{-1}\,\delta\langle A_r\rangle \\ &= (\text{variance of } A_r)^{-1}\,\delta\langle A_r\rangle\end{aligned} \tag{3.23}$$

or

$$\delta\langle A_r\rangle\delta\lambda_r \leqq \sum_i p_i(\delta p_i/p_i)^2 . \tag{3.24}$$

If m > 1, one needs to invert the matrix $\partial^2\lambda_0/\partial\lambda_r\partial\lambda_s$ to compute $\partial\lambda_r/\partial\langle A_s\rangle$.

Does it then follow that as the data becomes more accurate we know the λ_r's perfectly? The answer is yes but only in a limited sense. Yes, we can know precisely the numerical value of λ_r but it will always have a finite variance [9]. Indeed, the theoretical inherent variance of λ_r can be shown to be (m=1)

$$\Delta(\lambda_r) = -\partial\lambda_r/\partial\langle A_r\rangle$$

and, since for m=1 the variance of A_r is

$$\Delta(A_r) = -\partial\langle A_r\rangle/\partial\lambda_r$$

it follows that [8,9]

$$\Delta(\lambda_r)\Delta(A_r) = 1. \tag{3.25}$$

For a detailed derivation (also for the case m>1) we refer to Ref. 9.

4 APPLICATIONS

Many of our applications (for reviews see Refs. 15-18) have been to a situation which is as removed from classical thermodynamics as one can possibly get in the physical sciences. This is to collisions of composite projectiles (be they molecules or nuclei). These collisions are studied in the laboratory where each collision event is isolated and hence independent of all others. Yet to get good counting statistics many collisions are allowed to happen. The point is that when the colliding particles have internal structure there are very many possible final states and the experiment is to measure the distribution over these final states. But because there are very many accessible final states, the computation of their distribution using a conventional dynamical theory is exceedingly time consuming even on our fastest computers and has only been implemented for the simplest problems.

The experimental setup is thus precisely that described in Section 2. Yet if you hold the view that entropy is relevant only for a system of very many degrees of freedom (and some will further say, and at equilibrium) then you will clearly not apply the formalism for such problems. First, while the particles may be composite the number of degrees of freedom is very small. Three, in our first example below and several hundreds in the next one. Both are very finite. We also show a prediction for a problem with a few tens of degrees of freedom. Then, the system is very far removed from equilibrium however you choose to define the term. Yet if the discussion of Section 2 makes sense then we have an ideal testing ground: very many (in a sense that $\bar{A}_r \simeq \langle A_r \rangle$, cf. (2.26), i.e., $N \ggg n$) independent repetitions.

4.1 The prior distribution

The key to our successful application is the recognition (which essentially goes back to Planck [33] and later to von Neumann [34]) that in physics entropy is always the entropy of the distribution over quantum states. In a technical language their work identified the measure to be used (see also [35] and references therein). If the experimentalist at his convenience groups quantum states together into more coarse resolution then one is to use the grouping property of the entropy (see [36] or, in the present context, [16-18]) to determine the form of the entropy. The technical argument is, in brief, as follows. Let γ be a group of quantum states and p_γ the probability of the group. Then the entropy of the distribution of quantum states is

$$S = -\sum_\gamma p_\gamma \ln p_\gamma + \sum_\gamma p_\gamma S_\gamma \quad . \tag{4.1}$$

Here S_γ is the entropy of the distribution of quantum states within the group γ. Introducing the 'degeneracy' of the group γ, g_γ by

$$\ln g_\gamma = S_\gamma \geq 0 \tag{4.2}$$

we can write (4.1) as

$$S = -\sum_{\gamma} p_{\gamma} \ln(p_{\gamma}/g_{\gamma}) \,. \tag{4.3}$$

It proves convenient to rewrite (4.3) in terms of the normalized prior distribution p^{o}_{γ},

$$p^{o}_{\gamma} = g_{\gamma}/N \tag{4.4}$$

$$N = \sum_{\gamma} g_{\gamma} \,. \tag{4.5}$$

It then follows that

$$S = \ln N - DS \tag{4.6}$$

where DS,

$$DS = \sum_{\gamma} p_{\gamma} \ln(p_{\gamma}/p^{o}_{\gamma}) \tag{4.7}$$

is referred to, in our work, as 'the entropy deficiency'. The reason is that for a fixed values of the S_{γ}'s, the maximal value of S is $\ln N$. (Maximum over all possible variations in the p_{γ}'s.) Since $DS \geq 0$ (with equality if and only if $p_{\gamma} = p^{o}_{\gamma}$), DS measures how far is the entropy below its maximal value.

The prior distribution enters the present problem on physical grounds, but on the basis of symmetry considerations. (This is particularly obvious in von Neumann's work where the rationale was 'invariance under all unitary transformations.' For the question of why the 'classical' form of a sum rather than the quantum mechanical trace is appropriate in collision experiments, see [18,37]).

Note that (4.3) or (4.6) is to be used for the entropy even when all that was experimentally measured is the distribution p_{γ}.

4.2 Surprisal analysis

Surprisal analysis has been discussed in principle in Section 3.4. Consider first the simplest case where there is only one constraint (m=1). The Lagrangian to be minimized, (3.18), can then be written as

$$L = -[-\sum_{\gamma} p_{\gamma} \ln(p_{\gamma}/g_{\gamma})] + \lambda'_0 + \lambda' \sum_{\gamma} A_{\gamma} p_{\gamma}$$

$$= \sum_{\gamma} p_{\gamma} [\ln(p_{\gamma}/g_{\gamma}) + \lambda'_0 + \lambda' A_{\gamma}] \,. \tag{4.8}$$

It follows that the minimization of L (with respect to λ') can be done by graphical means. Plot $-\ell n(p_\gamma/g_\gamma)$ vs. A_γ. If the plot is linear L = 0 and the slope of the line is λ'. If due to experimental scatter (see Figure 1) the line is not quite straight - do not try a simple least square routine. The reason is that L is a weighted (by the p_γ's) sum of

Figure 1. Surprisal plot ($-\ell n[p(v,J)/p^0(v,J)]$ vs. the constraint, g_R) for the rotational state distribution of HD in the v=0 (ground) vibrational manifold of states in the single collision $H+D_2 \rightarrow HD+D$. The plot and data are from D.P. Gerrity and J.J. Valentini, [40]. The constraint, g_R, is the fraction of the available energy which is in the rotation of HD. By definition therefore the range of g_R is [0,1].

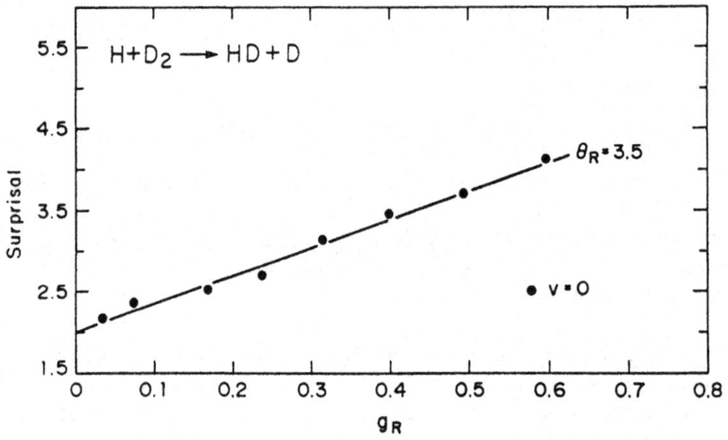

the deviations of $-\ell n(p_\gamma/g_\gamma)$ from the straight line. It is imperative to use weighted least squares, where points of higher p_γ's are to be closer to the line.

If one constraint does not suffice, then λ_r''s are to be determined numerically as was discussed in Section 3.4. (Of course, for the present problem, if summation in (3.18) is over γ then ℓnp_i is to be replaced by $\ell n(p_\gamma/g_\gamma)$.)

Our first example is for a 'state of the art' experiment [38,39,40]. Measured is the vibrational and rotational state distribution of the HD molecule following the rearrangement collision $H + D_2 \rightarrow HD + D$. (D is the heavier isotope of hydrogen, H is the lighter one.) This is the simplest chemical reaction and earlier quantum mechanical computations [41] have already demonstrated that a single constraint suffices to describe the rotational state distribution. The experiment was at a higher energy where since many more states are accessible, the exact quantal computation is prohibitive. Figure 1 shows a surprisal plot (adapted from the experimental paper) for the rotational state distribution in the ground vibrational state v=0. The plot, as implied by (4.8), is against the constraint (which is the fraction g of the total energy present as rotational energy of HD). Figure 2 is the fit to the actual distribution. It is for an experiment at a higher energy so that more states are accessible. The solid line is the distribution of maximal entropy determined by two constraints: one on the rotational state distribution and one of the vibrational one. To within the experimental error bars (also shown) two numbers (e.g., the two Lagrange multipliers) suffice to characterize the distribution.

Figure 2. The measured [40] rotational state distribution of HD in three different vibrational manifold (points with error bars) vs. the rotational quantum number J of HD. The solid line is the fit by a linear vibrational and rotational surprisal. Plot and data from D.P. Gerrity and J.J. Valentini, [40].

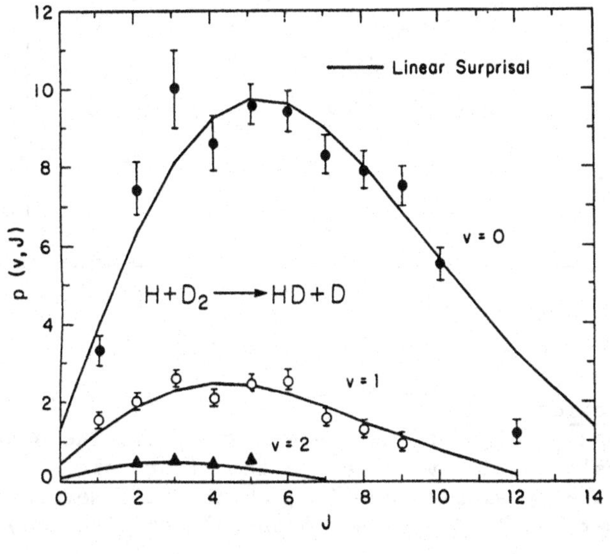

As already mentioned, computer solutions of the Schrödinger equation also lead to rotational state distributions which can be well characterized as having a linear surprisal [41]. That only a single constraint is relevant is not due therefore to any 'relaxation' (by secondary collisions) of the real nascent experimental data. Many many additional applications to molecular collisions are available [15-18]. But could the simplicity be due to molecules being more amenable to the concept of entropy? Let's raise the energy by eight orders of magnitude.

Our next example is nucleon transfer in a heavy ions collision at 105 MeV [42,43]. Shown in the top panel of Figure 3 is the surprisal plot for the distribution of the kinetic energy in the nucleon transfer process. The bottom panel shows the fit to the observed data. Note that the scale is 'counts' which is precisely the N_i's of Section 2.

Many more examples will be found in [42,43]. One can also consider the branching ratios amongst different transfer processes [44].

Figure 3. Surprisal plot (top panel) and fit (continuous line) to the observed (histogram) distribution of the ejectile kinetic energy in the nuclear transfer reaction at 105 MeV. Adapted from [42].

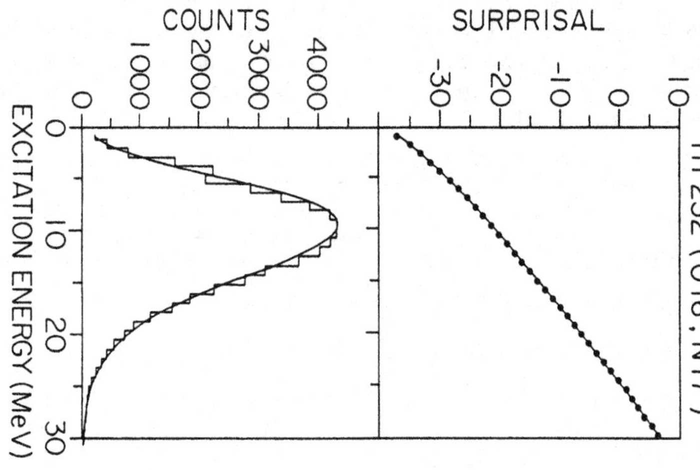

4.3 Prediction

Can one predict and not merely analyze? The answer is yes and for the general question we refer to the literature [17-20]. Here we consider only a very simple case [45] which did however suggest a number of new experiments (see Refs. 46 and 47 and also 48.)

Using a high power laser an isolated molecule can be made to absorb several photons in the visible/UV wavelengths. Such photons carry a significant amount of energy. Hence the molecule is very rich in energy and fragments, often to small pieces. What constraints govern the fragmentation pattern? The remarkable and, at the time [45], unexpected result is that these are very few in number, being just the conserved quantities: energy and chemical elements. But the number of elements in a given molecule is known (that is, after all, its chemical formula). Hence one can readily compute the fragmentation pattern as a function of the energy, $\langle E\rangle$, absorbed per molecule. A typical result, including a comparison with experiment is shown in Figure 4. Note that this is a strict prediction. Of course, now that one knows the constraints one can try to devise ways of experimentally introducing additional constraints. There is no reason however why such additional constraints cannot be incorporated while maximizing the entropy. When this is done [48], agreement with experiment can again be established.

Figure 4. Computed (via maximal entropy) fragmentation pattern of C_6H_6 at several energies, ⟨E⟩. The middle panel shows a comparison with experiment. The molecule is pumped by several (over a dozen) photons at a wavelength of λ = 504 nm. It both ionizes and fragments. Shown is the distribution of ionic fragments vs. mass. Adapted from [45].

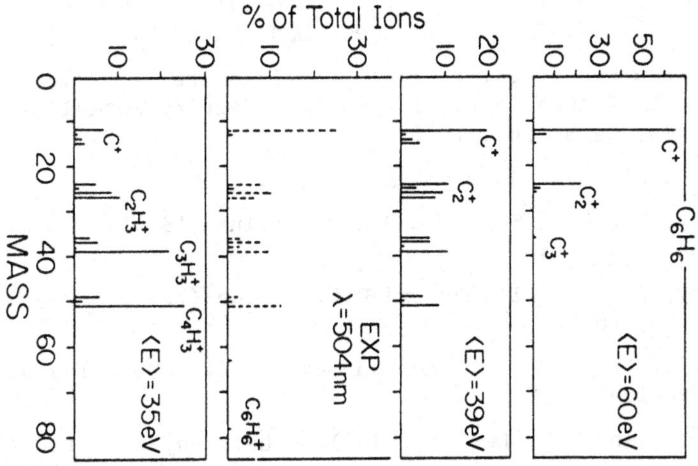

5 CONCLUDING REMARKS

The maximum entropy formalism has been discussed as a method for inducing a probability distribution for reproducible experiments. The resulting probabilities are well defined and independent of the number of times N the experiment has been repeated. Of course, for any finite N the observed (the 'sample averaged') frequency of event i need not numerically equal the probability p_i. In the limit $N \to \infty$ the probability, the observed frequency and the 'most probable' frequency all have the same value, but the probabilities induced by the maximum entropy formalism have a well defined meaning for any value of N.

It is time therefore to stop arguing about 'why' maximum entropy and address instead the really open question: why, for a wide variety of circumstances, do very few constraints suffice to determine the probability distribution.

ACKNOWLEDGEMENT

I have had the benefit of discussions with many colleagues. Special thanks are due to Y. Alhassid, H. Reiss, Y. Tikochinsky and N.Z. Tishby. Our work was supported by the U.S. Air Force Office of Scientific Research under Grant AFOSR-81-0030. This report was written while I was a visitor at the Department of Chemistry, UCLA. I thank the members of the department for their interest in this work and their warm hospitality.

BIBLIOGRAPHY

1 The maximum entropy formalism has been most actively championed by E.T. Jaynes. See, Papers on Probability Statistics and Statistical Physics. Reidel, Dordrecht, 1983.

2 For other early references see, for example, W.M. Elsasser, Phys. Rev. 52, 987 (1937); R.L. Stratonovich, Sov. Phys. J.E.T.P. 1, 426 (1955); U. Fano, Rev. Mod. Phys. 29, 74 (1957).

3 For statements by several current practitioners, see R.D. Levine and M. Tribus, eds., The Maximum Entropy Formalism. M.I.T. Press, Cambridge, 1979.

4 Levine, R.D. (1980). J. Phys. A. 13, 91.

5 Tikochinsky, Y., N.Z. Tishby and R.D. Levine (1984). Phys. Rev. Letts. 52, 1357.

6 Tikochinsky, Y., N.Z. Tishby and R.D. Levine (1984). Phys. Rev. A. 30, 2638.

7 Boltzmann, L. (1964). Lectures on Gas Theory. Berkeley U. Press, Berkeley.

8 Alhassid, Y., and R.D. Levine (1980). Chem. Phys. Letts. 73, 16.

9 Tikochinsky, Y. and R.D. Levine (1984). J. Math. Phys. 25, 2160.

10 Shannon, C.E. (1948). Bell System Tech. J. 27, 379.

11 Shore, J.E. and R.W. Johnson. IEEE Trans. Inform. Theory IT-26, 26 (1980), ibid. IT-29, 942 (1983).

12 Tribus, M. (1961). Thermostatics and Thermodynamics. Van Nostrand, Princeton.

13 Baierlein, R. (1971). Atoms and Information Theory. Freeman, San Francisco.

14 Landau, L.D. and E.M. Lifshitz (1977). Statistical Physics. Pergamon Press, Oxford.

15 Levine, R.D. and R.B. Bernstein (1974). Accts. Chem. Res. 7, 393.

16 Levine, R.D. and A. Ben-Shaul (1976). In Chemical and Biochemical Applications of Lasers, C.B. Moore, ed. Academic Press, N.Y.

17 Levine, R.D. (1981). Adv. Chem. Phys. 47, 239.

18 Levine, R.D. (1985). In Theory of Molecular Collisions, M. Baer, ed. C.R.C. Press, N.Y.

19 Alhassid, Y. and R.D. Levine (1977). Phys. Rev. A.18, 89.

20 Pfeifer, P. and R.D. Levine (1983). J. Chem. Phys. 78, 4938.

21 Jaynes, E.T. in Ref. 3, Section D in particular.

22 Arnold, V.I. and A. Avez (1978). Ergodic Problems of Classical Mechanics. Benjamin, N.Y.

23 Tishby, N.Z. and R.D. Levine (1984). Phys. Rev. A.30, 1477.

24 Levine, R.D. (1985). J. Phys. Chem. 89, 2122.

25 See, for example, Hogg, R.V. and A.T. Craig (1978). Introduction to Mathematical Statistics. Macmillan, N.Y.

26 van Campenhout, J. and T.M. Cover (1981). IEEE Trans. Inform. Theory IT-27, 483.

27 In this connection, see also Jaynes, E.T. (1982). Proc. IEEE 70, 939.

28 Levine, R.D. (1976). J. Chem. Phys. 65, 3302.

29 Procaccia, I. and R.D. Levine (1976). J. Chem. Phys. 65, 3357.

30 Agmon, N., Y. Alhassid and R.D. Levine (1979) (and in Ref. 3). J. Comput. Phys. 30, 250.

31 Alhassid, Y., N. Agmon and R.D. Levine (1978). Chem. Phys. Letts. 53, 22. See also Manoyan, J.M. and Y. Alhassid (1983). Chem. Phys. Letts. 101, 265.

32 Kinsey, J.L. and R.D. Levine (1979). Chem. Phys. Letts. 65, 413.

33 Planck, M. (1914). Theory of Heat Radiation. Blakiston, N.Y.

34 von Neumann, J. (1955). Mathematical Foundations of Quantum Mechanics. Princeton U. Press, Princeton.

35 Dinur, U., and R.D. Levine (1975). Chem. Phys. 9, 17.

36 Ash, R. (1956). Information Theory. Wiley, N.Y.

37 Levine, R.D. (1985). Phys. Lett.

38 Gerrity, D.P. and J.J. Valentini (1984). J. Chem. Phys. 81, 1298; Marinero, E.E., C.T. Rettner and R.N. Zare (1984). J. Chem. Phys. 80, 4142.

39 Zamir, E., R.D. Levine and R.B. Bernstein (1984). Chem. Phys. Letts. 107, 217.

40 Gerrity, D.P. and J.J. Valentini (1985). J. Chem. Phys. 82, 1323.

41 Wyatt, R.E. (1975). Chem. Phys. Letts. 34, 167. Schatz, G.C. and A. Kuppermann (1976). J. Chem. Phys. 65, 4668. Clary, D.C. and R.K. Nesbet (1979). J. Chem. Phys. 71, 1101.

42 Alhassid, Y., R.D. Levine, J.S. Karp and S.G. Steadman (1979). Phys. Rev. C.20, 1789.

43 Karp, J.S., S.G. Steadman, S.B. Gazes, R. Ledoux and F. Videback (1982). Phys. Rev. C.25, 1838.

44 Engel, Y.M. and R.D. Levine (1983). Phys. Rev. C.28, 2321.

45 Silberstein, J. and R.D. Levine (1980). Chem. Phys. Letts. 74, 6.

46 Lichtin, D.A., R.B. Bernstein and K.R. Newton (1981). J. Chem. Phys. 75, 5728.

47 Lubman, D.M. (1981). J. Phys. Chem. 85, 3752.

48 Silberstein, J., N. Ohmichi and R.D. Levine (1984). J. Phys. Chem. 88, 4952.

BAYESIAN NON-PARAMETRIC STATISTICS

Stephen F. Gull
Mullard Radio Astronomy Observatory
Cavendish Laboratory
Cambridge CB3 OH3
U.K.

John Fielden
Maximum Entropy Data Consultants Ltd.
33 North End
Meldreth
Royston SG8 6NR
U.K.

ABSTRACT

We consider the application of Bayes' theorem and the principle of maximum entropy to some problems that fall into the branch of statistics normally described as "non-parametric". We show how to estimate the quantiles and moments of a probability distribution, making only minimal assumptions about its form.

1 INTRODUCTION

In all experimental sciences there occur problems that involve interpreting and comparing samples of data from a probability distribution function (p.d.f.) whose form is unknown. Examples (taken from observational cosmology) might be to estimate the redshift distribution of quasars from a sample, or to compare the redshift distributions of two samples. We do not yet know enough about the physics of quasars or of the Universe to be able to predict the theoretical form of these distributions and there is certainly no reason to assume the distribution has some simple form, such as a Gaussian. When dealing with problems of this type there is a need for "non-parametric" statistical methods (see, for example, Siegel, 1956), which do not depend on the functional form of the underlying p.d.f. This independence is usually achieved by working with the cumulative probability distribution, q, where $q(x) = \int_a^x dx' \, \mathrm{prob}(x')$. This has given rise to some powerful tests (e.g., the Kolmogorov-Smirnov test), but there are pathological distributions for which these tests are not suited and they do not generalize easily to multivariate distributions.

Another approach that is often used in such problems is that of binning the data. This throws away information, especially for small samples. Binning also introduces problems of deciding where to place the bin boundaries; the results will clearly depend on such decisions.

This paper forms part of a programme to apply Bayesian methods to problems of this type, without either introducing unjustified assumptions about the p.d.f. or ignoring relevant information such as the ordering of a variable. As a modest step towards these general problems we consider here some simple examples where we make only minimal assumptions about the nature of the underlying p.d.f.

2 ESTIMATING THE POSITION OF THE MEDIAN

We will first consider the apparently simple problem of estimating the position of a quantile (e.g., the median of a p.d.f.). We emphasize that we are talking about estimating properties of the p.d.f. prob(x) such as:

$$x \text{ where } \int_a^x dx' \text{ prob}(x') = q,$$

and not just computing properties of the sample (e.g., the sample median).

Suppose that we have N samples, $\{x_i\}$, taken from a probability distribution prob(x), where $x \in [a, b]$. Denote by M_x the hypothesis that the median lies at x:

$$M_x : \int_a^x dx' \text{ prob}(x') = 1/2$$

Our object is to use Bayes' theorem to calculate the posterior probability of M_x:

$$\text{prob}(M_x \mid \{x_i\}) \propto \text{prob}(M_x) \text{ prob}(\{x_i\} \mid M_x).$$

The hypotheses M_x (for different x) are exclusive and exhaustive, since the median certainly lies in the interval [a, b]. We will therefore take the prior prob (M_x) to be uniform for all x in this interval. In any case our results depend only very weakly upon this assumption.

We are now face to face with the real problem: the likelihood $\text{prob}(\{x_i\} \mid M_x)$ is not uniquely determined by the position of the median. We merely know that equal amounts of probability lie to either side of the position x. But hypothesis M_x does constitute TESTABLE information (Jaynes, 1968) about prob(x); i.e., given any prob(x) we can decide immediately whether it is consistent with information M_x. Further, if the samples are EXCHANGEABLE, then M_x is testable information about the joint p.d.f. prob $(\{x_i\})$. If we seek a p.d.f. that incorporates the information available, yet is maximally non-committal about other parameters of the distribution, then we should use the distribution that has MAXIMUM ENTROPY under the constraint M_x. Hence we maximize:

$$S = - \int d^N x \text{ prob}(\{x\}) \log(\text{prob}(\{x\})/m(\{x\}))$$

under the constraints of normalization and the conditions (for all I):

$$\int_{D_I} d^N x \text{ prob}(x) = 1/2, \quad \text{where } D_I \text{ is the domain: } x_I \in [a,x].$$

For one sample and a uniform measure on [a, b] this has the simple form

$$\text{prob}(x') = \frac{1}{2(x-a)} \quad \text{for } x' < x \text{ and}$$

$$\text{prob}(x') = \frac{1}{2(b-x)} \quad \text{for } x' > x.$$

For multiple, exchangeable samples, again with uniform measure m({x}), the MAXENT distribution is necessarily independent because the constraints all take the form of separate equalities on each of the marginal distributions. We thus obtain the likelihood:

$$p(\{x_i\} \mid M_x) = \left[\frac{1}{2(x-a)}\right]^{N<} \left[\frac{1}{2(b-x)}\right]^{N>}$$

where $N_<$ is the number of data points $x_i < x$, and $N_>$ is the number of data points $x_i > x$. Using Bayes' theorem above, this is also the posterior $\text{prob}(M_x \mid \{x\})$ when viewed as a function of x.

The argument is easily generalized to other quantiles; if the proposition Q_x means that quantile q lies at position x:

$$\text{prob}(Q_x \mid \{x\}) \propto \left[\frac{q}{(x-a)}\right]^{N<} \left[\frac{(1-q)}{(b-x)}\right]^{N>}$$

Before giving examples, it is worth examining again the principles that underlie our calculation. We wish to estimate, from a sample, the position of the median making only minimal assumptions about the form of the distribution. Apart from our desire for a sensible average estimate, we would expect our uncertainties about the nature of the p.d.f. to be reflected as a relatively large uncertainty in the position of the median, that is, in the width of the posterior distribution. The MAXENT likelihood function we have calculated is, in a sense, the broadest p.d.f. that satisfies all the constraints that are available, and so helps us achieve our goal of making minimum assumptions. On the other hand, we are NOT necessarily claiming that the MAXENT distribution for some x is in fact the underlying p.d.f. Indeed, we are not even assuming that the joint p.d.f. is independent, though the MAXENT distribution happens to have this property.

2.1 Examples

For our first example we created 39 samples uniformly spaced between -1 and +1, with a prior range [-1.5,1.5]. Figure 1 shows the posterior p.d.f. of the lower quartile, the median and the upper decile. The distributions are extremely plausible, being centred about the appropriate positions and of widths that are comparable with $(\text{variance} / N)^{1/2}$ and also in agreement with the values obtained by previous binning methods (e.g., Yule & Kendall, 1946). The detailed structure, however, has an unfamiliar appearance, because the distributions are piece-wise inverse polynomials, with discontinuities at the sample points. These discontinuities may cause some surprise, and other workers have gone to considerable lengths to produce

Figure 1. Posterior p.d.f. of quantiles q=0.25, 0.5 and 0.9 for 39 uniform samples.

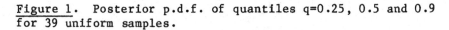

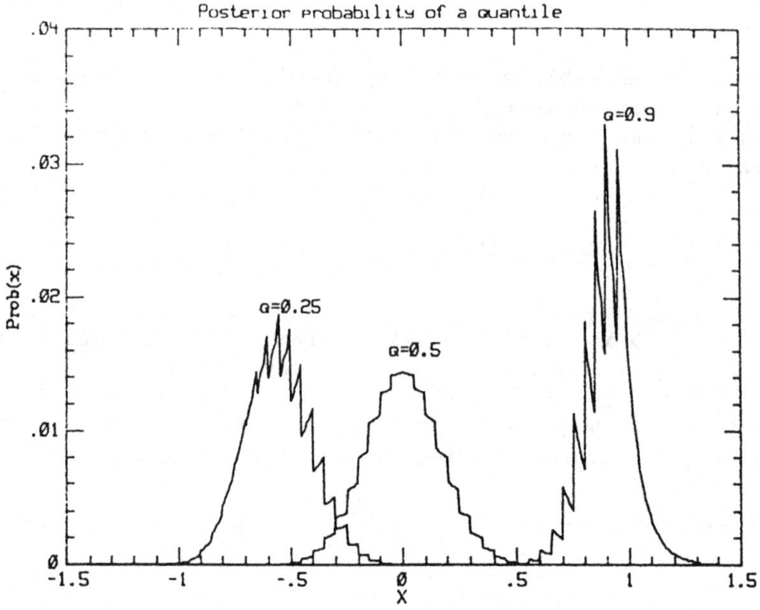

continuous p.d.f.s (even sometimes to the extent of tolerating a negative prob(x)!). We do not in fact find these discontinuities unaesthetic: it is the job of a p.d.f. to tell us how much probability falls into any interval of x, and hence it must certainly be integrable. But we cannot see any good reason why a p.d.f. should also be continuous.

An interesting feature of our solution is that the prior range [a, b] still appears in the posterior distribution no matter how many samples one has. In order to investigate this, Figure 2 shows the posterior p.d.f. of the median position for prior ranges of ± 1, ± 2 and ± 5. As we expect, the distribution gets wider as the range is increased, but not proportionally. Again, we feel that this behaviour is entirely reasonable when we consider how little has been assumed about the distribution.

Pathological multi-modal p.d.f.s cause problems for many statistical tests, and in Figure 3 we show an example where 36 samples are concentrated into 2 equal groups near ± 0.95. The p.d.f. of the median is itself bi-modal, falling sharply towards the middle. In this case, we believe that our answer is superior to that given by any previous method and to conform completely to common sense. We note in passing that whilst the median must almost certainly lie in a region where there are many samples, the mean does not have this property (see Section 3). For this case the p.d.f. of the mean is uni-modal and centred at x=0.

Figure 2. Median p.d.f. for prior ranges ±1, ±2 and ±5.

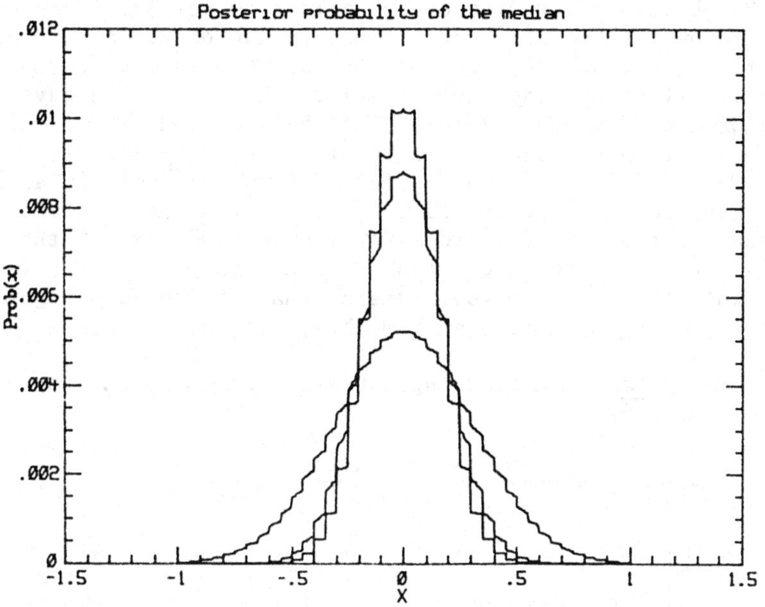

Figure 3. Median p.d.f. for bi-modal distribution of 36 samples.

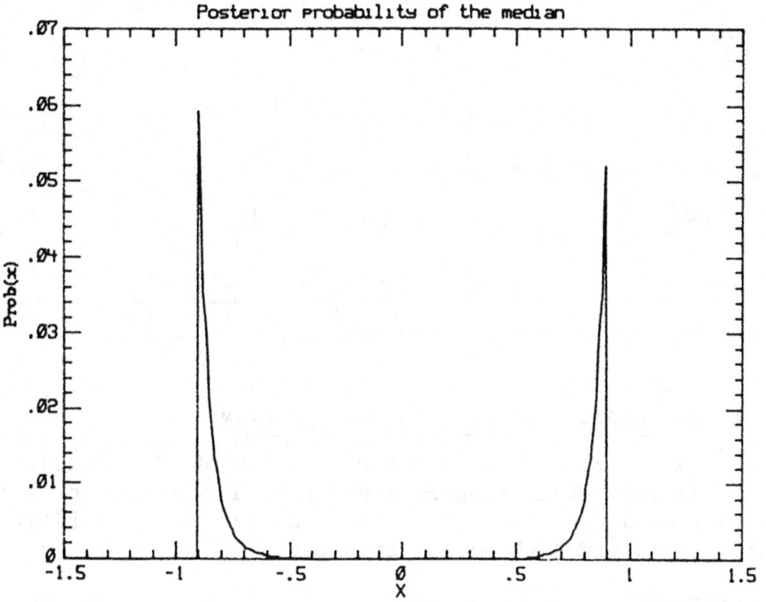

As a final, operational test we turn to observational cosmology. The median angular size of extragalactic radio sources in various luminosity classes has been much studied over the last 20 years. It may provide information on the evolution of the population of radio sources or, perhaps more interestingly, if the population of radio sources is much the same at all epochs then the apparent size of sources may indicate geometry of the Universe. Figure 4 is the distribution of the median angular size of 40 sources from a sample studied by Fielden et al. (1980). For this case we use a prior uniform in log (angular size) for small angles, varying between 10 milliarcsec (the approximate size of jets emanating from quasars) to a few arcmin (the size of the largest sources). The p.d.f. of the median for this sample is very reasonable, but its width (a factor of 10) warns us that we should not jump to hasty conclusions about the Universe from these data alone!

Figure 4. Median angular size for the 5C12 sample of radio sources.

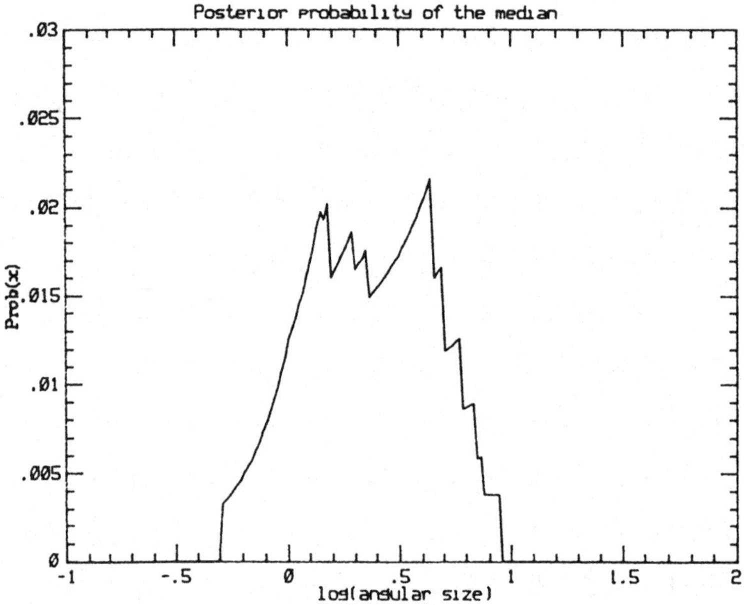

3 ESTIMATING THE POSITION OF THE MEAN

In a similar way to the estimation of the position of the median, we can also estimate parameters such as the mean or variance without assuming a detailed functional form for the probability distribution. Consider these two problems:

1) Proposition μ: "The mean lies at μ".

2) Proposition $\mu\sigma$: "The mean lies at μ and the variance is σ".

3.1 Estimation of the mean only

The posterior probability distribution is

$$\text{prob}(\mu \mid \{x_i\}) \propto \text{prob}(\mu)\ \text{prob}(\{x_i\} \mid \mu).$$

The solution has a strong dependence on the prior because of our weak assumptions and because the mean is very sensitive to small variations and asymmetries in the wings of the distribution. For this reason the problem is not soluble for a flat prior on $(-\infty,+\infty)$. If we restrict ourselves to a finite interval with a uniform prior, we get transcendental equations which can be solved numerically. However, for the purposes of illustration we will choose a prior that allows a simple analytic form for the solution. We shall take the prior on μ to be a Gaussian with variance M^2 and zero mean:

$$\text{prob}(\mu) = (2\pi\ M^2)^{-\frac{1}{2}} \exp - (\mu^2 / 2\ M^2)\ .$$

Again we derive the likelihood by maximizing entropy, taking as measure $m(\{x\})$ an independent Gaussian form:

$$m(\{x\}) \propto \exp - (\Sigma x_i^2 \ / \ 2\ M^2)\ .$$

We see that our prior can also be viewed as the MAXENT distribution without extra constraints. The likelihood is then:

$$\text{prob}(\{x_i\} \mid \mu) \propto \exp -(\Sigma_i (x_i - \mu)^2 \ / \ 2\ M^2)\ .$$

We see again that the MAXENT solution is independent. Hence the posterior p.d.f. is:

$$\text{prob}(\mu \mid \{x_i\}) \propto \exp -(\ N\ (\mu - \bar{x})^2 \ / \ 2\ M^2)\ .$$

where $\bar{x} = \Sigma_i\ x_i \ / \ (N+1)$.

The factor N+1 appears in the denominator of $\bar{x}$ (rather than N) because the prior chosen is equivalent to one zero sample. The distribution is a Gaussian of width $M \ / \ N^{1/2}$. Notice that this width depends only on the prior range and on the number of samples. In particular, it does not depend at all on the sample values! The reason for this is that the mean is very sensitive to outliers and, without the extra information that the variance exists, only the prior range gives a reliable estimate of the width.

3.2 Estimation of the Mean and Variance Together

The joint posterior probability distribution for the two parameters is:

$$\text{prop}(\mu\sigma \mid \{x_i\}) \propto \text{prob}(\mu\ \sigma)\ \text{prob}(\{x_i\} \mid \mu\ \sigma)\ .$$

The additional assumption of finite variance makes this problem better behaved than the previous one. We take as prior the form derived by Jeffreys (1939):

$$\text{prob}(\mu\ \sigma)\ d\mu\ d\sigma\ = d\mu\ d\sigma\ / \sigma;\ \mu \in (-\infty, +\infty),\ \sigma \in (0, +\infty).$$

Once more we compute the likelihood by the use of maximum entropy:

$$\text{prob}(\{x_i\} \mid \mu\sigma) \propto \sigma^{-N} \exp - \Sigma_i\ (x_i - \mu)^2 / 2\ \sigma^2.$$

and yet again the maximum entropy solution is independent. Hence the posterior probability becomes:

$$\text{prob}(\mu\sigma \mid \{x_i\}) \propto \sigma^{-(N+1)} \exp -N\ (\mu - \bar{x})^2 / 2\ \sigma^2.$$

where $\bar{x} = \Sigma_i\ x_i / N$ (the sample mean).

For comparison with the results of the previous section we can integrate out σ to get an estimate for μ alone, which has the form of Student's "t" distribution:

$$\text{prob}(\mu \mid \{x_i\}) \propto (\ V + N\ (\mu - \bar{x})^2\ )^{-N/2}.$$

where $V = \Sigma_i\ (x_i - \bar{x})^2$.

We should not be surprised to see that this is different from the estimation of μ alone, because in this later problem we have included the additional information that the variance exists.

If we instead integrate out μ we find that V/σ^2 is distributed as χ^2 with N-1 degrees of freedom.

4 TURNING THE PROBLEM AROUND

We have shown above how we can compute the posterior p.d.f. of a set of parameters $\{\lambda\}$ from some observations of exchangeable quantities $\{x_i\}$. Can we now use our estimated parameters to compute the probability of future samples from the distribution? We can certainly write:

$$\text{prob}(x \mid \{x_i\}) = \int d\{\lambda\}\ \text{prob}(x,\{\lambda\} \mid \{x_i\})$$

$$= \int d\{\lambda\}\ \text{prob}(x \mid \{\lambda\},\{x_i\})\ \text{prob}(\{\lambda\} \mid \{x_i\}).$$

We cannot go further without making assumptions about the meaning of the parameters. After all, why were we interested in the parameters $\{\lambda\}$ in the first place? We presumably consider them relevant to the p.d.f. and that they contain the information necessary to make REPRODUCIBLE predictions of x. If this is indeed the case, then the parameters $\{\lambda\}$ subsume all the detailed information contained in the sample $\{x_i\}$, which can then be dropped from the conditioning statement:

$$\text{prob}(x \mid \{x_i\}) = \int d\{\lambda\}\ \text{prob}(x \mid \{\lambda\})\ \text{prob}(\{\lambda\} \mid \{x_i\}).$$

The term prob($x \mid \{\lambda\}$) is found by MAXENT as before. This step is, of course, a "leap in the dark", but as Ed Jaynes has often pointed out, we have nothing to lose by going on with the calculation. If our predictions are useful, then so much the better; if they are completely wrong, then we have learnt that our assumptions about the parameters governing the probability distribution were incorrect, and this should help us to reformulate the problem.

This topic will be discussed further by Gull later in this conference, but we conclude by showing the result of assuming that the mean and variance are the relevant parameters. It should come as no surprise to statisticians or to physicists, being distributed about the sample mean as Student's "t" with N d.f. and having a width about equal to the sample standard deviation:

$$\text{prob}(x) \propto ((x - \bar{x})^2 + V)^{-N/2}.$$

5 CONCLUSIONS

We have come only a very small way towards our goal of providing Bayesian equivalents for non-parametric statistical tests. Our one new result is the estimation of the position of a quantile of a one-dimensional p.d.f. To achieve even this modest result we have had to use Bayes' theorem to estimate parameters from data and the principle of maximum entropy to assign probability distributions (likelihoods) given those parameters. When used together in this way we see clearly that MAXENT is not just a limiting form of Bayes' theorem, nor vice-versa; the two tools are complementary and the choice of which to use depends on the type of information available.

Our exercise has, however, shown where the real problem lies; there are no really "non-parametric" distributions, rather, it is our job to identify the parameters (or testable information) that are relevant to making reproducible predictions in any given case. This choice of a hypothesis space in which to work is our most difficult task and we have at present no systematic way of generating it. We can summarize our present understanding as follows:

1) We use Bayes' theorem to manipulate probabilities given conditioning data once those probabilities have been assigned.

2) It is the job of the MAXENT principle to assign probabilities given testable information once the appropriate hypothesis space has been defined.

3) It is for the moment left to our own information and common sense to determine which hypothesis space or parameters to use. We expect (Jaynes, 1985) that this process will involve an iterative improvement as we gradually learn the deficiencies of our earlier, simpler choices.

ACKNOWLEDGEMENTS

We thank the members of the Cambridge maximum entropy group, for their continued interest in these esoteric matters, particularly John Skilling, Geoff Daniell and Ed Jaynes.

REFERENCES

1 Fielden, J., A.J.B. Downes, J.R. Allington-Smith, C.R. Benn, M.S. Longair and M.A.C. Perryman (1983). Further High-Resolution Observations of Faint Radio Sources and the Angular Size-Flux Density Relation. Mon. Not. R. Astr. Soc. 204, 289-315.

2 Jaynes, E.T. (1968). Prior Probabilities. IEEE Trans. SSC-4, 227-241. Also reprinted in Jaynes (1983).

3 Jaynes, E.T. (1983). Papers on Probability, Statistics and Statistical Physics. Synthese Library Vol. 158, R.D. Rosenkrantz and D. Reidel (eds.).

4 Jaynes, E.T. (1985). Monkeys, Kangaroos and N. Paper presented at the 4th Workshop on Maximum Entropy and Bayesian Methods in Applied Statistics, Calgary.

5 Jeffreys, H. (1939). Theory of Probability. Oxford Univ. Press.

6 Siegel, S. (1956). Nonparametric Statistics for the Behavioral Sciences. McGraw-Hill.

7 Yule, G. and M.G. Kendall (1946). An Introduction to the Theory of Statistics. Charles Griffin, London.

GENERALIZED ENTROPIES AND THE MAXIMUM ENTROPY PRINCIPLE

J. Aczél and B. Forte

Apart from Physics, the first entropy measure was introduced by Hartley [8]. Here, as in what follows, we consider the entropy as measure of uncertainty about which of n events occurs from a partition $E_1, E_2, \ldots, E_N$ of the sure event. Hartley's entropy is just

$$\log_2 N \tag{1}$$

which looks pretty primitive, but is the only reasonable measure if we know only the number of events. Hartley has indeed consciously refused to work with probabilities. If we know at least how many (say, n) events have nonzero probabilities, we get the modification

$$\log_2 n \tag{2}$$

of Hartley's entropy used by Aczél, Forte and Ng [4].

If the probabilities $p_1, p_2, \ldots, p_N$ of $E_1, E_2, \ldots, E_N$ are known, we arrive at the Shannon entropy [11]

$$-\sum_{k=1}^{N} p_k \log_2 p_k \tag{3}$$

(extended by $0.\log 0 = 0$ by def. if some $p_k = 0$ is permitted)

or at its generalizations, for instance the Rényi entropies [10]

$$\frac{1}{1-\alpha} \log_2 \left(\sum_{k=1}^{N} p_k^{\alpha} \right) \tag{4}$$

(with $0^{\alpha}=0$ by def. if $p_k = 0$ permitted), where $\alpha \neq 1$

(they have the Shannon entropy as limit when $\alpha \to 1$ and have other desirable properties). The Hartley entropy log N is the special case of $p_1 = p_2 = \ldots = p_N = 1/N$ of both (3) and (4) without 0 probabilities, as is log n if 0 probabilities are permitted. Each is also the maximum of such Shannon and Rényi ($\alpha > 0$) entropies.

The Shannon and Rényi entropies for $\alpha>0$ ($\alpha \neq 1$) are (in a sense exact) lower bounds for average codeword lengths

$$\sum_{k=1}^{N} p_k n_k \tag{5}$$

or for exponential mean codeword lengths

$$\frac{\alpha}{1-\alpha} \log_2 \sum_{k=1}^{N} p_k 2^{(1-\alpha)n_k/\alpha},$$

respectively, (where the individual codeword lengths n_k are positive integers and

$$\sum_{k=1}^{N} 2^{-n_k} \leq 1)$$

These facts are related to other <u>inequalities</u> associated with these and other entropies, for instance the "how to keep the expert honest" inequality (see, e.g., [3]): An expert gives $q_1, q_2, \ldots, q_N$ as probabilities of the events (weather, market situations, etc.) E_1, $E_2, \ldots, E_N$ which in reality (or to the best of his knowledge) are $p_1, p_2, \ldots, p_N$. It is agreed that he gets paid the amount $f(q_k)$ if E_k happens. So his expected gain is

$$\sum_{k=1}^{N} p_k f(q_k).$$

How should the 'payoff function' f be chosen so that the expert's expected gain be maximal if he told the truth? Clearly f should then satisfy the inequality

$$\sum_{k=1}^{N} p_k f(q_k) \leq \sum_{k=1}^{N} p_k f(p_k). \tag{6}$$

It turns out that, without any further supposition on f, this inequality is satisfied for variable N or for fixed $N>2$ if, and only if, $f(p) = c \log_2 p+b$ ($c \geq 0$) so that the right hand side of (6) will be

$$c \sum_{k=1}^{N} p_k \log_2 p_k + b,$$

linking the subject to the Shannon entropy. Indeed, the "if" part of the above statement is equivalent to what is sometimes called the Shannon inequality:

$$-\sum_{k=1}^{N} p_k \log_2 q_k \geq -\sum_{k=1}^{N} p_k \log_2 p_k \quad (\sum_{k=1}^{N} p_k = \sum_{k=1}^{N} q_k = 1),$$

which in turn implies the above result that the Shannon entropy is a lower bound of the average codeword length

$$\sum_{k=1}^{N} p_k n_k.$$

There exist similar results with regard to the Rényi entropies.

If now we know about the events $E_1, E_2, \ldots, E_N$ more than just their probabilities, we arrive at inset entropies

$$c \sum_{k=1}^{N} p_k \log p_k + \sum_{k=1}^{N} p_k g(E_k) \tag{7}$$

where g is an arbitrary real valued function of the events, while c is an arbitrary constant (also the Hartley, Shannon and Rényi entropies could be multiplied by constants for most purposes). We can get to (7) among others again by the "how to keep the expert honest" method [1]: If we allow the payoff functions f to depend also upon the events E_k (not just their probabilities), then the previous 'keeping the expert honest' inequality is replaced by

$$\sum_{k=1}^{N} p_k f(q_k, E_k) \leq \sum_{k=1}^{N} p_k f(p_k, E_k) \tag{8}$$

and the general solution (again also for fixed $N>2$ and without further suppositions on f) will be $f(p,E) = c \log p + g(E)$ so that the right hand side of (8) becomes exactly (7).

Examples of other applications of inset entropies can be found, for instance, in the theory of gambling where Meginnis [9] interprets the second sum in

$$c \sum_{k=1}^{N} p_k \log p_k + \sum_{k=1}^{N} p_k g(E_k) \tag{7}$$

as the expected gain and the first as the "joy in gambling". Also, for the so-called continuous (partial) analogue of the Shannon entropy

$$-\sum_{k=1}^{N} p_k \log p_k, \text{ that is , for } -\int_{\alpha}^{\beta} \rho(t) \log \rho(t)\, dt,$$

(where ρ is the probability density function), the approximating sums of the integral are not Shannon entropies but inset entropies (7):

$$
\begin{aligned}
&-\sum_{k=1}^{N} \rho(\tau_k)\log \rho(\tau_k)(t_k-t_{k-1}) = \\
&-\sum_{k=1}^{N} \frac{F(t_k)-F(t_{k-1})}{t_k-t_{k-1}} \log \frac{F(t_k)-F(t_{k-1})}{t_k-t_{k-1}}(t_k-t_{k-1}) = \qquad (8)\\
&-\sum_{k=1}^{N} p_k \log p_k + \sum_{k=1}^{N} p_k \log \ell(E_k),
\end{aligned}
$$

where F is the probability distribution function, so $p_k = F(t_k) - F(t_{k-1})$ $(k = 1,2,\ldots,n)$ are the probabilities, $E_k = (t_{k-1}, t_k)$ and $\ell(E_k) = t_k - t_{k-1}$ [2]. For applications of (8) to geographical and economical analysis, see for instance [5, 6].

Further generalizations to entropies associated with random variables have been made by Forte (for instance, [7]).

Here we draw two consequences from the above:

(i) All (above) entropies are conditional on what we know about the events, the entropies being the measures of the remaining uncertainty about which of the events will happen. It is remarkable that, while each of the Hartley, modified Hartley, Shannon and inset entropies contains the previous ones as special cases, also each corresponds to more knowledge, that is, more conditions on the events.

(ii) All these entropies are connected to inequalities (and equations, see for instance [3]).

In another sense, the maximum entropy principle, of course, also relies on inequalities: we are looking among probability distributions $(p_1, p_2, \ldots, p_N)$, satisfying certain conditions (equations) for the one which maximizes a "suitable" entropy (makes it the largest, hence satisfying an inequality).

Perhaps the best known example is that the normal distribution maximizes the Shannon entropy

$$
-\sum_{k=1}^{N} p_k \log p_k
$$

under the conditions $p_1 + p_2 + \ldots + p_N = 1$, $a_1 p_1 + a_2 p_2 + \ldots + a_N p_N = 0$,

$$
a_1^2 p_1 + a_2^2 p_2 + \ldots + a_N^2 p_N = \sigma^2, \qquad (10)
$$

where $a_1, a_2, \ldots, a_N$ are the possible (real) values of a random variable, while σ^2 is its variance (also given). So, the Shannon entropy is a "suitable" entropy. Equations like (10) are again conditions representing our (partial) knowledge, this time about the otherwise unknown probabilities $p_1, p_2, \ldots, p_N$. In other words, there are two interpretations of the above argument: (i) It is usually considered to 'justify' the normal distribution, because the normal distribution maximizes the Shannon entropy (under appropriate conditions). (ii) We say that it can be interpreted also as 'justification' of the Shannon entropy, because the Shannon entropy is maximized by the normal distribution which is what we should get (under the same conditions), based on experience and usefulness.

We propose that "suitable" entropies should be introduced preferably as expressions, the maximization of which gives "useful" probability distributions. Entropies as measures of conditional uncertainty must take into account all kinds of informations provided by the problem, be they mathematical, scientific or "real life". The maximum entropy principle can be used to define some of those entropies. There remains much to do in this respect, even with regard to the above and other more or less generally used entropies.

This research has been supported in part by Natural Sciences and Engineering Research Council of Canada grants.

REFERENCES

1 Aczél, J. (1980). A Mixed Theory of Information. V. How to Keep the (Inset) Expert Honest. Journal of Math. Analysis and Applications 75, 447-453.

2 Aczél, J. (1978-80). A Mixed Theory of Information. VI. An Example at Last: A Proper Discrete Analogue of the Continuous Shannon Measure of Information (and its Characterization). Univ. Beograd. Publ. Elektrotehn. Fak. Ser. Mats. Fiz. Nr. 602-633, 65-72.

3 Aczél, J., and Z. Daróczy (1975). On Measures of Information and Their Characterizations. Academic Press, New York-San Francisco-London.

4 Aczél, J., B. Forte, and C.T. Ng (1974). Why the Shannon and Hartley Entropies are 'Natural'. Advances in Appl. Probab. 6, 131-146.

5 Batten, D.F. (1983). Spatial Analysis of Interacting Economies. Kluwer-Nijhoft, Boston-Hague-London.

6 Batty, M. (1978). Speculations on an Information Theoretical Approach to Spatial Representation. In Spatial Representation and Spatial Interaction. Nijhoft, Leiden-Boston, 115-147.

7 Forte, B. (1977). Subadditive Entropies for a Random Variable. Bol. Un. Mat. Ital. (5) 14B, 118-133.

8 Hartley, R.V. (1928). Transmission of Information. Bell Systems Technical Journal 7, 535-563.

9 Meginnis, J.R. (1976). A New Class of Symmetric Utility Rules for Gambles, Subjective Marginal Probability Functions, and a Generalized Bayes Rule. Business and Econ. Stat. Sec. Proc. Amer. Stat. Assoc., 471-476.

10 Rényi, A. (1961). On Measures of Entropy and Information. Proc. 4th Berkeley Symp. Math. Stat. and Prob. 1960, Vol. I, Univ. of Calif. Press, Berkeley, CA, 547-561.

11 Shannon, C.E. (1948). A Mathematical Theory of Communication. Bell Systems Technical Journal 27, 379-423, 623-656.

THE PROBABILITY OF A PROBABILITY

John F. Cyranski
Physics Department
Rockhurst College
Kansas City, Missouri

ABSTRACT

MAXENT (MAXimum ENTropy principle) is a general method of statistical inference derived from and intrinsic to statistical mechanics. The probabilities it produces are "logical probabilities" -- measures of the logical relationship between hypothesis and evidence. We consider the significance and applications of the "logical probability" of such probabilities. The probability of a "logical probability" is shown to be the probability of the evidence used for the "logical probability". This suggests a hierarchy of logics, with "evidences" defined as sets of probabilities on the preceding "logic". Applications to reliability theory are described. We also clarify the meaning of MAXENT and examine arguments in a recent article in which temperature fluctuations are introduced in thermal physics.

1 INTRODUCTION

A method fundamental to statistical physics is the maximization of entropy. In recent years, this method has been recognized as a general procedure for statistical inference based on the fact that "entropy" is essentially a measure of information uncertainty [1]. The probabilities one obtains using MAXENT (as the "Maximum Entropy Principle" is now called) have a natural interpretation which has not been generally recognized, even by advocates of the procedure. This is the "degree of belief" (DOB) interpretation [2] -- that "probability" is a measure of the logical relationship between two propositions: $p(H \mid E)$ expresses a (normalized) "degree of belief" (DOB) in the relationship of hypothesis H to evidence E. Indeed, MAXENT asserts precisely the (statistical) consequences of assumed evidence since it is based on the idea that one should choose as probability one which maximizes "uncertainty" consistent with the evidence. (See below.)

Within the DOB interpretation, it is meaningless to proclaim a "probability of a probability": "A question about the probability of a probability has no more point than a question about the probability of the statement that 2+2=4 or that 2+2=5, because a probability statement is, like an arithmetical statement, either (logically) true or (logically) false; therefore, its probability with respect to any evidence is either 1 or 0." [2]

While it is possible to interpret "probability of a probability" to mean the DOB of a relative frequency (RF), there are reasons to do otherwise. First, this does not seem to be the intent of Bayesians whose probabilities are exclusively DOB's or subjective [3]. Second, there is neither necessity for nor any reality to relative frequencies. For, it is known that given evidence that defines a RF (over a finite number of trials, naturally; a true RF demands an infinite number of trials which is why no such thing exists) MAXENT yields as DOB based on this evidence precisely the given RF [4,5]. Thus, the DOB concept is more general than RF, including the latter as a special case. It follows from this brief review that exclusive use of the DOB interpretation with MAXENT as DOB calculus permits greater generality and flexibility, and is logically more consistent than the so-called "objective probabilities". (Lest the reader mistake DOB's for "subjective probabilities", we need only point out that given the same evidence, all will calculate the same DOB using MAXENT. "Logical probabilities" are not subjective!)

The purpose of this paper is to extend the understanding of the deeper implications of this MAXENT-DOB scheme of statistical inference. Our focus on the DOB of a DOB issue reveals a novel interpretation for "evidence" (entirely consistent with MAXENT usage) which at the same time suggests how "logic" can be built from the bottom up (as opposed to the abstract formal logic, which is difficult to relate to concrete realizations -- semantics). Our approach sets the stage for more rigorous generalizations of MAXENT. We indicate some possibilities in passing. What is perhaps more of interest is the immediate application of the abstraction to problems related to the "reliability" of DOB assignments. In particular, we apply our approach to examine issues raised in Ref. [6], wherein a combination of MAXENT and Bayesian parameter estimation is used to derive some surprising results. (While many practitioners of MAXENT believe that MAXENT is an intimate part of Bayesian methodology, we feel that MAXENT is logically more consistent than Bayesian methods and can reproduce the correct results of Bayesian statistics: Indeed, Bayes conditionalization is a special case of MAXENT [7].) It is only, however, by examining MAXENT in the abstract context of measure theory -- rather than by looking at discrete case examples -- that one begins to appreciate the generality of the method. Thus, rather than following the tradition of trying to convince skeptics that MAXENT can be derived from game theory [8a] or as a consequence of Laplace's Principle of Indifference [8b], or from combinatorics [9a], or as an "optimal algorithm" [9b] -- among others! -- we will proceed from a relatively general perspective. The reader is asked to set aside all prejudices and allow our arguments if not to persuade, at least to stimulate fresh thinking.

Probability functions have as domain some type of "logic" -- usually a Borel algebra of subsets of a set [10,11]. While formally one may consider a logic as including all statements of ordinary language, a prerequisite of Bayesian methods [12], when one makes inferences it is typically about a particular class of objects! Thus, one may be concerned about the attributes of a gas. The set of attributes of a such class of objects is a "sublogic" in the sense that propositions about the color of your eyes are not in this logic. In physics the attributes necessarily form a Borel algebra of subsets of the class of all the objects because physical attributes must concern measurable properties of the objects. Thus, if we define a "scale" -- a function **T**:$0 \rightarrow R$ (random variables=measurable functions are distinguished throughout in boldface print) where 0 is the object class and R the real line -- then every Borel set Δ of reals (corresponding to precision limits) must be the image of an "attribute" H in the logic. (While many physicists ignore this, one can never measure a real number, only an interval [13].) This induces a Borel structure on the logic, which we denote by B(0). Much more general possibilities exist [14], but in (classical) physical applications ordinary measure theory suffices for our arguments. Note that if one has a list of measurable functions that are independent and characterize the physical system (such as position, momentum, energy, and time characterize a "free particle"), the "hypotheses" or propositions in the logic define subsets of particles sharing common value ranges of these variables. In the infinitely precise limit (classical mechanics), specific values of position, time, momentum, and energy, uniquely select a singleton -- that "free particle" having precisely these values of the variables. These, then, are the types of hypotheses relevant to physics.

Clearly, in order to develop a complete "natural language", the starting point of Bayesians, one must develop the individual logics of different object types and find a consistent scheme of imbedding these sub-logics in the larger scheme, where different objects can interact. This is usually done (in physics) using some form of "tensor product" [15], but a complete "bottoms up" approach to logic remains to be done. One step in the construction is to determine the relation of the logic of attributes of a system to the logic of evidence statements about the system. In particular, if a hypothesis is realized as a set of objects (that satisfy the hypothesis), what is an "evidence"?

Our answer to this question evolves from the MAXENT procedure. Typically, evidence for MAXENT is a statement that the expected value of a random variable (RV) takes on a certain value. (The expected energy of a gas molecule is specified, for example.) THIS EVIDENCE DEFINES A SET OF PROBABILITIES ON B(0) consistent with the constraint defined by the statement. Note that the "object class" of such evidence statements is never the same as that defined by the hypotheses. Thus, although formally -- as we show below -- it is possible to consider the set of evidence statements as a Borel algebra, just as is the set of hypotheses, these algebras are distinct. Since statements such as "H&E" -- the conjunction of hypothesis H with evidence E -- cannot be interpreted isomorphically as a set of objects $H \cap E$, such statements are nonsense in Boolean logic [16a]. Another way to see the problem is to note that evidence E is a semantic statement and thus is necessarily in a "metalanguage" (outside the language containing the hypotheses) [16b].

One consequence of this observation is that MAXENT is carefully distinguishing between hypothesis and evidence -- in perfect agreement with the DOB concept of probability [18]. However, the fact that "H&E" has no interpretation (in the sense of realization as a set of objects) suggests that Bayesian statistics rests on extremely insecure foundations for it is based entirely on the formal equality of all logical propositions. In particular, Bayes' Theorem (which is perfectly correct if applied to attributes from a single Borel algebra of subsets [11]) states that $p(H\&E) = p(H|E)p(E) = p(E|H)p(H)$; the validity of this basic relation depends on the meaningfulness of "H&E"! [19].

Thus, MAXENT clearly is doing something different from Bayesian statistics. Moreover, it is much more careful. In the next section we will analyze in more detail the structure of "evidences" -- including "prior evidence". In effect, we identify evidence with a subset of measures. Just as not every set of objects is "measurable" (Borel), not every evidence statement is meaningful to MAXENT. We suggest, however, that only evidence that can be expressed as a Borel set of probability measures is valid evidence for inference. This, it turns out, is not as restrictive as it appears, as MAXENT can be applied to a wide variety of evidence sets. In Section 3 we will consider some applications of the DOB of a DOB (reliability).

2 EVIDENCES AND INFORMATIONS

Let $M(O)$ represent the space of all sigma-finite measures on $B(O)$. One subset of $M(O)$ is the class of measures m which satisfy certain prior constraints: e.g., $m(A)=0$ for certain A's in $B(O)$ (a priori "impossible" properties: A physical example would be a statement that a particle can travel faster than light). Also, invariance properties of m under certain a priori "degrees of freedom" can be imposed [20]. One thus finds that prior evidence effectively defines a sub-class $M_0(O)\subseteq M(O)$. This has a most important consequence, namely that the class $M_0(O)$ consists of mutually equivalent ($m \ll m'$ and $m' \ll m$--"$\ll$" means absolutely continuous with respect to) measures [21]. Thus, any m in $M_0(O)$ asserts the same basic prior knowledge. Note that "There exists o satisfying H" corresponds to $m(H)>0$. Prior knowledge effectively determines existential quantifiers for the logic $B(O)$.

We assume that a posteriori evidence provides us with a DOB consistent with our prior assertions (otherwise, these prior arguments must be reconsidered). This means that we must consider $M_1(O)\subseteq M(O)$ defined by

$$M_1(O)= \{ p\varepsilon\ M(O) \mid p(O)=1, p \ll m, \text{ any } m \text{ in } M_0(O)\} \tag{1}$$

The potential DOB's thus are normalized measures which vanish for a priori "impossible" hypotheses. Beyond this we note that the effect of evidence is to narrow-down the class of admissable representative probabilities. For example, if we consider a hypothesis H then if E is the statement "H is definitely true", then E requires a DOB satisfying

$p(H|E)=1$. However, unless only H and its negation are in the logic, many probabilities satisfy this condition -- in fact, the set

$$E = \{p \,\varepsilon M_1(0) \mid p(H) = \int dm[dp/dm]C_H(o) = 1\} \quad (2)$$

where $C_H(o) = 1$ if o is in H, $= 0$ otherwise.

We are thus led to identify the evidence statement with the subset of $M_1(0)$ containing all probability measures consistent with the statement. In effect, then, the "object class" of (posterior) evidence statements is the set $M_1(0)$, and the statements form an algebra (not necessarily Boolean) defined by the subsets $E \subseteq M_1(0)$ that are allowed. As $M_1(0)$ can be identified with the space of functions $\{f: 0 \rightarrow [0,\infty] \mid \int dm\, f=1\}$ and this space has a natural topology [22] which can be used to generate a Borel algebra, we assume the logic of evidences to be Borel -- $B[M_1(0)]$. (One often uses the "weak topology", applicable when 0 is a metric space [33].)

Note that the construction can continue: Let $M[M_1(0)]$ be the class of all sigma-finite measures on $B[M_1(0)]$. This includes "probabilities" of sets of probabilities. One can impose prior conditions on $M[M_1(0)]$ to obtain $M_0[M_1(0)]$ and $M_1[M_1(0)]$ (probabilities in $M[M_1(0)]$ consistent with the prior). Evidence at this level is a subset $E' \subseteq M_1[M_1(0)]$ and represents, in words, evidence about the original evidences (sets of probabilities on B(0)). In this sense, "probability of a probability" has meaning: Explicitly it means a DOB assigned to the evidences (Borel sets of probability measures) on which the original DOB's are based. This is not quite the same as a DOB of a DOB: However, $dp'(p)/dm'$, the Radon-Nikodym probability density, has precisely the interpretation of a probability function on the space of probability measures (not necessarily DOB's).

Schematically we thus are led to a logical hierarchy generated by $L_0 = B(0)$: L_0, L_1 $(=B[M_1(0)])$, L_2 $(=B[M_1[M_1(0)]])$,... Here L_0 is the original logic, L_1 is a class of evidences for L_0, L_2 is a class of evidences about the evidences in L_1 (reliability of the evidences in L_1 based on "external" knowledge), and so on. We shall not pursue this provocative connection with the theory of logic here.

At this point we note that given a "prior measure" $m: B(0) \rightarrow [0,\infty]$, the information embodied in a "probability" measure p on B(0) is defined by

$$S(p,m) = \int dp \ln(dp/dm), \quad p \ll m$$
$$= \infty \quad \text{otherwise} \quad (3)$$

This "relative entropy" or "cross-entropy" (among numerous aliases) has been justified in various ways as an appropriate measure of information [23] and we shall accept it without question. Note that we only require that m be a sigma-finite measure in general. The negative of (3) is "uncertainty" (entropy).

What is the information content, I(E), of the evidence? Clearly, if $E \subset E'$, $I(E) \geq I(E')$, since E constrains the probabilities more than does E' and is thus "more informative". The extreme case is $I(\{\}) = \infty$. If one further requires that $I(E) = F[\{S(p,m) \mid p \in E\}]$, then one functional form is $I(E) = \inf \{S(p,m) \mid p \in E\}$. A consistent and reasonable way to define the measure $p(H|E)$ is to identify it with that p in E (if it exists) for which $I(E) = S(p,m)$. Note that (a) this method avoids the questionable Bayesian method and (b) this method (essentially MAXENT, as "information", eq.(3), is the negative of the usual entropy [23]) is extremely conservative, for precisely the information content of the evidence -- no more, no less, is assumed [4]. In effect, MAXENT is a method for obtaining the logical consequences of the "evidence", assuming it is "total" and that no irrelevant or extraneous information is included [24a].

We remark that MAXENT makes no claim to obtain "the probability underlying the process". Indeed, there is no such thing, according to the DOB viewpoint. Rather, MAXENT tells us which probability measure most conservatively represents the actual evidence we used. It is therefore a kind of "evidence tester", for the MAXENT predictions based on E can be compared with experimental data (new E). If there is good agreement between predictions and experiment, this means that the original E is all the evidence needed to make accurate predictions. (This is a remarkable feature of thermodynamics, where evidence about very few variables is sufficient to determine all the macroscopic features of the system.) More often, the new evidence refines or contradicts the original evidence, in which case one must recalculate the DOB using MAXENT and the new evidence [24b].

We have suggested above that the evidences form the Borel algebra $B[M_1(O)]$. Do MAXENT solutions exist on arbitrary Borel sets of measures? Relatively little work has been expended on characterizing the existence of MAXENT solutions [23,25,26,27]. One of the most general criteria, perhaps, is the following: If $M_1(O)$ is a metric space, suppose E is closed in the weak topology and that $I(E) > -\infty$. Then a MAXENT solution exists in E [26]. (E must be convex to ensure uniqueness.) Typically, evidence has the form $E = \{p \mid \int dp\mathbf{T} = \tau\}$ where $\mathbf{T}$ is bounded and continuous. If $I(E) > -\infty$, such evidences have MAXENT solution. More generally, if G(p) is convex and lower semicontinuous (weak topology) then $E = \{p \mid G(p) \leq \gamma\}$ is convex and closed, admitting MAXENT solution when $I(E) > -\infty$. MAXENT clearly imposes minimal restriction on the evidences. Also, since the closed sets generate the algebra, appropriate limiting properties should permit extension of MAXENT to open sets. Thus, the union of N closed sets is open in the limit $N \to \infty$, and if the sequence of MAXENT solutions thus defined has a limit, then this limit is reasonably the DOB for the open set. On the other hand, the intersection (conjunction) of infinitely many closed sets may be a singleton, $\{p^*\}$. In this case we say we have total knowledge since any other evidence either changes nothing or contradicts the total knowledge. An interesting example of total knowledge is given by the sequence $E_N = \{p \mid \int dp\mathbf{X}_N(o) = (1/N)\Sigma\langle\mathbf{X}(o_i)\rangle\}$, where $\mathbf{X}(o_i)$ = 1 if hypothesis H occurs in the ith trial (o_i), 0 otherwise; where N is the number of samples, and $\mathbf{X}_N(o)$ is the sample average [11]. Then MAXENT yields [4]

$$p(H|E_N)=\langle X_N\rangle=\text{No. of times H occurred in N trials} \tag{4}$$

Provided the sequence converges to $\{p^*\}$, this limit p^* defines the "objective probability" as "total knowledge" based on counting the frequency of H in $N\to\infty$ trials.

Finally, we note that the MAXENT DOB explicitly depends on the prior measure m. This suggests that we generalize MAXENT to seek $(p^*,m^*)\varepsilon ExM_0(0)$ such that $S(p^*,m^*)=I'(E)=\inf\{S(p,m)|(p,m)\varepsilon ExM_0(0)\}$. We conjecture that the usual arguments for the prior select m that satisfies $I'(E) \approx \inf\{S(p,m)|p\varepsilon E\}$. Not only would this explain the success of the usual MAXENT applications, but this would also explain why MAXENT solutions are relatively insensitive to "reasonable choices of prior [28]. (This could also happen if E is "close to" total knowledge.)

3 APPLICATIONS TO "RELIABILITY"

As illustration of the consequences of the logical hierarchy we have established, we consider application of MAXENT at the secondary level: Find $p'(E|E')$, where E' is in L_2, for all E in L_1. Since $p(H|E)$ is presumably given (by applying MAXENT on E in L_1), this (effectively) amounts to calculating the DOB of a DOB. Assume m' is a fixed prior over L_2 and let p' be any probability in $M_1[M_1(0)]$. Then $E'\subseteq M_1[M_1(0)]=\{p' \text{ on } L_1|p'\ll m'\}$.

To be explicit, we first treat a case suggested by S. Gull. Assume as evidence E' in L_2 that the relative information is no greater than W (with $m(0)=1$ so $S(p,m)\geq 0$ for all p):

$$E' = \{p'\varepsilon\, M_1[M_1(0)]\,|\int S(p,m)\, dp'(p)\leq W\} \tag{1}$$

MAXENT yields for p' on L_1 the density:

$$dp'[p]/dm' = \exp\{-\, bS(p,m)\}/Z(b) \qquad \text{a.e.}[m'] \tag{2}$$

$$Z(b) = \int dm' \exp\{-\, bS(p,m)\} \tag{3}$$

$$W = -\,d\ln(Z[b])/db \qquad [\text{condition for b}] \tag{4}$$

from which we see that

$$S(p',m) = -\, bW - \ln(Z[b]) \tag{5}$$

Suppose that the bound W is lowered by $\Delta W<0$. Then, as $\Delta[\ln Z]=-W\Delta b$ and the functional form of S is fixed, we obtain

$$\Delta I(E') = -\, b\Delta W \tag{6}$$

so that provided $b>0$, the minimal information about the evidences is increased. If m' is itself a probability, then this clearly indicates that a decrease in the average information content of the evidences increases the minimum information about the evidences. Put another way, our uncertainty about the original set of evidences is smaller than our uncertainty about the new set, defined by the lower average cross-entropy. The lower intrinsic average uncertainty of the evidences is compensated by an increase in the uncertainty about the choice of evidence. This suggests a kind of complementarity between the types of evidence.

Let us now consider a potentially practical application of "probability of a probability": What is the DOB that a weather forecaster is "correct" in his predictions? (This problem was suggested in different form by A. Russell.) Let us assume for simplicity that there are two mutually exclusive alternatives only one of which is valid on a given day -- e.g., "it rains" = H and "it does not rain = H^c. The forecaster attempts to classify the days o O as to whether they are in H or not. This can be viewed as a communication problem between the sub-algebra $K=\{\ \{\}, H, H^c, O\}\subseteq B(O)$ and $B(O)$ (generated by the atoms $\{o\}$), wherein the communication channel linking days with outcomes is a probability measure $p:K\times B(O)\to[0,1]$ [30]. Thus, $M_1(O\times O)$ is the space of such "channels". The forecaster supplies, based on his evidence E, the <u>conditional</u> probability measure $p[(H\,|\,|o\ )]$. While how he determines this is irrelevant to the evaluation of his "reliability", we note that the process ultimately involves a kind of "regression" analysis [29].

Now an error in classification defines <u>distortion</u> [30]. A natural measure of this is

$$d(H'),o)=C_{H'^c}(o);\ H'\varepsilon K \tag{7}$$

which, for example, assigns unit distortion if $o\varepsilon H$ is misclassified as being in "not H", and zero distortion if o is correctly classified. If one calculates the expectation of (7) based on any "channel" p one obtains the <u>probability of mis-classification using p</u>:

$$\mathbf{x}(p)=\Sigma_{H'\varepsilon K,o\varepsilon O}p(H',\{o\})d(H',o)=p(H,H^c)+p(H^c,H):\quad M_1(O\times O)\to[0,1] \tag{8}$$

which, as we have noted, is a linear functional on $M_1(O\times O)$, assumed to be measurable. This expectation is a natural criterion for evaluating the forecaster.

Let us assume that we observe the forecaster on a set $O'\subseteq O$ of N days. To <u>estimate</u> the probability of mis-classification we employ the predictions, $p[(H'|\,|o)]$ $(o\varepsilon O,H'\varepsilon K)$ and assume that all days in O' are equiprobable: $p[\{o\}]=1/N$ for $o\varepsilon O'$. <u>Post facto</u> we know which $o\varepsilon O'$ are in H or not. Thus, we readily calculate

$$D=\Sigma_{o\varepsilon O'}p(H',\{o\})d(H',o)=(1/N)\{\Sigma_{o\varepsilon O'\cap H^c} \quad p[(H||o)] + \Sigma_{o\varepsilon O'\cap H} \quad p[(H^c||o)] \tag{8}$$

With perhaps some kindness, we finally assume that the expected probability of misclassification not exceed this estimate. Thus, our evidence for "reliability" is:

$$E'=\{p'\varepsilon M_1[M_1(O)]|\int dp'(p)\ \mathbf{x}(p) \leq D \quad \text{a.e.}[m'] \tag{9}$$

Applying MAXENT to this we obtain

$$dp'(p)/dm' = \exp[-b\mathbf{x}(p)]/Z(b) \tag{10}$$

$$Z(b)=\int dm'(p)\ \exp[-b\mathbf{x}(p)] \tag{11}$$

$$D=-d(\ln[Z(b)])/db \tag{12}$$

It is reasonable to assume that the prior m' defines a distribution function for $\mathbf{x}$ that has uniform density on [0,1]:

$$F_{m'}(x)=m'\{p\varepsilon M_1(OxO)|\mathbf{x}(p)\leq x\}= \int_{-\infty}^{x} dv\ C_{[0,1]}(v)\} \tag{13}$$

Using this, (11) becomes

$$Z(b) = \int_0^1 dx\ \exp[-bx]=[1-b]/b \tag{11'}$$

and using (12) we find

$$D=1/b - \exp(-b)/[1-\exp(-b)] \tag{14}$$

We can now determine the probability that the probability of misclassification not exceed q:

$$F(q) = p'\{p|\mathbf{x}(p)\leq q\}= \int_0^q dx\ e^{-bx}/Z(b)=[1-e^{-bq}]/[1-e^{-b}] \tag{15}$$

Using these formulae we can calculate for a given D the confidence level [1-F(q)] x 100% that the probability of a correct forecast be at least 1-q. For example, if we wish to determine our confidence that the probability of a correct forecast is at least 1-q=.8, we obtain the following confidence levels for various estimates of the distortion (D):

D =	.98	.90	.80	.72	.58	.46	.34	.23	.19	.05
% Confidence =	0	.003	1	4	13	24	38	56	64	98

In particular, if during the test period the sample probability of misclassification, D, is .05, then we are 98% confident that the probability that this forecaster will be correct is .8 (loosely, that the forecaster is right at least "80% of the time"). If instead we are only interested in the confidence that the forecast is right at least "half the time", q=.5 and we get

D =	.98	.90	.80	.72	.58	.46	.34	.23	.19	.05
% Confidence =	0	.6	8	18	38	56	73	88	96	98

Thus, if the sampled probability of forecast error during the test period is .46, we have 56% confidence that the forecasts are right at least "half the time".

The above can easily be extended to the case of N alternative (but exclusive) hypotheses: Simply replace $\mathbf{x}(p)$ by the sum of $p(H_n, H_n^c | E)$ over n. With H_n defining a precision interval for a physical quantity (such as the concentration of chromium in a sample as determined by a laboratory [31]) one can determine the "reliability" of laboratories or instruments by obtaining D from a sequence of tests with known values of the quantity.

As a final illustration, we examine the arguments of Ref. 6 from our perspective. In thermodynamics, extensive and intensive variables are on an equal footing, thanks to the Legendre transformations. However, in statistical physics, some variables (such as energy $\mathbf{E}$) can fluctuate (have non-zero variance $\sigma^2 = \langle[\mathbf{E} - \langle\mathbf{E}\rangle]^2\rangle$), but the conjugate variables (inverse temperature $\boldsymbol{\beta}$ in this case) are determined without dispersion. The stated goal of [6] is to achieve "democracy" at the statistical level by deriving a dual distribution (on temperature) that reflects reliability of the assignment $\langle\mathbf{E}\rangle, \beta$ ($\beta = 1/kT$, where k is Boltmann's constant and T is absolute temperature). An intriguing consequence of the arguments of [6] is the existence of an "inherent" uncertainty relation, "$\sigma_E^2 \sigma_{\boldsymbol{\beta}}^2 = 1$".

Since "temperature" is a quantity that characterizes the "equilibrium ensemble" [32] it appears that [6] is seeking a way to treat the probability of the ensembles which, as each ensemble defines a Maxwell-Boltzmann distribution of molecular energies, amounts to the DOB of a DOB. Unfortunately, the authors of [6] do not recognize the unique thermodynamic significance of the Boltzmann distribution: Derived from maximum entropy given fixed expected energy, it is necessarily a DOB that defines a logical relation between hypotheses H (Borel sets of energies) and evidence E (defined by the Lagrange parameter β). One cannot use parameter estimation methods for β because these require that $p(H|E)$ be either a joint probability or a conditional probability measure (as assumed in [6]). In other words, while it is possible (the arguments are far from being simple) to formally estimate β under the assumption that the ensemble is drawn from a Maxwell-Boltzmann population, such an approach is inconsistent with thermodynamics [24a]. In particular, the original problem of thermodynamic "democracy" may not be addressed by this altered theory of thermodynamics.

It is actually easy to see that "democracy" already exists in statistical thermodynamics without recourse to Bayesian arguments. In what follows, O is the set of micro-systems (molecules, etc.) whose thermodynamics is being considered.

To begin with, we note that at the thermodynamic level all variables are measurable functions on $M_1[O]$. For example thermodynamic energy is the expected energy $\langle \mathbf{E}\rangle = \mathbf{F}_{\mathbf{E}}(p) = \int dp\ \mathbf{E}(o)$ -- clearly a functional of the probability measures. Also, the inverse temperature, β, is defined by the average energy (using MAXENT). The evidence is, with $\varepsilon = \langle \mathbf{E}\rangle$

$$E[\mathbf{E},\varepsilon]=\{p \mid \int dp\ \mathbf{E} = \varepsilon\} \tag{16}$$

and $\beta=\beta(\varepsilon)$ is defined via

$$\varepsilon=-\partial \ln Z[\beta]/\partial\beta ; Z[\beta]=\int dm\{\exp - \beta E\} \tag{17}$$

The Boltzmann distribution that results is simply

$$dp(o)/dm=\exp\{-\beta E(o)\}/Z(\beta) \quad \text{a.e.}[m] \tag{17b}$$

This suggests that as a random variable,

$$\boldsymbol{\beta}_{E}[p) = \beta(\varepsilon) \text{ for } p\varepsilon E[\mathbf{E},\varepsilon] \tag{18}$$

which defines $\boldsymbol{\beta}: M_1(O)\rightarrow R$ provided $M_1(O)$ equals the union of all $E[\mathbf{E}, \varepsilon]$ for ε in the range of $\mathbf{E}$. In other words, temperature depends on the probabilities through its dependence on the evidence $\langle\mathbf{E}\rangle$; and MAXENT relates all probabilities defining the same $\langle\mathbf{E}\rangle$ to the same value $\beta(\varepsilon)$. From this viewpoint, there is no asymmetry between ensemble energy $\mathbf{F}_{\mathbf{E}}(p)$ and "temperature" $\boldsymbol{\beta}_{\mathbf{E}}(p)$, but there is an essential asymmetry between $\mathbf{E}(o)$ and $\boldsymbol{\beta}_{\mathbf{E}}(p)$ -- the energy random variable is a function of the objects (molecules in a gas) while the temperature is a function of the probabilities over the Borel sets of the objects! Thermodynamics deals with the variables $\mathbf{F}_{\mathbf{E}}$ and $\boldsymbol{\beta}_{\mathbf{E}}$, not $\mathbf{E}$ and $\boldsymbol{\beta}_{\mathbf{E}}$.

Fluctuations in "temperature" clearly refer to the statistics of the equilibrium systems: i.e., we may not know $\varepsilon=\langle\mathbf{E}\rangle$ (as opposed to $\mathbf{E}$) accurately. (In practice, we seek β directly with a thermometer and use (17) to estimate ε.) Ref. 6 employs a subtle combination of MAXENT (over $M_1[O]$, although not stated) and Bayesian parameter estimation (which requires a "joint probability" on $B(O)xB(M_1[O])$ to be consistent! [19]) in order to ascertain the "best" value for β based on the average energy of a small sample of the gas molecules. A key step is the (traditional Bayesian) requirement that this choice minimize the "expected loss" for a loss function (shown by consistency arguments to be quadratic in β). Such a requirement is, we feel, alien to the conservative spirit of MAXENT in that it imposes optimism on physics. Why, in other words, should the physical temperature minimize "average loss"? (The same criticism applies in general to Bayesian methods -- another argument in favor of exclusive use of MAXENT!) While ordinarily "reliability" can only be tested by sampling many such estimates to see the variation in $\langle \mathbf{E} \rangle$, apparently "optimism" in [6] supplants such sampling.

Let us consider I trials in which $\varepsilon_i = (1/N)\Sigma\langle \mathbf{E}(o_{k(i)})\rangle$ is the sample average on the ith trial and $\langle \mathbf{E}(o_{k(i)})\rangle$ is the value of the estimate (=observed value) of $\mathbf{E}(o)$ for the kth molecule on trial i. (We follow the prescription of [6] in estimating average energy by observing the energy of a small number N of molecules allowed to escape through a hole in the vessel holding the gas.) On each trial, MAXENT determines the estimate $\langle \beta(p)\rangle_i = \beta(\varepsilon_i)$ (see above), so we may use the average of these estimates as evidence:

$$E'[\boldsymbol{\beta},b]=\{p'\varepsilon M_1[M_1(0)] \mid \int dp'(p) \quad \boldsymbol{\beta}_{\mathbf{E}}(p)=b\} \tag{19}$$

where $b=(1/I)\Sigma\rho(\varepsilon_i)$. Using MAXENT we find

$$dp'(p)/dm'=\exp\{-\gamma(b)\beta_{\mathbf{E}}(p)\}/Z[\gamma(b)] \quad \text{a.e.}[m'] \tag{20a}$$

$$Z[\gamma(b)]= \int dm'(p)\ \exp\{-\gamma(b)\boldsymbol{\beta}_{\mathbf{E}}(p)\ \} \tag{20b}$$

$$b=-\partial \ln Z(\gamma)/\partial\gamma \quad [\text{for } \gamma=\gamma(b)] \tag{20c}$$

Using (20) one can calculate (in principle)

$$\sigma^2_{\mathbf{F}_{\mathbf{E}}}= \int dp'(p)\ [\mathbf{F}_{\mathbf{E}}(p) - \langle\mathbf{F}_{\mathbf{E}}\rangle]^2 \tag{21a}$$

$$\langle\mathbf{F}_{\mathbf{E}}\rangle = \int dp'(p)\ \mathbf{F}_{\mathbf{E}}(p) \tag{21b}$$

which define the variance and the expected average energy (= thermodynamic energy) and

$$\sigma^2_{\boldsymbol{\beta}_{\mathbf{E}}} = \int dp'(p)\ [\boldsymbol{\beta}_{\mathbf{E}}(p) - b]^2 \tag{22}$$

which defines the temperature fluctuation. It appears from the formalism of [6] that one must formally define

$$dp(o)/dm = \exp\{-\langle\boldsymbol{\beta}_{\mathbf{E}}\rangle\mathbf{E}(o)\}/Z[\langle\boldsymbol{\beta}_{\mathbf{E}}\rangle] \tag{23a}$$

$$dp'(p)/dm'=\exp\{-\langle\mathbf{F}_{\mathbf{E}}\rangle\boldsymbol{\beta}_{\mathbf{E}}(p)\}/Z[\langle\mathbf{F}_{\mathbf{E}}\rangle\} \tag{23b}$$

in order to obtain $\sigma^2_{\mathbf{E}}\sigma^2_{\boldsymbol{\beta}_{\mathbf{E}}}= 1$ of [6]. More feasible than (23a) is to replace the usual MAXENT-Boltzmann ensemble by

$$\langle dp/dm\rangle(o)=\int dp'(p)[\exp\{-\beta_E(p)E(o)\}/Z(\beta_E(p))] \qquad (24)$$

However, neither option is required in order to understand the nature of thermodynamics, "fluctuations" and reliability.

4 CONCLUSIONS

In this paper we have attempted to motivate interest in the deeper implications of the MAXENT principle of inference. Realizing that the method remains controversial, we believe that part of the controversy is due to the absence of a reasonable exposition of the entire inference process. By this we mean a careful explanation of the domain of the method (a Borel algebra of subsets of a particular class of objects), the nature of "evidence" in (at least) statistical inference (subset of probability measures consistent with the verbally expressed constraint), and a motivation for the procedure divorced from both Bayesian and frequentist concepts. What the method is really trying to accomplish simply cannot be discovered by examining special cases and discrete examples -- welcome as these are to indicate the usefulness of the procedure. There remains a need for a more general mathematical analysis of issues such as the existence of MAXENT solutions based on evidences that are not simply linear, convex sets. We have suggested some of the possibilities in this paper -- possibilities that indicate even wider application of the method. In particular, we considered how the DOB of a DOB -- an apparently philosophical "pseudo-problem" to some -- actually leads to simple and general ways of discovering the reliability of DOB assignments. In so doing, we have had to criticize an article by proponents of MAXENT for the casual mixing of Bayesian and MAXENT formalisms. We hope that our results will stimulate further careful investigation of this most promising method of inference.

REFERENCES AND NOTES

1a Jaynes, E.T. (1957). Phys. Rev., Vol. 106, 620.

1b Jaynes, E.T. (1957). Phys. Rev., Vol. 108, 171.

2 Carnap, R. (1950). Logical Foundations of Probability. Chicago U. Press, Chicago.

3 Fine, T. (1973). Theories of Probability. Academic Press, New York.

4 Cyranski, J.F. (1978). Found. Phys., Vol. 8, 493.

5 Cyranski, J.F. (1982). Inform. Sci., 26, 257.

6 Tikochinsky, R. and R.D. Levine (1984). J. Math. Phys., Vol. 25, 2160.

7 Williams, P.M. (1980). Brit. J. Phil. Sci., 31, 131.

8a Topsoe, F. (1979). Kybernetika, 15, 8.

8b Hadjisavvas, N. (1981). J. Stat. Phys., 26, 807.

9a Jaynes, E.T. (1979). The Maximum Entropy Formalism, p.15, R.D. Levine and M. Tribus (eds.), MIT Press, Cambridge, MA.

9b Tikochinsky, Y., N.S. Tishby and R.D. Levine (1984). Phys. Rev. Letts., 52, 1357.

10 Renyi, A. (1970). Probability Theory. North-Holland, Amsterdam.

11 Papoulis, A. (1965). Probability, Random Variables, and Stochastic Processes. McGraw-Hill, New York.

12 Cox, R.T. (1979). The Maximum Entropy Formalism, R.D. Levine and M. Tribus (eds.), p.119. MIT Press, Cambridge.

13 Birkhoff, G. and J. von Neumann (1936). Annals. Math., 37, 823.

14 Cyranski, J.F. (1981). J. Math. Phys., 22, 1467.

15 Cyranski, J.F. (1978). Found. Phys., 9, 641.

16a Nilsson, N.J. (1971). Problem-Solving Methods In Artificial Intelligence, Chap. 6. McGraw-Hill, New York.

16b Curry, H.B. (1977). Foundations of Mathematical Logic. Dover, New York.

17 Halmos, P.R. (1950). Measure Theory. Van Nostrand, New York.

18 Strictly speaking, a MAXENT probability measure is not a DOB in Carnap's sense. Carnap allows H and E to coexist in a language and construes p(H E) as a generalization of the rules of logical deduction. Moreover, he allows p(H E) = m(H&E)/m(E) (a "conditional" measure) where m(A) is determined (by additional axioms) on the sentences that logically imply A. While MAXENT hypotheses and evidences cannot easily be incorporated in a single language, MAXENT defines the "logical relationship" between H and E as the inference about H that assumes only the information contained in E, in particular the minimal information asserted by E. (See Section 2.) It is for this reason that we consider MAXENT as a calculus of "logical relationships" -- a sense perhaps weaker (more general) than Carnap's. For arguments that MAXENT is a special type of DOB (in Carnap's sense) see: Dias, P.M.C. and A. Shimony (1981), Adv. App. Math., Vol. 2, p. 172.

19 One can still obtain Bayes' Theorem with H and E in different logics provided one formally takes "H&E" to mean (H,E) in the logical product $L_{objects} \times L_{evidences}$. Such a generalization is prerequisite to Bayesian "parameter estimation" methods. See, for example: Fraser, D.A.S. (1976), Probability and Statistics, Duxbury Press, North Scituate, MA.

20 Jaynes, E.T. (1968). IEEE Trans. Syst. Sci. & Cyber., Vol. SSC-4, 227.

21 Varadajaran, V.S. (1970). Geometry of Quantum Theory, Vol. 2. Van Nostrand Reinhold, NY.

22 Kuratowski, K. (1968). Topology. Trans. A. Kirkor. Academic, NY.

23a Shore, J. and R. Johnson (1980). IEEE Trans. Inf. Th., Vol. IT-26, 26.

23b Ochs, W. (1975). Reps. Math. Phys., Vol. 8, 109.

23c Ochs, W. (1975). Reps. Math. Phys., Vol. 9, 135.

23d Ochs, W. (1975). Reps. Math. Phys., Vol. 9, 331.

24a It follows that MAXENT DOB's cannot be viewed as conditional measures, for these are ordinarily defined by $m(A||B) = m(A\&B)/m(B)$ (where "A&B" is the intersection of the sets corresponding to the propositions [10,11,17]). Even in terms of measures on $L_0 \times L_1$, MAXENT yields measures that are not conditionals. For, according to [30], a conditional probability measure is a function $q(o,E)$ on $O \times L_1$ which is L_0-measurable for all E and a probability measure on L_1 for all o in O. In particular, $q(o,EUE') = q(o,E) + q(o,E')$ when $E\&E' = \{\}$. However, in this event MAXENT yields as $p \ll m$ that measure which satisfies $S(p,m) = I(EUE') = \min\{I(E),I(E')\}$. Thus, if $I(E)<I(E')$, $dp(o,EUE')/dm = dp(o,E)/dm$ so that MAXENT yields a function that does not satisfy the basic criterion of "conditional probability". The same conclusion obtains if one attempts to define a conditional probability as $m(H|E) = m(H,E)/m(E) = m[(H,M_1(O))\&(O,E)]/m[O,E]$, where $m(E) = m(O,E)$ is the marginal of m on L_1.

24b This is actually a form of "reliability" study (Section 3): In effect, one tries to find the evidences (i.e., variables whose expectation values define elements of L_1) that best agree with experiments (=evidence about the evidences). For practical applications of this use of MAXENT (called "surprisal analysis") see: Levine, R.D. (1979), in The Maximum Entropy Formalism, R.D. Levine and M. Tribus (eds.), p.247, MIT Press, Cambridge, MA.

25 Posner, E.D. (1975). IEEE Trans. Inf. Th., IT-21, 388.

26 Csiszar, I. (1975). Ann. Prob., 3, 146. One can easily generalize the arguments of Theorem 2.1 to show that even for a sigma-finite prior distribution m, S(p,m) achieves its minimum on any set of probabilities that is closed in the weak topology, provided I(E) is finite and $M_1(O)$ is a metric space.

27 Kullback, S. (1967). Information Theory and Statistics. Dover, New York.

28 Tzannes, N.S. Personal communication.

29 The forecaster requires an estimating (Boolean) function in F = {f:O->K}, K={ {},H,H^c,O}, so that given o in O, f(o)=H ("it rains on day o") or f(o)=H^c ("it does not rain on day o) can be computed. At least part of the "evidence" for inferring f in F can consist of a finite sample {(o_i,B_i)|B_i=H or H^c}, i=1,2,...I, so that the "regression" analogue is apparent. If we assume (or define F so) that sets {f in F|f(A)=B; A in B(O), B in K} are measurable (in B(F)), then we induce p[(B,A)|E] = p[(f|f(A)=B}|E] for A in B(O), B={},H,H^c, or O. In particular, p[(H||o)|E] = p[{f|f(o)=H}|E] defines the (conditional) probability of rain on day o in terms of the probability for the set of estimating functions that yield rain on day o.

30 Berger, T. (1971). Rate Distortion Theory. Prentice-Hall, Englewood Cliffs, NY.

31 Grubbs, F.E. (1982). JTEVA, Vol. 10, 133.

32 Adkins, C.J. (1968). Equilibrium Thermodynamics. McGraw-Hill, London.

33 Parthasarathy, K.C. (1967). Probability Measures on Metric Spaces. Academic Press, New York.

PRIOR PROBABILITIES REVISITED

N.C. Dalkey
Cognitive Systems Laboratory
School of Engineering and Applied Science
University of California, Los Angeles

ABSTRACT

Unknown prior probabilities can be treated as intervening variables in the determination of a posterior distribution. In essence this involves determining the minimally informative information system with a given likelihood matrix.

Some of the consequences of this approach are non-intuitive. In particular, the computed prior is not invariant for different sample sizes in random sampling with unknown prior.

1 GENERALITIES

The role of prior probabilities in inductive inference has been a lively issue since the posthumous publication of the works of Thomas Bayes at the close of the 18th century. Attitudes on the topic have ranged all the way from complete rejection of the notion of prior probabilities (Fisher, 1949) to an insistence by contemporary Bayesians that they are essential (de Finetti, 1975). A careful examination of some of the basics is contained in a seminal paper by E.T. Jaynes, the title of which in part suggested the title of the present essay (Jaynes, 1968).

The theorem of Bayes, around which the controversy swirls, is itself non-controversial. It is, in fact, hardly more than a statement of the law of the product for probabilities, plus the commutativity of the logical product. Equally straightforward is the fact that situations can be found for which representation by Bayes theorem is unassailable. The classic classroom two-urn experiment is neatly tailored for this purpose. Thus, the issue is not so much a conceptual one, involving the "epistemological status" of prior probabilities, as it is a practical one. In practice, the required prior probabilities are often unknown, or poorly known.

The present paper presents an approach to the estimation of prior probabilities when these are unknown. The approach is a generalization of maximum entropy methods. It was derived with a quite different rationale, and thus represents a convergence of two different streams of thought.

2 FIGURES OF MERIT

As a foundation for a theory of estimation, it is necessary to introduce a figure of merit, a measure of the excellence of an estimate.

Figures of merit are commonly some form of discrepancy measure, e.g., if I am asked to guess the height of a distant tree, the excellence of my guess is determined by comparing it with the actual height. In the measurement literature a wide variety of scores can be found -- absolute difference, squared difference, percentage difference, and the like.

Estimates of probabilities have the difficulty that the true or actual probability is rarely available for comparison. An ingenious way to sidestep this difficulty has been found in the theory of proper scores (Savage, 1971). Let E be a partition on an event space, and e an unspecified member of E. Let R be an estimate of the probability distribution on E. Finally, let S(R,e) be a function which assigns the score (rating, reward, payoff, etc.) if R is the estimated probability distribution on E and the event e occurs. If P(E) is the actual probability distribution on E, the expected score for the estimate R is $\sum_E P(e)S(R,e)$. Notice that the score S(R,e) can be assigned knowing only the estimate R and the event e that actually occurs, without knowing the actual probability P.

A score rule is called proper (reproducing, honesty- promoting, admissable, etc.) if it fulfills the condition

$$\sum_E P(e)S(R,e) \leqq \sum_E P(e)S(P,e) \qquad (1)$$

i.e., a score is called proper if the expected score is a maximum when the estimate is the same as the distribution which determines the expectation. (1) is analogous to the requirement for a discrepancy score that the "error" be a minimum when the estimate is precisely the same as the actual quantity.

It is convenient to introduce some definitions:

$$G(P,R) = \sum_E P(e)S(R,e)$$

$$H(P) = G(P,P) = \sum_E P(e)S(P,e)$$

$$N(P,R) = H(P) - G(P,R)$$

G(P,R) is the expected (discrepancy) score if R is the estimate and P is the actual distribution. H(P) plays a special role for probabilistic scores. For error measures that are analogous to a distance, e.g., the absolute difference, $H(x) = |x-x| = 0$ for all x. However, for proper scores, H(P) represents a measure of the excellence of a distribution P on its own so to speak. N(P,R) is the net score if R is estimated and P obtains. Note that from (1) N(P,R) is always non negative.

Since G(P,R) is an expectation, it is linear in P. An important property of H(P) is that it is convex (Dalkey, 1982).

There is a very large family of scores that fulfill (1). They range from scores derived from decisional payoff matrices to scores appropriate primarily for scientific contexts (Dalkey, 1980). The most widely used of the latter is the logarithmic score, S(R,e) = log R(e). Note that -H(P) for the logarithmic score is precisely the Shannon entropy for the distribution P.

Proper scores can play the same role in inductive logic that truth-value plays in traditional logic. In fact, the truth-value is a form of proper score. If an individual believes a given statement is true, but asserts the negation, his expected "score" is false, clearly less excellent than if he had asserted what he believed.

Proper scores enable the verification of statements of the probability of a single case, a possibility usually denied in the literature of probability theory. If an estimator asserts "P(e) = p", where e is a specific event such as "rain tomorrow", one need only wait until tomorrow and (for the logarithmic score) award the prediction with the score log p if it rains, or log (1-p) if it doesn't. The dependence upon the occurrance of a specific event gives the requisite tie to reality needed for a verification procedure, and the dependence on the asserted probability furnishes the requisite dependence on the content of the assertion.

3 MIN-SCORE INDUCTION

Given an appropriate figure of merit, it is feasible to formulate an inductive logic that is an extension of classical logic. A general structure for a logic is a collection of rules which transform a set of premises into a conclusion. In the classical case, if the premises are true and the inference is valid (i.e., follows the rules), then the conclusion must be true. That simple guarantee is, of course, precisely what makes classical logic useful in inquiries. As might be expected, the nature of the guarantee is somewhat more complex in inductive logic.

In the most elementary case, consider a partition E on an event space, where E represents the events of interest, i.e., E specifies the events for which a probability distribution is desired. We assume that there is a probability distribution on the event space, and thus, in particular, there is a distribution P(E) on the partition E. In the relevant case, P(E) is unknown, but there may be some (partial) information concerning P(E). Suppose the partial information consists of knowing that P(E) is in some class K of distributions on E. In the extreme case of no information, K is the set of all possible distributions on n events, where n is the number of events in E; i.e., K is just the simplex Z_n of all probability distributions on n events. If K is a unit class, then P(E) is completely known. In intermediate cases, K is some subset of Z_n.

We can take the specification of K as the premises of an inference. What is desired as a conclusion is some estimate R(E) of the distribution on E. Since by assumption the actual distribution P is in K, it might be supposed that R must be selected from K. However, there is no formal constraint that R be in K; it could be any distribution in Z. Assuming that a score rule S has been adopted, the actual expectation is G(P,R). The inductive rule to be employed in this paper is derived from two postulates: P1 - the selection rule should guarantee at least the expected score of R, i.e., it should guarantee H(R). Formally this requires $G(P,R) \geq H(R)$, for any P in K; P2 - the selection rule should assure the positive value of information, i.e., if additional information is obtained, then the expected score should not decrease. Formally, if $K' \subset K$, then $H(R') \geq H(R)$.

These two postulates lead to a specific selection rule which could be called the min-score rule: select the R in the closure of the convex hull of K that minimizes H(R) (Dalkey, 1982). If K is convex and closed, then R will be in K; if K is not convex and closed, then R may not be in K, but will be in the closure of the convex hull of K (Dalkey, 1985).

P1 appears to be essential for any kind of inference. The user of the conclusion must be confident that he will achieve at least as high an expected score as the conclusion promises. P2 is more germaine to induction. In the case of complete information, the positive value of information is a theorem (Lavalle, 1978). It appears a-fortiori plausible that additional information should be constructive in the case of incomplete information.

If the score rule adopted is the logarithmic score, then for the elementary case under consideration, the min-score rule is precisely the maximum entropy procedure. As noted above, the expected log score is just the negative of the entropy. Minimizing the log score is equivalent to maximizing the entropy. For more highly structured problems, the min-score approach may lead to a different analysis than current practice with maximum entropy methods. This divergence will show up in the analysis of unknown prior probabilities.

4 PRIOR PROBABILITIES

The elementary min-score rule does not involve the distinction between prior and posterior probabilities. The information class K does represent "prior information", but is not expressed as a probability.

Historically, the notion of prior probability has been employed in the context of "updating". A probability distribution is known for an event set E. New evidence I, either planned as in an experiment, or fortuitous as in casual observation, comes to attention, and the problem arises of revising the probability distribution on E to reflect the new evidence. In this case, the old distribution P(E) is the prior and the new distribution $P(E|I)$ is the "posterior". Of course, $P(E|I)$ can operate as a new prior if further evidence arises. The distinction is significant only for a given instance of updating. Another way of putting the same point is that the distinction between primary events and evidence is not a formal aspect of the calculus of probabilities.

As long as the updating is conducted with complete information (all the relevant probabilities known), there is no conceptual difficulty. A variety of updating procedures is available, depending on what is known concerning the relationships between the evidence and the primary events. The one most frequently employed is the theorem of Bayes, $P(E|I) = P(E)P(I|E)/P(I)$.

Difficulties do arise, of course, if the relevant probabilities are not completely known. Essentially, what the analyst needs to know for the updating step is the joint distribution P(E.I). A frequent situation is that in which the likelihoods $P(I|E)$ are known, but not the joint distribution. In the context of min-score inference, the class K can be taken to be a set of joint distributions, constrained by the requirement that they generate the known likelihoods, i.e., P(E.I) is in K if $P(E.I)/\sum_{I} P(E.I) = P(I|E)$.

In the given instance, the class K can be characterized equivalently by the set of joint distributions $P(E.I) = P(E)P(I|E)$ where P(E) can be any distribution on n events (since P(E) is totally unknown). However, it is clearly incorrect to select the min-score distribution in Z_n for P(E) since this ignores the role of the score rule. The score for the updating problem is related to the posterior probability $P(E|I)$, not to the prior. In colloquial terms, the analyst is not being paid to estimate the prior, or, from the standpoint of the decisionmaker, his payoff will be determined by implementing the posterior, not the prior.

A further complication arises in imposing the score rule for the case of incomplete information. With complete information, it is legitimate to ignore all potential evidence except the specific item that actually obtains. Considering I as a set of possible items of evidence (observations, data, signals, etc.), and i as a member of I, then in practice what is wanted is $P(E|i)$ when i is known. This feature has been elevated to the status of a principle by some writers -- the posterior determined by an item of evidence i should be a function solely of i and not of any other potential evidence that might have been observed.

That principle cannot be maintained in the case of incomplete information. For the illustrative case where the likelihoods, but not the prior, are known, the information in the likelihood matrix concerning potential (but not observed) evidence is relevant to the assessment of the observed evidence. As a simple example, consider the case of two events e and $\bar{e}$ (the bar indicating negation). Suppose there are two possible pieces of evidence, i and $\bar{i}$. Let $P(i|e) = q$ and $P(i|\bar{e}) = r$. Without loss of generality, we can let $q>r$. (If $q = r$, the evidence is trivial.) Set $P(e) = p$. With $p = r/(q+r)$, $P(e|i) = 1/2$. Thus, whatever q and r, a prior probability can be assigned that makes the evidence completely uninformative (at least for any symmetrical score rule).

The example clearly generalizes to several events and several potential items of evidence. Thus, for the assessment of evidence in the case of incomplete information, it is necessary to treat the evidence and the events of interest as an information system, and the selection of a prior probability as the design of a min-score information system. For the logarithmic score, this requirement can be restated as designing a minimally informative information system (Dalkey, 1980).

Summarizing: for the updating problem, the probabilities of interest are the posterior conditional probabilities $P(E|I)$; it is the expected score of these probabilities which determines the value of the new evidence. However, there is a separate posterior for each potential item of evidence; thus, the complete assessment consists of the average of these expectations over the potential items of evidence. Denoting the average expected score by $H(E|I)$, we have

$$H(E|I) = \sum_{I} P(i) \sum_{E} P(e|i)S(i,e) \tag{2}$$

where $S(i,e)$ is shorthand for "the score given that $P(E|i)$ is the estimate, and e occurs".

For the logarithmic score, (2) can be unpacked in the form of a well-known formula in information theory

$$H(E|I) = H(E) + H(I|E) - H(I) \tag{3}$$

That is, the average information furnished by an information system (E,I) is the information contained in the prior distribution $P(E)$, plus the average information in the likelihood matrix $P(I|E)$ minus the information in the initial distribution on the evidence $P(I)$. Notice that there is a simple duality between events and evidence. From (3) $H(I|E) = H(I) + H(E|I) - H(E)$.

If the prior probabilities $P(E)$ are not known, the min-score inference rule prescribes minimizing $H(E|I)$ as a function of the distribution $P(E)$ over the class K of joint distributions $P(E.I)$ constrained by the likelihood matrix $P(I|E)$. The maximum entropy rule (for the log score) is now extended to a maximum expected entropy rule.

For the illustrative case of two events described above: $H(E) = p\log p + (1-p)\log(1-p)$, $H(I|E) = p(q\log q + (1-q)\log(1-q)) + (1-p)(r\log r + (1-r)\log(1-r))$, $H(I) = (pq + (1-p)r)\log(pq + (1-p)r) + (p(1-q) + (1-p)(1-r))\log(p(1-q) + (1-p)(1-r))$. (I've expanded this elementary case in somewhat tedious detail because the role of the prior probability p is different from the usual form of max entropy analysis). $H(E|I)$ can be minimized as a function of p by elementary differentiation and setting the result equal to 0. The solution is obtained by solving for p the implicit equation

$$\frac{p}{(1-p)} e^{H(q)-H(r)} = \left(\frac{pq + (1-p)r}{p(1-q)+(1-p)(1-r)}\right)^{q-r} \qquad (4)$$

The solution is not particularly intuitive. If q and r are symmetric, i.e., $r = 1-q$, the min-score p is the classic uniform distribution, $p = 1/2$. However, if q and r are not symmetric, and each is rather far from 1/2, the min-score prior is not uniform. For example, if $q = .9$ and $r = .025$, the min-score prior is about .63. Roughly speaking, the min-score solution puts greater weight on the "less informative" prior event.

An even less intuitive result is obtained if the observation is iterated, e.g., if two independent observations are made. The min-score prior computed from the extension of (4) to two observations is not the same as the prior computed for one observation; e.g., the min-score prior for $q = .8$, $r = 0$, is .625 for one observation and is .69 for two independent observations. The "discounting" of the more informative event is more drastic for the two-observation case; the difference between q and r has a more pronounced effect on the likelihoods for two observations.

In the classic calculus of probabilities, the effect of an additional observation can be computed by "updating", i.e., by using the posterior probability for one observation as the new prior for an additional observation. This procedure is not valid for the case of an unknown prior. One way of expressing what is going on is to note that in the min-score analysis, the solution is sensitive not only to the inputs, but also to the precise question being asked. As remarked above, the question being asked in the case of additional evidence is the posterior probability given the evidence. If the evidence changes, then a new prior must be computed. Another way of saying the same thing is that the relevant K for the case of one observation is a set of joint distributions of the form $P(E.I)$; for two observations the relevant K is a subset of distributions of the form $P(E.I_1.I_2)$. This characteristic of the min-score rule has serious implications for general purpose inference mechanisms, e.g., expert systems. In a medical expert system, for example, there is a basic difference between the diagnostic and the prognostic use of data from the min-score point of view. A system could not use the same set of "best-guess priors" for both types of estimate.

Some readers may find this dependence on the specific question being asked a serious drawback to min-score procedures. There is no question but that it is a serious practical complication. A single prior distribution cannot be computed and then plugged into each new problem. However, the "difficulty" serves to emphasize the basic difference between complete and partial information. In the case of partial information and updating on new evidence, the prior probabilities are "intervening variables", serving to complete the analysis, not to advance knowledge. The new knowledge is contained in the posterior estimates.

5 RANDOM SAMPLING WITH UNKNOWN PRIOR

The classroom example of the previous section has a highly structured frame of reference. In practice most problems are not so neatly packaged. A case in point is random sampling with unknown prior. An elementary example is an exotic coin with unknown probability of heads. Another example is the case of the possibly loaded die treated by Jaynes (Jaynes, 1968). In the classroom example, there are two well-defined "states of nature" and a fairly clear interpretation of the prior probability -- someone presumably selected one of the two states, e.g., one of two urns, according to a specific probability distribution. In the case of the exotic coin, the states of nature are not "given" and a mechanism to incarnate a prior probability is even less apparent.

A frame of reference for such problems was devised by Laplace. For the exotic coin, each potential probability of heads is considered to be a separate state. For the loaded die, each potential probability distribution on the six faces is a state. The prior probability, then, is a distribution on these states. If it is assumed for the coin that any probability of heads from 0 to 1 is possible, then a prior probability would be a density on the interval 0--1. For the die, with similar freedom, a prior probability is a density on the simplex of distributions for six events.

The model is illustrated in Fig. 1 for the coin. There is a continuum of states, labelled by the probability p of heads. The prior is a density D(p) whose integral is 1; and the likelihoods are just the probabilities p for a single flip of the coin. For multiple flips, assuming independence, the likelihoods are the Bernoulli probabilities $P(p,n,m) = \binom{n}{m} p^m (1-p)^{n-m}$ for the case of m heads in n flips of the coin.

For the logarithmic score, the continuous version of (3) holds, where

$$H(E) = \int_0^1 D(p)\log D(p)dp$$

$$H(I|E) = \int_0^1 D(p)\left[\sum_{m=0}^{n} P(p,n,m)\log P(p,n,m)\right]dp$$

$$H(I) = \sum_{m=0}^{n} \int_0^1 D(p)P(p,n,m)dp \log \int_0^1 D(p)P(p,n,m)dp$$

Figure 1. Laplace model for binary-event random sampling with unknown prior.

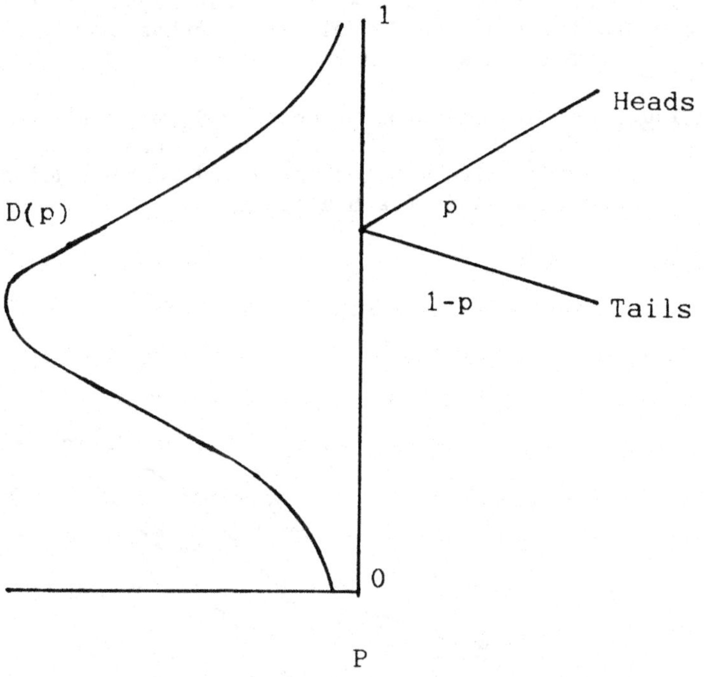

The min-score problem, then, is to find $D^o(p)$ which minimizes $H(E|I)$.

Because of the symmetry of the state space -- for every state p there is an antisymmetric state 1-p -- we can expect the min-score $D^o(p)$ to be symmetrical in p. For one observation, symmetry implies that P(I) is uniform, i.e., P(heads) = 1/2. Thus, the term H(I) is invariant under changes in D(p), and we have

$$D^{\circ}(P) \propto e^{-H(p)} \tag{5}$$

and since - H(p) = Entropy(p), equivalently

$$D^{\circ}(p) \propto e^{\mathrm{Ent}(p)} \tag{5'}$$

The posterior density of p, given the observation i, is then

$$D^{\circ}(p|i) = \frac{pe^{-H(p)}}{\int_0^i pe^{-H(p)}\,dp} \tag{6}$$

In Fig. 2, the prior and posterior densities are drawn for the case of a single observation. Also shown in dashed lines are the uniform prior and the posterior density for the uniform prior. If we take the average of p for the posterior distribution as the "best guess" p, then, for the uniform prior it is 2/3, and for the min-score prior it is .64. The difference is not large for a single observation.

Figure 2. Min-score prior density $D^o(p)$, and min-score posterior density $D^o(p|i)$ for single observation (solid lines), with uniform prior $D_u(p)$ and corresponding posterior density $D_u(p|i)$ (dashed lines).

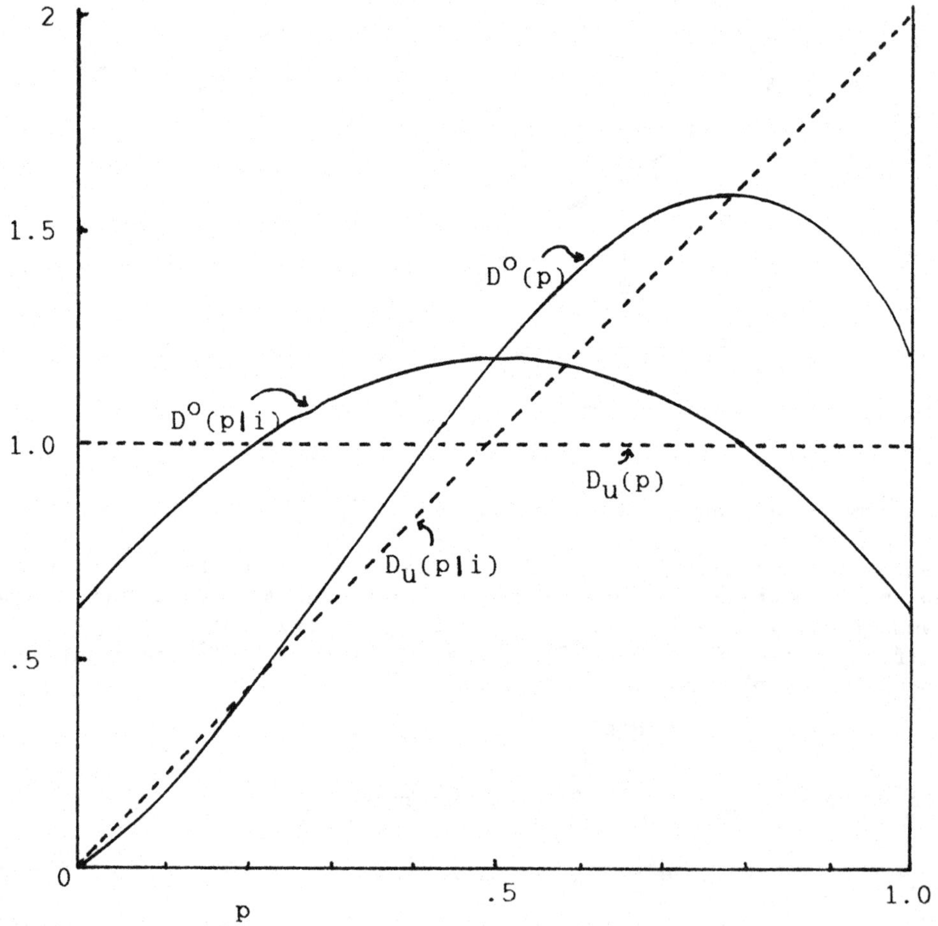

For multiple observations, it is no longer the case that H(I) is invariant under changes in D(P). It is instructive, however, to glance at the result if H(I) is assumed to be invariant. In that case we would have

$$D^{o}(p) \propto e^{-nH(p)} \tag{7}$$

i.e., the prior density becomes increasingly concentrated around 1/2 with increasing n, where n is the number of observations. Fig. 3 shows this "first approximation" $D^0(p)$ for several n. It is clear that for this approximation, $D^0(p)$ converges to a distribution concentrated at $p = 1/2$ as $n \to \infty$.

<u>Figure 3</u>. Approximate min-score priors, $ke^{-nH(p)}$, for various sample sizes n.

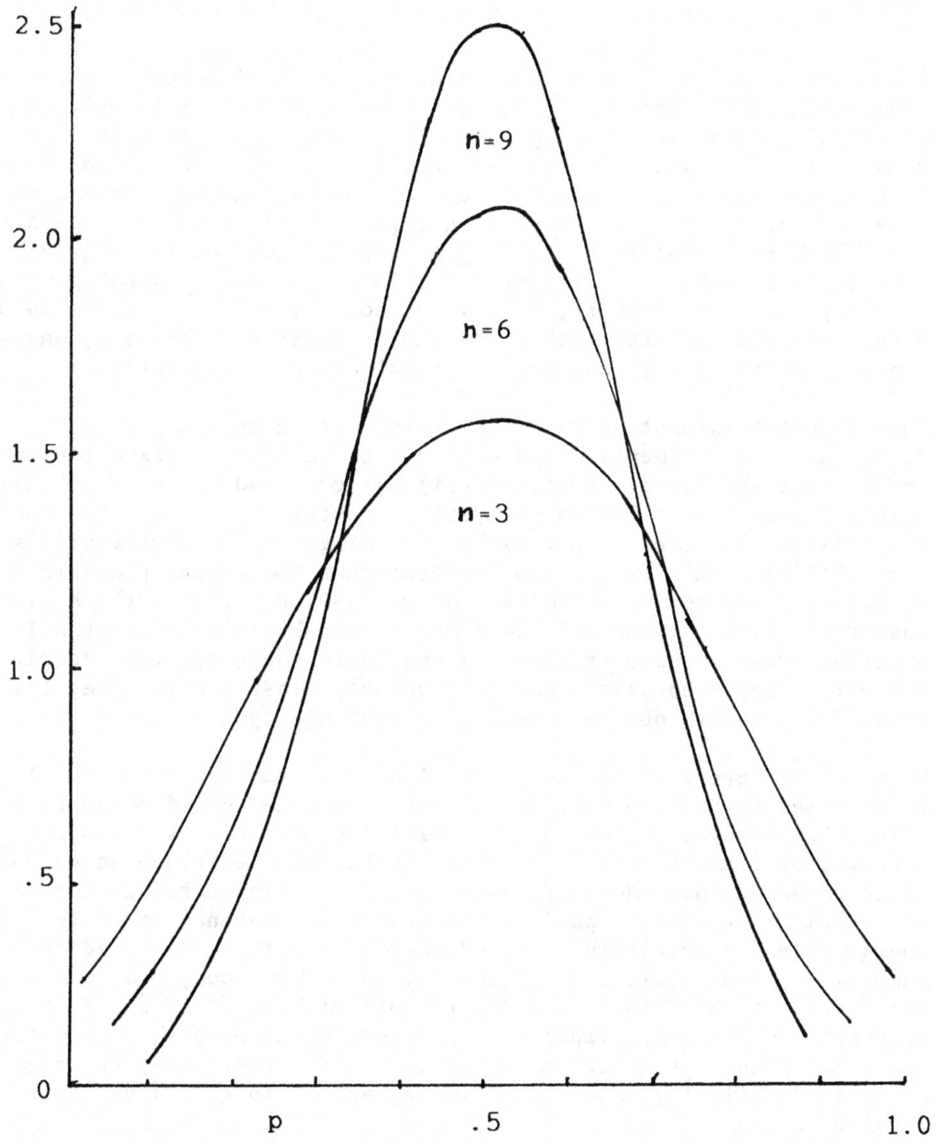

I do not have an exact solution for the case of $n>1$. If we consider two extreme distributions, the uniform distribution $D_u(p) = 1$, and the distribution $D_{1/2}(p)$ concentrated at $p = 1/2$, we can say that $H(I|E) = H(I)$ for $D_{1/2}(p)$, i.e., $H(I|E) - H(I) = 0$, and $H(E) = \infty$. At the other extreme, $D_u(p)$, $H(E) = 0$, $H(I|E) = -n/2$ and $H(I) = \log 1/(n+1) - 1/(n+1) \sum_{m=0}^{n} \log \binom{n}{m}$. Since H is convex, and $H(I|E)$ is an average while $H(I)$ is an H of averages, $H(I|E)$ is always greater than $H(I)$, but the difference is concave. Thus D^o is intermediate between D_u and $D_{1/2}$. $H_u(I|E) - H_u(I) \to \infty$ as $n \to \infty$, thus $D^o \to D_{\frac{1}{2}}$. as $n \to \infty$.

Even without an exact solution, then, we can conclude that as $n \to \infty$, the asymptotic D^o is massed at 1/2. Qualitatively, this implies that the amount of information in large samples with unknown prior is less than classical theory would imply. For binary events, this implies that the best guess is closer to 1/2 than the classic m/n guess, where m is the observed frequency of an event in n trials. This result is compatible, e.g., with the observation that opinion polls are, on the average, too extreme, i.e., they tend to predict a larger margin of winning than is actually observed (Dembert, 1984). Of course, political polls involve potential forms of error other than the statistical analysis, and thus the compatibility with our present result is only suggestive.

The present treatment of random sampling with unknown prior deals with fixed-sample experiments. The number n of samples is fixed beforehand, and a posterior distribution, $P^o(E|i)$, is computed for each potential sampling pattern i. Furthermore, the computation is conducted under the supposition that the score will be determined by the posterior distribution $P^o(E|i)$. In effect, this requires that the actual probability p be observed before the score can be awarded. For the textbook two-urn case there is no difficulty; determining which urn was selected is straightforward. However, in more realistic contexts, this requirement may not be implementable. For the possibly biassed coin case, there is no way to directly observe the actual probability.

What can be observed (in theory, at least) is further samples. It might be supposed that an operational scoring mechanism could be devised by using the computed posterior to predict the probability of further observations, and base the score on the occurrance or non-occurrance of these observations. However, as we have seen, to introduce further observations requires expanding the frame of reference to $(E.I_1.I_2)$, where I_1 is the initially observed sample, and I_2 is the predicted sample. This analysis can, of course, be carried out, and is a legitimate application of the min-score formalism; but, as seen earlier, this requires, in essence, computing a new best guess prior. In the former case, the figure of merit is $H(E|I_1)$; in the latter case, the figure of merit is $H(I_2|I_1)$, and these two may appear to give divergent results.

6 CONSTRAINTS AND STATISTICS

The justification of the min-score rule assumes that the unknown probability function P is contained in the knowledge set K. In other words, it is assumed that whatever constraints delimit K are categorical. In contrast, it is common in applications of maxent methods to introduce constraints that do not have this property. A frequently used form of constraint is one derived from an observed statistic, i.e., given a statistic S, with observed value s, it is assumed that the class K consists of those probability distributions whose expectation for S is the observed value s. As an example, in the case of the biased die analyzed by Jaynes, he assumes that the probability distribution under investigation has a theoretical average equal to the observed sample average.

It is clear that such statistical constraints are not categorical. For the die example, any distribution within the interior of the simplex of all distributions on six events could give rise to the observed statistic. Many of these would have very small likelihoods of generating a sample with the observed average, but that is a fact to be exploited by the analysis, not ignored. The justification of the step from sample data to theoretical expectations is somewhat obscure in the literature. Jaynes uses terms such as "compatible with the given information" (Jaynes, 1968) or "allowed by the data" (Jaynes, 1982), but in light of the compatibility of the observed statistic with any underlying distribution, it is not clear how these terms are being used.

If the complete frequency table arising from a random experiment is available to the analyst then the maxent procedure becomes irrelevant. The constraint convention -- expectation = observed value -- leads to $P_i = f_i/N$ where f_i is the observed frequency for event i, and N is the total number of sample points. Some obscurity arises in this regard concerning the question whether the justification of the procedure is intended to be asymptotic (infinite sample) or not. But in practice, it seems clear that the procedure is intended to apply to finite samples.

In the min-score approach, observations are not considered as constraints on the knowledge class K, but rather are elements of an information system. In the case of the loaded die, what is sought is the best-guess posterior density on probability distributions on the faces of the die given the observed average. The class K is all joint densities D(P.A), where P is a probability distribution on six events, and A is a potential observed average over N tosses of the die. Analysis consists of finding the minimally informative information system with this structure. As in the case of the binary event sampling analyzed in the previous section, there is no constraint on the possible probability distributions P.

If it is presumed that uncertainty arises in a given experiment, not from "error" but from the fact that the expectation of a statistic need not be the same as the observed value, then the min-score procedure is a way to deal with uncertainty without the addition of an error term.

7 COMMENTS

The analysis of unknown prior probabilities presented above leaves a great deal to be desired as far as mathematical implementation is concerned; but there does not appear to be any deep mathematical issues involved.

The same cannot be said for logical issues. One that appears particularly critical is the fact that a min-score estimate is simply a best guess that depends on the score rule and on the specific question being asked. This characteristic seems to deny the possibility that min-score inference can be used to add to the store of knowledge. In a sense, this result is inherent in the formalism. By definition, all that is known is the class K and observations I.

This issue will be explored at somewhat greater depth in an upcoming paper (Dalkey, 1985).

REFERENCES

1 Dalkey, N.C. (1980). The Aggregation of Probability Estimates. UCLA-ENG-CSL-8025.

2 Dalkey, N.C. (1982). Inductive Inference and the Maximum Entropy Principle. UCLA-ENG-CSL-8270, presented at the 2nd Workshop on Maximum Entropy. U. of Wyoming, August, 1982.

3 Dalkey, N.C. (1985). The Representation of Uncertainty. (In prep.)

4 de Finetti, B. (1975). Theory of Probability. Vol. II, John Wiley and Sons, New York.

5 Dembert, L. (1984). Roper Cites Flaws in Political Polls. Los Angeles Times, May 26, p.16.

6 Fisher, R.A. (1949). The Design of Experiments. Oliver and Boyd, London.

7 Jaynes, E.T. (1968). Prior Probabilities. IEEE Transactions on Systems Science and Cybernetics, SSC-4, 227-241.

8 Jaynes, E.T. (1982). On the Rationale of Maximum-Entropy Methods. Proceedings of the IEEE, 70, 939-952.

9 Lavalle, I.H. (1978). Fundamentals of Decision Analysis. Holt, Rinehart & Winston, New York.

11 Savage, L.J. (1971). Elicitation of Personal Probabilities and Expectations. Jour. Amer. Stat. Assoc., 66, 783-801.

BAND EXTENSIONS, MAXIMUM ENTROPY AND THE PERMANENCE PRINCIPLE

Robert L. Ellis, University of Maryland
Israel Gohberg*, Tel Aviv University
David Lay, University of Maryland

*partially supported by an NSF grant

ABSTRACT

This paper explains the connection between the basic theorem of the well-known maximum entropy method and the following extension problem for band matrices. Let R be an nxn band matrix with bandwidth m. We seek an nxn positive definite matrix H whose band is the same as that of R and whose inverse is a band matrix with bandwidth m. The matrix H has the property that its determinant is larger than the determinant of any other nxn positive definite matrix that agrees with R in the band. An important role is played by a permanence principle that allows us to reduce the size of the matrix in the band extension problem.

INTRODUCTION

The maximum entropy method was introduced by J. Burg in connection with problems in signal processing and is extensively used in applications. It is based on the following theorem [2]:

Theorem A. Let $r_{-m}, r_{-m+1}, \cdots, r_m$ be given complex numbers such that the corresponding (m+1) x (m+1) Toeplitz matrix

$$\{r_{j-k}\}_{j,k=0}^{m}$$

is positive definite. There is a unique function

$$h(e^{i\omega}) = \sum_{k=-\infty}^{\infty} h_k e^{ik\omega}$$

with

$$\sum_{k=-\infty}^{\infty} |h_k| < \infty$$

and with the following properties:

(i) $h_k = r_k$ for $|k| \leq m$.

(ii) h is a positive function and 1/h has the form

$$1/h(e^{i\omega}) = \sum_{k=-m}^{m} b_k e^{ik\omega}.$$

(iii) For any positive function g whose Fourier coefficients (g_k) satisfy

$$\sum_{k=-\infty}^{\infty} |g_k| < \infty$$

and $g_k = r_k$ for $|k| \leq m$, we have

$$\frac{1}{2\pi}\int_{-\pi}^{\pi} \log g(e^{i\omega})d\omega \leq \frac{1}{2\pi}\int_{-\pi}^{\pi} \log h(e^{i\omega})d\omega \tag{1}$$

with equality if and only if g = h.

The inequality in (1) expresses the fact that h has maximum entropy. Generalizations of this theorem exist for the block, discrete and continuous cases [3,4,5]. Recently a theorem that may be viewed as an analog of the preceding theorem for finite matrices was proved [6]. The simplest version of the theorem for finite matrices is the following:

Theorem B. Let

$$R = \{r_{j-k}\}_{j,k=0}^{n}$$

be a Toeplitz matrix with $r_k = 0$ for $|k| > m$. Suppose the principal submatrix

$$\{r_{j-k}\}_{j,k=0}^{m}$$

is positive definite. There is a unique positive definite Toeplitz matrix

$$H = \{h_{j-k}\}_{j,k=0}^{n}$$

with the following properties:

(i) $h_k = r_k$ for $|k| \leq m$.

(ii) Let $H^{-1} = \{s_{jk}\}_{j,k=1}^{n}$. Then $s_{jk} = 0$ for $|j-k| > m$.

(iii) For any positive definite Toeplitz matrix

$G = \{g_{j-k}\}_{j,k=0}^{n}$ that satisfies the condition $g_k = r_k$ for $|k| \leq m$, we have

$$\det G \leq \det H \tag{2}$$

with equality if and only if $G = H$.

The matrix H is called the band extension of R.

In order to explain the connections between these two theorems, we show that the inequality in (2) can be replaced by the following condition:

$$\frac{\det G(1,\ldots,k)}{\det G(1,\ldots,k-1)} \leq \frac{\det H(1,\ldots,k)}{\det H(1,\ldots,k-1)} \quad (2 \leq k \leq n+1) \tag{3}$$

where $G(1,\ldots,k)$ denotes the kxk leading principal submatrix of G.

It turns out that Theorem A can be viewed as a restatement of Theorem B for infinite Toeplitz matrices with (2) replaced by (3). Let

$$R = \{r_{j-k}\}_{j,k=0}^{\infty}$$

be a semi-infinite Toeplitz matrix with $r_k = 0$ for $|k| > m$ and with the matrix

$$\{r_{j-k}\}_{j,k=0}^{m}$$

positive definite. We seek a positive definite Toeplitz matrix

$$H = \{h_{j-k}\}_{j,k=0}^{\infty}$$

with $h_k = r_k$ for $|k| \leq m$ such that (3) holds for $k \geq 1$ and for any other positive definite semi-infinite Toeplitz matrix G with $g_k = r_k$ for $|k| \leq m$. In this analog, the matrix H in Theorem B corresponds to the function h in Theorem A. To establish the connection between (1) and (3), we will first show that

$$\frac{\det H(1,\ldots,k)}{\det H(1,\ldots,k-1)} = \exp \frac{1}{2\pi} \int_{-\pi}^{\pi} \log h(e^{i\omega})d\omega \quad (k > m).$$

For any other positive function g as in Theorem A and its corresponding Toeplitz matrix G, a theorem of Szegö states that

$$\lim_{k\to\infty} \frac{\det G(1,\ldots,k)}{\det G(1,\ldots,k-1)} = \exp \frac{1}{2\pi} \int_{-\pi}^{\pi} \log g(e^{i\omega})d\omega.$$

The connections between Theorem A and Theorem B are established in the last section of this paper, where a complete proof of Theorem A based on its finite dimensional analog is presented. To make the paper self-contained, we prove Theorem B in the first section.

There is another connection between the finite and infinite cases that is used in the proofs. If a band matrix R admits a positive definite band extension H, then each principal section of R admits its own positive definite band extension, and the extension of a principal section of R coincides with the corresponding principal section of H, even though H is computed from the larger matrix. This permanence principle reduces the band extension problem for infinite matrices to that for finite matrices, and for finite matrices it reduces the problem to one-level extensions. Section 2 is devoted to the permanence principle, and Section 3 concerns one-level extensions.

Finally, although Toeplitz matrices are at the heart of the problems we consider, the basic results in the finite dimensional case are valid for all positive definite matrices.

1 FINITE-DIMENSIONAL BAND EXTENSIONS

An nxn matrix $R = (r_{jk})$ is called a band matrix with bandwidth m if $r_{jk} = 0$ for $|j-k| > m$. Here m satisfies $0 \leq m \leq n - 1$. The elements r_{jk} for $|j-k| \leq m$ form a "band" of $2m + 1$ diagonals surrounding the main diagonal of R. An nxn matrix $H = (h_{jk})$ is an extension of R if H agrees with R in the band, i.e., if $h_{jk} = r_{jk}$ for $|j-k| \leq m$. An extension H of R is a band extension if H is invertible and H^{-1} is a band matrix with bandwidth m. For any nxn matrix G, and for $1 \leq j \leq k \leq n$, we let $G(j,\ldots,k)$ denote the principal submatrix of G given by

$$G(j,\ldots,k) = \begin{bmatrix} G_{jj} & \cdots & G_{jk} \\ \vdots & & \vdots \\ G_{kj} & \cdots & G_{kk} \end{bmatrix}.$$

Theorem 1. Let R be an nxn band matrix with bandwidth m. Then R admits a positive definite band extension if and only if

$$R(j,\ldots,j+m) \text{ is positive definite for } 1 \leq j \leq m \tag{1.1}$$

In this case the positive definite band extension of R is unique.

Proof. Suppose that H is a positive definite band extension of R. Then $R(j,\ldots,j+m) = H(j,\ldots,j+m)$ for $1 \leq j \leq n - m$. These matrices are positive definite because H is. Using an LU factorization of H^{-1}, we have

$$H^{-1} = XX^* \tag{1.2}$$

where X is lower triangular with positive real entries on its diagonal. Then $X = H^{-1}X^{*-1}$, which shows that X is a lower triangular band matrix. Now let $X = (x_{ij})$ and take $1 \leq j \leq n - m$. Since H is an extension of R, X^{*-1} is upper triangular, and

$$HX = X^{*-1} \tag{1.3}$$

it follows that

$$R(j,\ldots,j+m)\begin{bmatrix} x_{jj} \\ x_{j+1,j} \\ \vdots \\ x_{j+m,j}\end{bmatrix} = \begin{bmatrix} x_{jj}^{-1} \\ 0 \\ \vdots \\ 0\end{bmatrix} \tag{1.4}$$

or

$$R(j,\ldots,j+m)\begin{bmatrix} x_{jj}^2 \\ x_{jj}x_{j+1,j} \\ \vdots \\ x_{jj}x_{j+m,j}\end{bmatrix} = \begin{bmatrix} 1 \\ 0 \\ \vdots \\ 0\end{bmatrix} \quad (1 \leq j \leq n-m). \tag{1.5}$$

But $R(j,\ldots,j+m)$ is positive definite, so $R(j,\ldots,j+m)$ is invertible and the upper left entry of its inverse is positive. Therefore (1.5) determines $x_{jj},\ldots,x_{j+m,j}$ uniquely. For $n - m < j \leq n$, it follows similarly from (1.3) that

$$R(j,\ldots,n)\begin{bmatrix} x_{jj}^2 \\ x_{jj}x_{j+1,j} \\ \vdots \\ x_{jj}x_{nj}\end{bmatrix} = \begin{bmatrix} 1 \\ 0 \\ \vdots \\ 0\end{bmatrix} \quad (n-m < j \leq n) \tag{1.6}$$

and this determines $x_{jj},\ldots,x_{nj}$. Thus the band of X and hence X itself is determined by (1.2) and the requirement that X be lower triangular and have positive entries on the diagonal. This shows that if R has a positive definite band extension, then the extension is unique.

For the converse, suppose that (1.1) holds. Then we may define an invertible lower triangular band matrix $X = (x_{ij})$ by (1.5) and (1.6). In order to define a positive definite extension of R, we introduce the notation $\{A\}_\ell$ to stand for the matrix in Ω that coincides with a matrix A in Ω below the band and whose entries inside and above the band are zero. Then we define an extension H of R by first defining

$$\{H\}_\ell = - \{RX\}_\ell X^{-1}$$

and then setting

$$H = \{H\}_\ell + R + \{H\}_\ell^* .$$

Clearly H is a selfadjoint extension of R. Next, we observe that

$$HX = -\{RX\}_\ell + RX + \{H\}_\ell^* X.$$

From the definition of X in (1.5) and (1.6), we see that the entries of RX inside the band are zero below the diagonal and coincide with the entries of X^{*-1} on the diagonal. Hence $-\{RX\}_\ell + RX$ has zero entries below the diagonal and agrees with X^{*-1} on the diagonal. Therefore HX is upper triangular and agrees with X^{*-1} on the diagonal. Since X^* is upper triangular, it follows that X^*HX is a selfadjoint upper triangular matrix with diagonal entries equal to one, so $X^*HX=I$, which is equivalent to (1.2). Hence H is positive definite and H^{-1} is a band matrix with bandwidth m.

From equation (1.3) it is possible to obtain recursion formulas for the elements of H outside the band. If $1 \leq j < i \leq n$ and $j < n - m$, then the product of row i of H and column j of X is zero because HX is upper triangular. Therefore

$$H_{ij}x_{jj} + H_{i,j+1}x_{j+1,j} + \ldots + H_{i,j+m}x_{j+m,j} = 0$$

so that

$$H_{ij} = -(H_{i,j+1}x_{j+1,j} + \ldots + H_{i,j+m}x_{j+m,j})x_{jj}^{-1} \quad (j < j \leq n-m). \tag{1.7}$$

Also, since H is selfadjoint,

$$H_{ij} = -x_{ii}^{-1}(\bar{x}_{i+1,j} + \ldots + \bar{x}_{i+m,i}H_{i+m,j}) \quad (1 < j,\ i \leq n-m). \tag{1.8}$$

<u>Theorem 2</u>. Let R be an nxn band matrix with bandwidth m that admits a positive definite band extension H, and let G be any positive definite extension of R. Then

$$\frac{\det G(1,\ldots,k)}{\det G(1,\ldots,k-1)} \leq \frac{\det H(1,\ldots,k)}{\det H(1,\ldots,k-1)} \quad (2 \leq k \leq n) \tag{1.9}$$

with equality for all k if and only if G = H.

<u>Proof</u>. We may write

$$H = LDL^* \text{ and } G = MEM^*$$

where L and M are lower triangular matrices with diagonal entries equal to one, and D and E are positive definite matrices. Then

$$\left.\begin{aligned} H(1,\ldots,k) &= L(1,\ldots,k)D(1,\ldots,k)L^*(1,\ldots,k) \\ G*1,\ldots,k) &= M(1,\ldots,k)E(1,\ldots,k)M^*(1,\ldots,k) \end{aligned}\right\} (1 \leq k \leq n) \tag{1.10}$$

and therefore

$$\det H(1,\ldots,k) = \det D(1,\ldots,k) = D_{11}\cdots D_{kk}$$
$$\det G(1,\ldots,k) = \det E(1,\ldots,k) = E_{11}\cdots E_{kk}$$

Consequently,

$$D_{11} = H_{11}, \quad E_{11} = G_{11},$$

and

$$D_{kk} = \frac{\det H(1,\ldots k)}{\det H(1,\ldots,k-1)}, \quad E_{kk} = \frac{\det G(1,\ldots k)}{\det G(1,\ldots k-1)} .$$

Thus (1.9) is equivalent to stating that $E_{kk} \leq D_{kk}$ for $2 \leq k \leq n$.

Since H^{-1} is a band matrix and $L^{-1} = (DL^*)H^{-1}$, it follows that L^{-1} is a (lower triangular) band matrix. Writing

$$G = H + (G-H)$$

we obtain

$$L^{-1}MEM^*L^{*-1} = D + L^{-1}(G-H)L^{*-1} .$$

Let

$$A = D + 1^{-1}(G-H)L^{*-1} . \tag{1.11}$$

Then A is positive definite because E is positive definite and $A = (L^{-1}M)E(L^{-1}M)^*$. Let us partition A and write

$$A = \begin{bmatrix} A_{11} & A_{12} \\ A_{12}^* & A_{22} \end{bmatrix}$$

where A_{11} is a positive definite (n-1)x(n-1) matrix and A_{22} is a (positive) scalar. In fact, since G - H has zeros in the band and since L^{-1} and L^{*-1} are triangular and banded, it follows that $L^{-1}(G-H)L^{*-1}$ has zeros on the diagonal. Thus the diagonal entries of A are just the entries of D, and so $A_{22} = D_{nn}$. Factoring A, we have

$$(L^{-1}M)E(L^{-1}M)^* = N \begin{bmatrix} A_{11} & 0 \\ 0 & D_{nn} - A_{12}^*A_{11}^{-1}A_{12} \end{bmatrix} N^* \tag{1.12}$$

where

$$N = \begin{bmatrix} I & 0 \\ A_{12}^*A_{11}^{-1} & I \end{bmatrix} .$$

Then

$$(N^{-1}L^{-1}M)E = \begin{bmatrix} A_{11} & 0 \\ 0 & D_{nn} - A_{12}^{*}A_{11}^{-1}A_{12} \end{bmatrix} (N^{-1}L^{-1}M)^{*-1} .$$

Since E is diagonal and $N^{-1}L^{-1}M$ is lower triagular with 1's on the diagonal, it follows that

$$E_{nn} = D_{nn} - A_{12}^{*}A_{11}^{-1}A_{12} .$$

Since A_{11} is positive definite, we conclude that $E_{nn} \leq D_{nn}$, with equality if and only if $A_{12} = 0$. From this and (1.12), it follows that $E_{nn} = D_{nn}$ only if

$$(L^{-1}(G-H)L^{*-1})_{jn} = 0 \qquad (1 \leq j \leq n-1) .$$

To verify (1.9) for $2 \leq k \leq n - 1$, we use (1.10) and repeat the argument above with $G(1,\ldots,k)$ and $H(1,\ldots,k)$ in place of G and H, respectively. Notice that equality holds in (1.9) for all k if and only if

$$(L^{-1}(G-H)L^{*-1})_{jk} = 0 \qquad (1 \leq j \leq k-1) .$$

We have already observed that the diagonal entries in $L^{-1}(G-H)L^{*-1}$ are zeros. Since $L^{-1}(G-H)L^{*-1}$ is selfadjoint and L is nonsingular, it follows that G = H if and only if equality holds in (1.9) for all k.

Corollary 3. Let R be an nxn band matrix with bandwidth m that admits a positive definite band extension H, and let G be any positive definite extension of R. Then det G $\leq$ det H, with equality holding if and only if G = H.

Proof. Note that det H is the product of H_{11} and the quotients on the right side of (1.9), with a similar formula for det G. By Theorem 2, each term in the product giving det H dominates the corresponding term in the product giving det G. Hence det G $\leq$ det H, and equality holds if and only if equality holds in (1.9) for all k, in which case G must equal H, by Theorem 2.

2 THE PERMANENCE PRINCIPLE

Let R be an nxn band matrix with bandwidth m that admits a positive definite band extension H. For any principal submatrix $R(j,\ldots,k)$ of R with $k - j > m$, it is possible to apply Theorem 1 with n replaced by $k - j + 1$ to obtain a $(k-j+1)x(k-j+1)$ band extension of $R(j,\ldots,k)$. Unexpectedly, it turns out that the entries of this extension coincide with the corresponding entries of H. Essentially, this is due to the fact that each entry $H_{j,k}$ in H is determined by the entries in the band of $R(j,\ldots,k)$, as follows from (1.7) and (1.8)

Theorem 4. (The Permanence Principle.) Let R be an nxn band matrix with bandwidth m that admits a positive definite band extension H. Then

$H(j,\ldots,k)$ is the unique positive definite band extension of $R(i,\ldots,j)$ for $1 \leq j \leq k \leq n$.

Proof. First suppose $1 < k \leq n$. Then $H(1,\ldots,k)$ is a positive definite extension of $R(1,\ldots,k)$. To prove $H(1,\ldots,k)^{-1}$ is a band matrix, we write

$$H^{-1} = YY^*$$

where Y is an invertible upper triangular band matrix with bandwidth m. Since Y^{*-1} is lower triangular and Y^{-1} is upper triangular, it follows that

$$\begin{aligned} H(1,\ldots,k) &= (YY^*)^{-1}(1,\ldots,k) \\ &= Y^{*-1}(1,\ldots,k)Y^{-1}(1,\ldots,k) \\ &= [Y(1,\ldots,k)]^{*-1}[Y(1,\ldots,k)]^{-1} . \end{aligned}$$

Consequently,

$$H(1,\ldots k)^{-1} = Y(1,\ldots,k)Y(1,\ldots,k)^*$$

so that $H(1,\ldots,k)^{-1}$ is a band matrix. Therefore, $H(1,\ldots,k)$ is the unique positive definite band extension of $R(1,\ldots,k)$, by Theorem 1.

Now suppose $1 \leq j < n$. Then $H(j,\ldots,n)$ is a positive definite extension of $R(j,\ldots,n)$. Using the factorization $H^{-1} = XX^*$, where X is an invertible lower triangular band matrix and proceeding as before, we find that

$$H(j,\ldots,n)^{-1} = [X(j,\ldots,n)][X(j,\ldots,n)]^* .$$

It follows that $H(j,\ldots,n)$ is the unique positive definite band extension of $R(j,\ldots,n)$.

Finally, suppose $1 \leq j < k \leq n$ and apply the second case discussed above with n replaced by k and R replaced by $R(1,\ldots,k)$. Since

$$R(j,\ldots,k) = [R(1,\ldots,k)](j,\ldots,k)$$

with a similar formula for $H(j,\ldots,k)$, we obtain the conclusion of the theorem.

Let R be an nxn band matrix with bandwidth $m < n - 1$. We define as follows an nxn selfadjoint band matrix R' that has bandwidth m+1 and extends R. For $1 \leq k \leq n - m - 1$, Theorem 1 guarantees an (m+2)x(m+2) positive definite band extension of $R(k,\ldots,k+m)$. The entry in the upper right corner of this matrix is taken as the entry in position k, m+k+1 of R'. In this way, we determine the entries of R' that lie on the diagonal just above the band of R. Since R' is to be selfadjoint, this also determines the entries of R' that lie on the diagonal just below the band of R. The extension R' is called the one-level extension of R. The next theorem reduces the problem of obtaining the positive definite band extension to constructing a series of one-level extensions.

Theorem 5. Let R be an nxn band matrix with bandwidth m that admits a positive definite band extension H. Then H is the result of n-m one-level extensions starting with R.

Proof. It follows from Theorem 4 that the entries in the band of R' coincide with the corresponding entries of H. Since H^{-1} is a band matrix with bandwidth m (and hence also bandwidth m+1), it follows from the uniqueness in Theorem 1 that H is the positive definite band extension of R'. Applying this result again, we see that H is the positive definite band extension of (R')', the one-level extension of R'. The theorem follows by finite induction.

Corollary 6. Let R be a Toeplitz band matrix that admits a positive definite band extension H. Then H is Toeplitz.

Proof. By Theorem 5, it suffices to prove that the one-level extension of a Toeplitz band matrix R is Toeplitz. But this is clear since R(k,...,k+m) is independent of K.

Although the extension H in Corollary 6 is Toeplitz, the matrix X in (1.2) need not be Toeplitz. However, it is "almost Toeplitz". It follows from (1.5) that the first m+1 entries in the first column of X are independent of n. (The rest of the entries in the first column are zero.) Also, since R(j,...,j+m) is independent of j for $1 \leq j \leq n - m$, the first n - m columns of X form a Toeplitz band matrix, say T(n). Therefore, X has the form

$$\left[\begin{array}{c|c} & 0 \\ T(n) & \\ & N \end{array}\right] \tag{2.1}$$

where N is an mxm invertible lower triangular matrix.

The permanence principle in its simplest form is implicitly contained in Burg's dissertation [2]. It is also a topic of discussion in a paper of van den Bos [9], who suggested that it does not hold in full generality. We also mention a preprint of Feder and Weinstein [7], who have a version of the permanence principle, but still doubt its full generality.

3 ONE-LEVEL EXTENSIONS

Let R be an nxn band matrix with bandwidth n-2 and assume that R(1,...,n-1) and R(2,...,n) are positive definite. For any complex number w let H(w) be the selfadjoint nxn matrix that agrees with R in the band and whose entry in the lower left corner is w. Write H(w) in the form

$$H(w) = \begin{pmatrix} A & d^* \\ d & R_{nn} \end{pmatrix}$$

where $A = R(1,\dots,n-1)$ and $d = [w, R_{n2}, \dots, R_{n,n-1}]$. Then

$$H(w) = \begin{pmatrix} I & 0 \\ dA^{-1} & 1 \end{pmatrix} \begin{pmatrix} A & 0 \\ 0 & v \end{pmatrix} \begin{pmatrix} I & A^{-1}d^* \\ 0 & 1 \end{pmatrix} \tag{3.1}$$

where

$$v = R_{nn} - dA^{-1}d^* .$$

We will first find the values of w for which H(w) is positive definite. Since R(1,...,n-1) is positive definite, H(w) is positive definite if and only if det H(w) > 0, which by (3.1) is equivalent to v > 0, i.e.,

$$dA^{-1}d^* < R_{nn} .$$

This inequality can be rewritten as

$$\langle A^{-1}d^*, d^* \rangle < R_{nn} . \tag{3.2}$$

Let us write

$$d^* = \bar{w}e_1 + b$$

where

$$e_1 = \begin{bmatrix} 1 \\ 0 \\ \vdots \\ 0 \end{bmatrix} \quad \text{and} \quad b = \begin{bmatrix} 0 \\ R_{2n} \\ \vdots \\ R_{n-1,n} \end{bmatrix} .$$

Then

$$\langle A^{-1}d^*, d^* \rangle = \alpha w\bar{w} + \beta w + \bar{\beta}\bar{w} + \gamma$$
$$= \alpha\left(w + \frac{\bar{\beta}}{\alpha}\right)\left(\bar{w} + \frac{\beta}{\alpha}\right) + \gamma - \frac{|\beta|^2}{\alpha}$$

where

$$\alpha = \langle A^{-1}e_1, e_1\rangle$$

$$\beta = \langle A^{-1}b, e_1\rangle$$

$$\gamma = \langle A^{-1}b, b\rangle \ .$$

Since A^{-1} is positive definite, we have $\alpha > 0$ and $\gamma \geq 0$. We can rewrite (3.2) in the form

$$\left|w - \frac{\bar{\beta}}{\alpha}\right|^2 < \frac{R_{nn}-\gamma}{\alpha} + \frac{|\beta|^2}{\alpha^2} \tag{3.3}$$

Therefore, H(w) is positive definite if and only if (3.3) is satisfied. By Theorem 1, there are solutions of (3.3), so the right side of (3.3) is positive. Thus the set of all w for which H(w) is positive definite is an open disk with center

$$w_0 = -\frac{\bar{\beta}}{\alpha} \ .$$

Let

$$A^{-1} = (s_{jk})_{j,k=1}^{n-1} \ .$$

Then

$$\alpha = s_{11} \quad \text{and} \quad \beta = \sum_{k=2}^{n-1} s_{1k}R_{kn} \ .$$

Therefore

$$w_0 = -\frac{1}{s_{11}}\sum_{k=2}^{n-1} R_{nk}s_{k1} \ . \tag{3.4}$$

Similarly, if we partition H(w) in the form

$$H(w) = \begin{pmatrix} R_{11} & g \\ g^* & R(2,\ldots,n) \end{pmatrix}$$

and let $R(2,\ldots,n)^{-1} = (t_{jk})_{j,k=2}^{n}$, we obtain

$$w_0 = \frac{1}{t_{nn}}\sum_{k=2}^{n-1} t_{nk}R_{k1} \ .$$

By comparing (3.4) with (1.7) with $j = 1$ and $i = n$ we see that $H(w_0)$ is the positive definite band extension of R. These results are summarized in the following theorem.

Theorem 7. Let R be an nxn band matrix with bandwidth n-2 such that $R(1,\dots,n-1)$ and $R(2,\dots,n)$ are positive definite. For any complex number w, let H(w) be the selfadjoint matrix that agrees with R in the band and whose lower left entry is w. Let

$$R(1,\dots,n-1)^{-1} = (s_{jk})_{j,k=1}^{n-1}$$

and

$$R(2,\dots,n)^{-1} = (t_{jk})_{j,k=2}^{n} \;.$$

(a) The values of w for which H(w) is positive definite form an open disk with radius ρ and center w_0 given by

$$\rho = \frac{\sqrt{\det R(1,\dots,n-1)\det R(2,\dots,n)}}{\det R(2,\dots,n-1)} \tag{3.5}$$

$$w_0 = -\frac{1}{s_{11}}\sum_{k=2}^{n-1} R_{nk}s_{k1} = -\frac{1}{t_{11}}\sum_{k=2}^{n-1} t_{nk}R_{k1} \;. \tag{3.6}$$

(b) $H(w_0)$ is the positive definite band extension of R.

Proof. All of the theorem has been proved except for (3.5). Since $R(1,\dots,n-1)$ and $R(2,\dots,n)$ are positive definite, there are unique numbers $x_2,\dots,x_{n-1}$, P_{n-1}, $w_2,\dots,w_{n-1}$, and Q_{n-1} with $P_{n-1} \neq 0$ and $Q_{n-1} \neq 0$ such that

$$R(1,\dots,n-1)\begin{bmatrix} 1 \\ x_2 \\ \vdots \\ x_{n-1} \end{bmatrix} = \begin{bmatrix} P_{n-1} \\ 0 \\ \vdots \\ 0 \end{bmatrix} \tag{3.7}$$

and

$$R(2,\dots,n)\begin{vmatrix} w_2 \\ \vdots \\ w_{n-1} \\ 1 \end{vmatrix} = \begin{vmatrix} 0 \\ \vdots \\ 0 \\ Q_{n-1} \end{vmatrix} \tag{3.8}$$

From (3.7) and (3.8) it follows that

$$P_{n-1} = \frac{\det R(1,\dots,n-1)}{\det R(2,\dots,n-1)} \tag{3.9}$$

$$Q_{n-1} = \frac{\det R(2,\dots,n)}{\det R(2,\dots,n-1)} \tag{3.10}$$

In particular, $P_{n-1} > 0$ and $Q_{n-1} > 0$. We define

$$\Delta'_n = w + \sum_{j=2}^{n-1} R_{nj}x_j \tag{3.11}$$

$$\Delta''_n = \sum_{j=2}^{n-1} R_{1j}w_j + \bar{w} \tag{3.12}$$

$$c'_n = -\frac{\Delta'_n}{Q_{n-1}} \tag{3.13}$$

$$c''_n = -\frac{\Delta''_n}{P_{n-1}} \tag{3.14}$$

$$P_n = P_{n-1} + c'_n\Delta''_n \tag{3.15}$$

$$Q_n = c''_n\Delta'_n + Q_{n-1} \tag{3.16}$$

From (3.11)-(3.16) it follows that

$$P_n = P_{n-1}(1-c'_nc''_n) \tag{3.17}$$

$$Q_n = Q_{n-1}(1-c'_nc''_n) \tag{3.18}$$

$$H(w)\left(\begin{bmatrix}1\\x_2\\\vdots\\x_{n-1}\\0\end{bmatrix}+c_n'\begin{bmatrix}0\\w_2\\\vdots\\w_{n-1}\\1\end{bmatrix}\right)=\begin{bmatrix}P_n\\0\\\vdots\\0\\0\end{bmatrix} \tag{3.19}$$

$$H(w)\left(c_n''\begin{bmatrix}1\\x_2\\\vdots\\x_{n-1}\\0\end{bmatrix}+\begin{bmatrix}0\\w_2\\\vdots\\w_{n-1}\\1\end{bmatrix}\right)=\begin{bmatrix}0\\0\\\vdots\\0\\Q_n\end{bmatrix}. \tag{3.20}$$

If $H(w)$ is positive definite, then it follows from (3.19) that P_n is not 0 and

$$P_n = \frac{\det H(w)}{\det R(2,\ldots,n)} \tag{3.21}$$

which implies that $P_n > 0$. On the other hand, if $P_n > 0$, then it follows from (3.19), and the fact that $R(2,\ldots,n)$ has rank $n-1$, that $H(w)$ has rank n. Then (3.21) holds, so $\det H(w) > 0$, which implies that $H(w)$ is positive definite. Therefore, $H(w)$ is positive definite if and only if $P_n > 0$, which by (3.17) is equivalent to

$$c_n'c_n'' < 1 . \tag{3.22}$$

Similarly, $H(w)$ is positive definite if and only if $Q_n > 0$ and in that case

$$Q_n = \frac{\det H(w)}{\det R(1,\ldots,n-1)} \tag{3.23}$$

Suppose again that $H(w)$ is positive definite. From (3.7) and (3.8) it follows that

$$x_j = \frac{s_{j1}}{s_{11}} \text{ and } w_j = \frac{t_{jn}}{t_{nn}} . \qquad (2 \le j \le n-1) .$$

Comparing (3.11) and (3.12) with (3.6), we find that

$$\Delta_n' = w + \sum_{j=2}^{n-1} R_{nj}x_j = w + \sum_{j=2}^{n-1} R_{nj}\frac{s_{j1}}{s_{11}} = w - w_0 \tag{3.24}$$

$$\Delta_n'' = \sum_{j=2}^{n-1} R_{1j}w_j + \bar{w} = \sum_{j=2}^{n-1} R_{1j}\frac{t_{jn}}{t_{nn}} + \bar{w} = \overline{w - w_0} \tag{3.25}$$

Therefore

$$\Delta_n'' = \bar{\Delta}_n' \tag{3.26}$$

and

$$\sum_{j=2}^{n-1} R_{1j}w_j = \overline{\sum_{j=2}^{n-1} R_{nj}x_j} \,. \tag{3.27}$$

From (3.13), (3.14) and (3.26) we find that

$$c_n'' = \overline{c_n'}\,\frac{Q_{n-1}}{P_{n-1}} \tag{3.28}$$

Combining (3.11)-(3.14), we have

$$w = -\sum_{j=2}^{n-1} R_{nj}x_j - c_n' Q_{n-1} \tag{3.29}$$

$$\bar{w} = -\sum_{j=2}^{n-1} R_{1j}w_j - c_n'' P_{n-1}. \tag{3.30}$$

Then by (3.27), (3.29) and (3.30),

$$\left|w + \sum_{j=2}^{n-1} R_{nj}x_j\right|^2 = c_n'c_n'' Q_{n-1}P_{n-1} \,. \tag{3.31}$$

Therefore if H(w) is positive definite, then by (3.31) and (3.22) w must satisfy

$$\left|w + \sum_{j=2}^{n-1} R_{nj}x_j\right| < \sqrt{Q_{n-1}P_{n-1}} \,. \tag{3.32}$$

Conversely, suppose w is a complex number that satisfies (3.32). Define

$$c_n' = -\frac{w + \sum_{j=2}^{n-1} R_{nj}x_j}{Q_{n-1}} \tag{3.33}$$

$$c''_n = - \frac{\overline{\bar{w} + \sum_{j=2}^{n-1} R_{nj} x_j}}{P_{n-1}} \tag{3.34}$$

Then (3.32)-(3.34) imply that $c'_n c''_n < 1$. If we now define Δ'_n, Δ''_n, P_n, and Q_n by (3.11), (3.12), (3.15) and (3.16), then (3.19) and (3.20) are satisfied. Since $c'_n c''_n < 1$, it follows from an earlier comment that $H(w)$ is positive definite. Thus we have proved that $H(w)$ is positive definite if and only if w satisfies (3.32), i.e., w lies in the open disk with center w_0 and radius ρ given by

$$\rho = \sqrt{Q_{n-1} P_{n-1}} = \frac{\sqrt{\det R(1,\ldots,n-1) \det R(2,\ldots,n)}}{\det R(2,\ldots,n-1)} .$$

This completes the proof of the theorem.

Suppose the matrix R in Theorem 7 is Toeplitz. Then $R(1,\ldots,n-1) = R(2,\ldots,n)$, so it follows from (3.9) and (3.10) that $P_{n-1} = Q_{n-1}$ and from (3.28) that $c''_n = \bar{c}'_n$. Then (3.17) and (3.18) become

$$P_n = P_{n-1}(1-|c'_n|^2).$$

Thus the numbers c'_n and c''_n may be regarded as generalizations of the reflection coefficients that appear in the modern version of the Levinson algorithm employed by Burg [2].

4 INFINITE-DIMENSIONAL BAND EXTENSIONS

In this section we give a new proof of Theorem A, the maximum entropy theorem. Our proof will clarify the connection between this theorem and the matrix result in Theorem B.

A doubly infinite Toeplitz matrix

$$R = (r_{j-k})_{j,k=-\infty}^{\infty}$$

is called a band matrix with bandwidth m if $r_k = 0$ for $|k| > m$. The following two theorems summarize our finite-dimensional results in the context of extensions of infinite Toeplitz band matrices.

Theorem 8. Let $R = (r_{j-k})_{j,k=-\infty}^{\infty}$ be a Toeplitz band matrix with bandwidth m, and suppose that the principal section $(r_{j-k})_{j,k=0}^{m}$ is positive definite. Then there exists a unique Toeplitz matrix $H = (h_{j-k})_{j,k=-\infty}^{\infty}$ with the property that $h_k = r_k$ for $|k| \leq m$, and for $n > m$ the principal section $(h_{j-k})_{j,k=0}^{n}$ is a positive definite Toeplitz band extension of $(r_{j-k})_{j,k=0}^{n}$.

Proof. For $n > m$, let H_n be the unique positive definite Toeplitz band extension of $(r_{j-k})_{j,k=0}^{n}$. Theorem 1 and Corollary 6 guarantee that H_n exists. By the permanence principle, there is a unique sequence $\{h_k : -\infty < k < \infty\}$ such that $H_n = (h_{j-k})_{j,k=0}^{n}$ for each $n > m$. Clearly the matrix $H = (h_{j-k})_{j,k=-}^{\infty}$ has the desired properties.

By Theorem 5, the matrix H_n in the proof above may be obtained as a one-level extension of H_{n-1}, if one so desires. In this way one may view H as constructed from R by a sequence of one-level extensions. (This approach to the maximum entropy extension was already taken by Burg [2, pages 31-36].)

The next theorem gives an extremal characterization of the matrix H constructed in Theorem 8. Later we shall show how this extremal property translates into the maximum entropy statement in Theorem A.

Theorem 9. Let R and H be as in Theorem 8, and let $H_n = (h_{j-k})_{j,k=0}^{n}$. Let $G = (g_{j-k})_{j,k=-\infty}^{\infty}$ be any Toeplitz matrix such that $g_k = r_k$ for $|k| \leq m$ and such that the principal sections $G_n = (g_{j-k})_{j,k=0}^{n}$ are positive definite. Then

$$\frac{\det G_n}{\det G_{n-1}} \leq \frac{\det H_n}{\det H_{n-1}} \qquad (n \geq 2) \tag{4.1}$$

with equality holding for all n if and only if G = H.

Proof. For $n \leq m$, $G_n = H_n$. For $n > m$, H_n is the positive definite band extension of $(r_{j-k})_{j,k=0}^{n}$, and G_n is any positive definite extension. Since G_n is an $(n+1)\text{x}(n+1)$ matrix, $G_n = G_n(1,\ldots,n+1)$ and $G_{n-1} = G_{n-1}(1,\ldots,n)$, in the notation of Section 1, with similar formulas for H_n and H_{n-1}. Thus (4.1) is just a restatement of (1.9).

We shall now derive Theorem A from Theorems 8 and 9. Given $r_{-m},\ldots,r_m$, as in Theorem A, let $r_k = 0$ for $k > m$, let $R = (r_{j-k})_{j,k=-\infty}^{\infty}$ and let $H = (h_{j-k})_{j,k=-\infty}^{\infty}$ be the extension of R described in Theorem 8. The proof will now consist of two steps. The first step is to show that $\Sigma|h_k| < \infty$ and that the function h defined by

$$h(z) = \sum_{k=-\infty}^{\infty} h_k x^k \qquad (|z| = 1)$$

has the properties listed in (ii) of Theorem A. The second step in the proof will be to derive the inequality (iii) of Theorem A.

Step 1. We begin by extending the principal sections H_n $(n>m)$ to be bounded invertible operators on $\ell^1(-\infty,\infty)$ as follows. It suffices to consider only odd values of n, say $n = 2k+1$. Let $\tilde{H}_n$ be the operator that acts as H_n on vectors in ℓ^1 of the form $(\ldots,0, \xi_{-k}, \ldots, \xi_0,\ldots,\xi_k,0,\ldots)$ and acts as the identity on vectors of the form

$(\ldots,\xi_{-k-1},0,\ldots,0,\xi_{k+1},\xi_{k+2},\ldots)$. Thus we may represent $\widetilde{H}_n$ by a block matrix of the form

$$\widetilde{H}_n = \begin{bmatrix} I & 0 & 0 \\ 0 & H_n & 0 \\ 0 & 0 & I \end{bmatrix} .$$

Next, as in the proof of Theorem 1, let X_n be the lower triangular band matrix with positive entries on the diagonal. Then $H_n^{-1} = X_n X_n^*$ and this factorization leads to

$$\widetilde{H}_n^{-1} = \widetilde{X}_n \widetilde{X}_n^* \tag{4.2}$$

where $\widetilde{X}_n$ is represented by a block matrix partitioned in the same way as the matrix for $\widetilde{H}_n$, namely,

$$\widetilde{X}_n = \begin{bmatrix} I & 0 & 0 \\ 0 & X_n & 0 \\ 0 & 0 & I \end{bmatrix} \tag{4.3}$$

From (2.1) we see that X_n itself has the form

$$X_n = \begin{bmatrix} T_n & 0 \\ M_n & N \end{bmatrix} \tag{4.4}$$

where T_n is an $(n-m+1)\mathrm{x}(n-m+1)$ lower triangular Toeplitz band matrix, N is a fixed lower triangular $m\mathrm{x}m$ matrix, and the nonzero columns of M_n are independent for n for $n \geq 2m$. Furthermore, if we denote the part of the first column of T_n that lies in the band by the vector

$$b = \begin{bmatrix} b_0 \\ b_1 \\ \vdots \\ b_m \end{bmatrix} \tag{4.5}$$

then this vector is independent of n for $n \geq 2m$. It follows from (4.3) and (4.4) that the operator $\widetilde{X}_n$ converges pointwise on $\ell^1(-\infty,\infty)$ to a bounded linear operator X. The matrix representing X is an infinite lower triangular Toeplitz band matrix whose nonzero entries in each column are given by the vector b in (4.5). Recall from the construction of X_n that b satisfies the equation

$$R_m b = \begin{bmatrix} b_0^{-1} \\ 0 \\ \vdots \\ 0 \end{bmatrix} .$$

where $R_m = (r_{j-k})_{j,k=0}^{m}$ is positive definite. We shall show that the zeros of the polynomial

$$b(z) = b_0 + b_1 z + \ldots + b_m z^m$$

all lie outside the unit disk $|z| \leq 1$. Observe that the vector $b_0 b$ is the first colum of R_m^{-1}. Using the standard formula for the entries in R_m^{-1}, we see that

$$b_0 b(z) = \frac{1}{\det R_m} \cdot \det \begin{bmatrix} 1 & z & \ldots & z^m \\ r_1 & r_0 & & r_{-m+1} \\ r_2 & r_1 & & r_{-m+2} \\ \vdots & & \ddots & \\ r_m & r_{m-1} & & r_0 \end{bmatrix}$$

Reversing the order of the rows and the order of the columns of the matrix above, we have

$$b(z) = \frac{b_0^{-1}}{\det R_m} \cdot \det \begin{bmatrix} r_0 & r_1 & \ldots & r_m \\ r_{-1} & r_0 & & r_{m-1} \\ \vdots & & \ddots & \\ r_{-m+1} & r_{-m+2} & & r_1 \\ z^m & z^{m-1} & & 1 \end{bmatrix}$$

Since R_m is positive definite, a result of Szegö shows that $b(z) \neq 0$ for $|z| \leq 1$. (See [1], pages 14-15.) Hence $1/b(z)$ is analytic in $|z| \leq 1$ and has the form

$$1/b(z) = \sum_{k=0}^{\infty} c_k z^k \qquad (|z| \leq 1) \tag{4.6}$$

for some sequence (c_k) such that

$$\sum_{k=0}^{\infty} |c_k| < \infty$$

Set $c_k = 0$ for $k < 0$. Then a direct calculation using (4.6) shows that X has the bounded inverse

$$X^{-1} = (c_{j-k})_{j,k=-\infty}^{\infty} \tag{4.7}$$

In particular, X^{-1} is a lower triangular Toeplitz matrix.

We now claim that $\widetilde{X}_n^{-1}$ converges pointwise on ℓ^1 to X^{-1}. Since we already have $\widetilde{X}_n$ converging pointwise to X, it suffices to show that the norms of the $\widetilde{X}_n^{-1}$, as operators on ℓ^1, are bounded as $n \to \infty$. By (4.3), it suffices to show that the ℓ^1 operator norms of the matrices X_n^{-1} are bounded. Observe from (4.4) that X_n^{-1} has the form

$$X_n^{-1} = \begin{bmatrix} T_n^{-1} & 0 \\ U_n^{-1} & N^{-1} \end{bmatrix}$$

where $U_n^{-1} = -N^{-1}M_nT_n^{-1}$. Since N^{-1} and the nonzero columns of M_n are independent of n, it suffices to show that the ℓ^1 operator norms of the matrices T_n^{-1} are bounded, for then the same will hold for U_n^{-1} and hence for X_n^{-1}. Clearly T_n is a principal section of X. Since X^{-1} is lower triangular, T_n^{-1} is a principal section of X^{-1}. It follows that $\|T_n^{-1}\| \le \|X^{-1}\|$ for $n > m$. Hence the sequence $\{\|X_n^{-1}\|\}$ is bounded. We conclude that $\widetilde{X}_n^{-1}$ converges pointwise to X^{-1}.

Since $\widetilde{X}_n$ and X correspond to band matrices, the conjugate transposes of these matrices determine bounded operators on ℓ^1, which we write as $(\widetilde{X}_n)^*$ and X^*, respectively. An argument similar to that given above will show that $(\widetilde{X}_n)^{*-1}$ converges pointwise on ℓ^1 to X^{*-1}. Finally, from the factorization (4.2) of $\widetilde{H}_n^{-1}$ we conclude that the $\widetilde{H}_n^{-1}$ converge pointwise on ℓ^1 to the bounded linear operator XX^*. From this and the fact that $\widetilde{H}_n$ converges to H, it follows that H is an invertible bounded operator on ℓ^1 and

$$H^{-1} = XX^*, \quad H = X^{*-1}X^{-1} . \tag{4.8}$$

The fact that $H = (h_{j-k})_{j,k=-\infty}^{\infty}$ is bounded as an operator on $\ell^1(-\infty,\infty)$ is equivalent to the condition

$$\sum_{k=-\infty}^{\infty} |h_k| < \infty .$$

Thus we may define

$$h(z) = \sum_{k=-\infty}^{\infty} h_k z^k \qquad (|z| = 1) .$$

A routine calculation using (4.8), (4.7) and (4.6) shows that

$$h(z) = \sum_{k=0}^{\infty} \Big(\sum_{j=0}^{\infty} \bar{c}_j c_{j+k} \Big) z^k$$

$$= [b(z)\overline{b(z)}]^{-1} = |b(z)|^{-2} > 0 \qquad (|z| = 1)$$

Hence h has the properties listed in (ii) of Theorem A. It is easily shown that H is positive definite as an operator on $\ell^2(-\infty,\infty)$, and hence it is appropriate to refer to H as the positive definite Toeplitz band extension of R.

Step 2. Now we shall use Theorem 9 to derive the extremal characterization of the function h in Theorem A. From (4.4) it is clear that

$$\frac{\det X_n}{\det X_{n-1}} = b_0 \qquad (n > m+1)$$

Since $H_n = X_n X_n$, it follows that

$$\frac{\det H_n}{\det H_{n-1}} = \frac{|\det X_{n-1}|^2}{|\det X_n|^2} = b_0^{-2} \qquad (n > m+1)$$

which shows that the right side of (4.1) is independent of n. Now let g be any function as in (iii) of Theorem A. That is, $g(z) > 0$ for $|z| = 1$ and there exists a sequence (g_k) such that $g_k = r_k$ for $|k| \leq m$, with $\Sigma |g_k| < \infty$ and

$$g(z) = \sum_{k=-\infty}^{\infty} g_k z^k \qquad (|z| = 1)$$

Then the matrix $G = (g_{j-k})_{j,k=-\infty}^{\infty}$ satisfies the conditions of Theorem 9. We shall prove that the left side of (4.1) is a decreasing function of n.

For $n > 1$, we apply a Hadamard-Fischer inequality to the $(n+1)x(n+1)$ positive definite matrix G_n and find that

$$\frac{\det G_n(1,\ldots,n) \cdot \det G_n(2,\ldots,n+1)}{\det G_n(2,\ldots,n)} \geq \det G_n(1,\ldots,n+1) .$$

Hence

$$\frac{\det G_n(1,\ldots,n)}{\det G_n(2,\ldots,n)} \geq \frac{\det G_n(1,\ldots,n+1)}{\det G_n(2,\ldots,n+1)}$$

that is,

$$\frac{\det G_{n-1}}{\det G_{n-2}} \geq \frac{\det G_n}{\det G_{n-1}}$$

because G_n is a Toeplitz matrix. Thus the left side of (4.1) is a decreasing function of n.

Suppose $g \neq h$. Then $G \neq H$, and so a strict inequality holds in (4.1) for some n. By what we have proved, it follows that

$$\lim_{n\to\infty} \frac{\det G_n}{\det G_{n-1}} < \lim_{n\to\infty} \frac{\det H_n}{\det H_{n-1}} .$$

The maximum entropy inequality (iii) in Theorem A now follows from a theorem of Szegö that says that

$$\lim_{n\to\infty} \frac{\det G_n}{\det G_{n-1}} = \exp\left\{\frac{1}{2\pi}\int_{-\pi}^{\pi} \log g(e^{i\omega})\, d\omega\right\} . \tag{4.9}$$

An analogous formula holds for the function f. A proof of this is in [8, pages 77-78]. We shall describe a variation of this proof in order to illustrate the connection between the finite- and infinite-dimensional situations.

Since $(g_{j-k})_{j,k=0}^{n}$ is positive definite for each n, the system of equations

$$\sum_{k=0}^{n} g_{j-k}x_k = \delta_{j0} \qquad (j = 0,1,\ldots,n) \tag{4.10}$$

has a unique solution. If $x_0^{(n)}$ denotes the first coordinate of the solution of (4.10), then

$$x_0^{(n)} = \frac{\det G_{n-1}}{\det G_n} .$$

The properties assumed for g imply that G is positive definite as an operator on $\ell^2(0,\infty)$. From this it is a standard fact that as $n \to \infty$, $x_0^{(n)}$ converges to the first coordinate of the solution $(x_k)_{k=0}^{\infty}$ in $\ell^1(0,\infty)$ of

$$\sum_{k=0}^{\infty} g_{j-k}x_k = \delta_{j0} \qquad (j = 0,1,\ldots) . \tag{4.11}$$

See [8, page 74] for a proof. Thus

$$x_0 = \lim_{n\to\infty} \frac{\det G_{n-1}}{\det G_n} \tag{4.12}$$

Since the solution (x_k) to (4.11) is in $\ell^1(0,\infty)$, we may define a function

$$x(z) = \sum_{k=0}^{\infty} x_k z^k \qquad (|z| \leq 1)$$

Then (4.11) is equivalent to

$$g(z)x(z) = 1 + y(z) \qquad (|z| = 1) \tag{4.13}$$

where

$$y(z) = \sum_{k=-\infty}^{-1} y_k z^k, \quad \sum_{k=-\infty}^{-1} |y_k| < \infty .$$

Observe that $\overline{x(z)g(z)} = \overline{x(z)g(z)} = 1 + \overline{y(z)}$, because $g(z) > 0$ for $|z| = 1$. hence

$$(1 + \overline{y(z)})x(z) = \overline{x(z)}g(z)x(z) = \overline{x(z)}(1 + y(z)), \quad (|z| = 1) \tag{4.14}$$

Since $\bar{z} = 1/z$ when $|z| = 1$, the left side of (4.14) is represented by an absolutely convergent series with nonnegative powers of z. Also, the right side of (4.14) is represented by an absolutely convergent series with nonpositive powers of z.

It follows that

$$(1 + \overline{y(z)})x(z) = x_0 \qquad (|z| = 1) \tag{4.15}$$

In fact, $\overline{y(z)}$ has an extension to an analytic function on $|z| < 1$, which we again denote by $\overline{y(z)}$. Then $(1 + \overline{y(z)})x(z)$ is identically equal to x_0 for $|z| \leq 1$. If x_0 were zero, then $x(z)$ would be identically zero for $|z| \leq 1$ (since $1 + \overline{y(z)}$ is not zero for z sufficiently close to $z = 0$). This is impossible, so $x_0 \neq 0$ and $x(z) \neq 0$ for $|z| \leq 1$. This implies that $1/x(z)$ is analytic for $|z| \leq 1$ and

$$1/x(z) = x_0^{-1}(1 + \overline{y(z)}) \qquad (|z| \leq 1).$$

Thus (4.13) may be written in the form

$$g(z) = (1 + \overline{y(z)})x_0^{-1}(1 + y(z)) \qquad |z| = 1. \tag{4.16}$$

We may take the logarithm of both sides, since $g(z) > 0$ for $|z| = 1$. Since $1 + \overline{y(z)}$ is analytic and does not vanish for $|z| < 1$, there is a branch of $\log(1 + \overline{y(z)})$ that is analytic for $|z| < 1$. Therefore it follows from (4.16) that

$$\frac{1}{2\pi}\int_{-\pi}^{\pi} \log g(e^{i\omega})d\omega = \log x_0^{-1} .$$

Combining this with (4.12), we obtain Szego's result (4.9).

BIBLIOGRAPHY

1 Ahiezer, N.I. and M. Krein (1962). Some Questions in the Theory of Moments. Translations of Mathematical Monographs, Volume 2. American Mathematical Society, Providence, Rhode Island.

2 Burg, J. (1975). Maximum Entropy Spectral Analysis. Ph.D. Dissertation, Stanford University, Stanford, California.

3 Dym, H. and I. Gohberg (1979). Extensions of Matrix Valued Functions with Rational Polynomial Inverses. Integ. Equations Operator Theory 2, 503-528.

4 Dym, H. and I. Gohberg (1980). On an Extension Problem, Generalized Fourier Analysis, and an Entropy Formula. Integ. Equations Operator Theory 3, 143-215.

5 Dym, H. and I Gohberg (1982/83). Extensions of Kernels of Fredholm Operators. Journal d'Analyse Mathematique 42, 51-97.

6 Dym, H. and I Gohberg (1981). Extensions of Band Matrices with Band Inverses. Linear Algebra and Appl. 36, 1-24.

7 Feder, M. and E. Weinstein. On the Finite Maximum Entropy Extrapolation. Preprint.

8 Gohberg, I. and I. Feldman (1974). Convolution Equations and Projection Methods for their Solution. Translations of Mathematical Monographs, Volume 41. American Mathematical Society, Providence, Rhode Island.

9 van den Bos, A. (1971). Alternative Interpretation of Maximum Entropy Spectral Analysis. IEEE Trans. On Information Theory IT-17, 493-494.

THEORY OF MAXIMUM ENTROPY IMAGE RECONSTRUCTION

John Skilling
Cambridge University
Department of Applied Mathematics and Theoretical Physics
Silver Street, Cambridge CB3 9EW, England

ABSTRACT

We address the problem of reconstructing an image from incomplete and noisy data.

Maximum entropy is to be preferred to the direct Bayesian approach, which leads to impossibly large problems of computation and display in the space of all images. The "monkey" approach to maximum entropy, in which N balls are thrown randomly at the cells of an image, invites us to interpret entropy as the logarithm of a prior probability. Such an interpretation leads to severe difficulties with the parameter N, as well as leaving unsolved the problem of selecting a single image for final display.

It is better to justify maximum entropy by the "kangaroo" argument. This shows that maximum entropy is the only generally applicable technique which will give the correct answer to certain simple problems.

As originally presented, maximum entropy fails to incorporate many types of prior knowledge. This defect can be remedied by a small change of interpretation, with great potential benefit for the quality of reconstructed images.

1 INTRODUCTION

In this paper, I shall discuss the theory of using maximum entropy to reconstruct probability distributions. This involves outlining the related Bayesian approach, and also involves developing a way of incorporating relevant prior knowledge. To make the arguments less abstract, I shall discuss the techniques in the light of one particular application, that of reconstruction of an image from incomplete and noisy data. Readers should be aware that my comments are necessarily biassed by my current thinking, and that not all the ideas are universally accepted. I am very conscious that the good ideas are due to other people, in particular Jaynes, Gull, Shore and Johnson: I shall be responsible for any wrong ones.

Let us start by setting up some formalism. An "image" p bears a very close relationship to a probability distribution function (pdf). It can be divided into cells i=1,2,... and described by assigning a number p_i to each cell. This number can be identified with the intensity of light relating to the ith cell. Because light intensities are positive and are additive between cells, the p_i are also positive and additive

$$p_i \geq 0 \quad , \quad p_{iUjU\ldots} = p_i + p_j + \ldots \tag{1}$$

If, furthermore, we decide that our results should be independent of the physical units of the image, we may normalise p to

$$\Sigma p_i = 1 \ , \tag{2}$$

Technically, p is now the set of proportions (of total flux) describing the normalised image, but it also obeys the Kolmogorov axioms and can be legitimately treated as a probability distribution in the space S of pdf's describing single samples. Proportions and probabilities are isomorphic. Specifically, p_i can be identified with the probability that "the next photon" to be received from the image would have come from cell i (Skilling and Gull, 1984a). This gives a very natural and convenient way of thinking about the abstract problem of drawing samples from a pdf, though in other applications such as spectral analysis, the identification can seem somewhat forced.

Later, we shall need to consider drawing many samples from the pdf p. Successive samples from any particular pdf would be independent (the iid assumption), so that the probability of drawing samples from cells i,j,k... successively would be

$$p_{ijk\ldots} = p_i p_j p_k \cdots \ , \tag{3}$$

which is a distribution from a larger space S^N of pdf's describing multiple samples. (Here one has to be a little careful with the photon model. Successive photons could physically interfere with each other in the receiver if the image were too bright. This would alter the response functions of the detectors and hence the numerical values of the data, but it would not alter the algebra of abstractly drawing samples from that pdf which is isomorphic to the image. An opposing view has been developed by Kikuchi and Soffer (1977) and by Frieden (1983).)

Individual samples may be combined to give the histogram n describing the frequency n_i of occurrences of cell i in the multiple sample.

When the number N of samples exceeds the number of cells, the histogram becomes a recognisable, though quantised, image, and as $N \to \infty$, we may expect the normalised histogram n_i/N to approach the original distribution p (Figure 1).

Figure 1. Sampling from an "image".

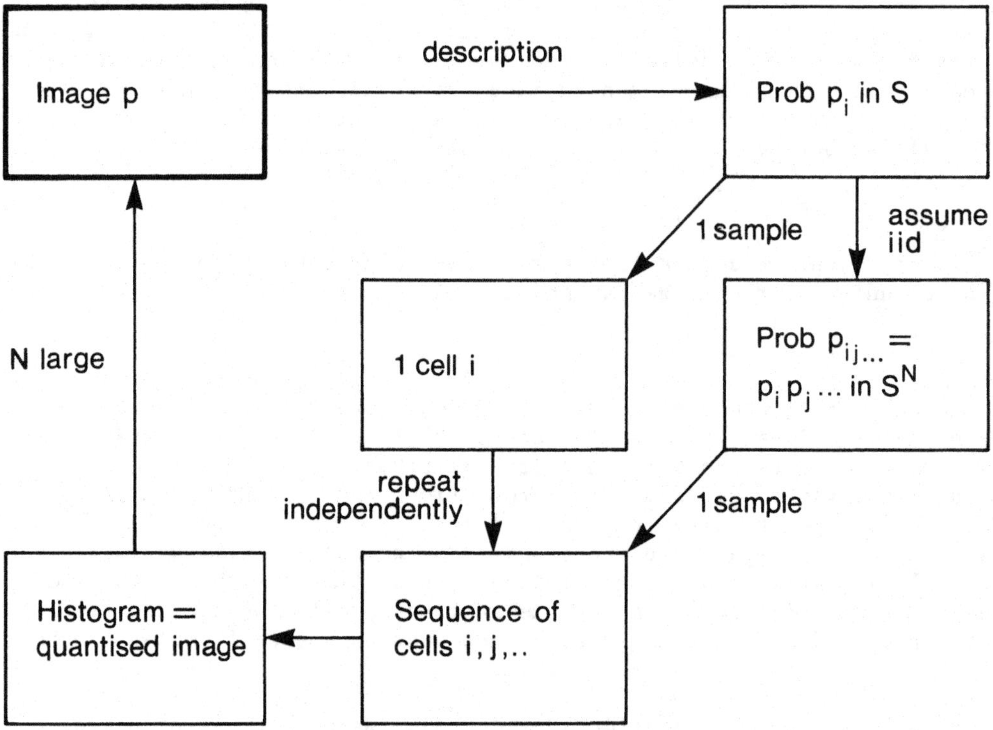

2 THE DIRECT BAYESIAN APPROACH

In practical image reconstruction, we do not have complete knowledge of the image p. All we have are constraints given by some form of data D_k (k=1,2,...), which may be Fourier transform components, line integrals, convolution averages or indeed almost anything else. Usually the constraints are "soft" being given to us in the "noisy" or probabilistic form of likelihoods

$$L_k(p) = \text{prob}(D_k \mid p)\ . \tag{4}$$

These are often taken to be of Gaussian form

$$L_k(p) = (2\pi\sigma_k)^{-1/2} \exp(-(M_k(p)-D_k)^2/2\sigma_k^2) \tag{5}$$

where $M_k(p)$ is the mock datum which would have been observed from p in the absence of noise (having standard deviation σ_k). "Hard" constraints, by contrast, are absolute restrictions on the mock data, such as definitive equality or inequality constraints. These constitute "testable" information in the sense of Jaynes (1978) on the space S of pdf's of single samples i, and also on the space S^N if we allow independent sampling. It is possible to determine unambiguously whether or not p is consistent with a hard constraint, but not with a soft constraint. Technically, soft constraints are not testable.

If, as in many practical cases, the "soft" noise residuals (M_k-D_k) are independent between successive measurements k, the probability of obtaining the entire dataset D from an image p is

$$L(p) = \text{prob}(\ D \mid p\ ) = \prod_k L_k\ (p) \propto \exp(-\ \text{chisquared}\ /2), \qquad (6)$$

$$\text{chisquared} = \sum_k (M_k\ (p) - D_k)^2/\sigma_k^2\ . \qquad (7)$$

Hard constraints can be subsumed in this treatment by requiring L(p) to be zero for images which would break the constraints.

As happens in inverse problems, the probability we have is the wrong way round. We want to draw conclusions about p, given D, but we are given prob(D | p) instead. Bayes' theorem is the standard tool for inverting probabilities, and it gives us the posterior probability distribution (Figure 2)

$$\text{post}(\ p \mid D\ ) \propto \text{prob}(\ D \mid p\ )\ \text{prior}(p) = L(p)\ \text{prior}(p) \qquad (8)$$

relative to some assigned prior distribution prior(p).

Figure 2. Bayesian procedure.

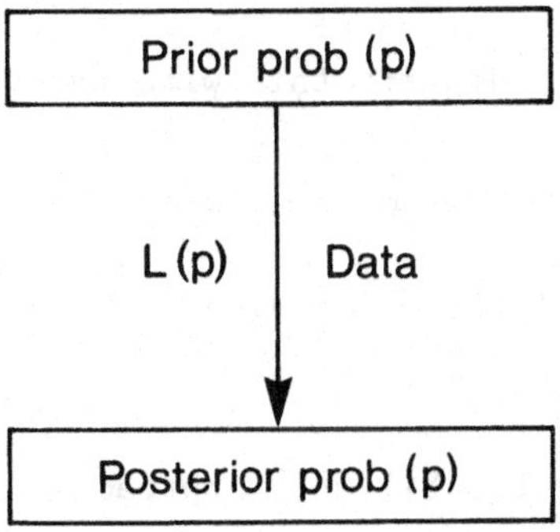

It may be difficult to assign a useful prior to the space of all possible images p, and we will return to this question later. The serious difficulty, seen as soon as practical computations are attempted, is that the space of all possible images is enormous, having a degree of freedom for each and every cell of the image. Correspondingly, we may expect an enormous number of images p to have substantial posterior probability. Remember that any collection of data is finite, whereas an image may be defined with an arbitrarily large number of cells.

As a matter of practical necessity, we must present just one image p (or at most a few) as "the result" of the experiment which gave us the data D. The straightforward Bayesian approach, however appealing it may be from an Olympian viewpoint, does not tell us which single p to select. This simple point is crucial.

3 THE MONKEY APPROACH

Here, we take the idea of sampling pdf's seriously, by considering a team of monkeys throwing balls (i.e., classical distinguishable particles) randomly (i.e., independently from some distribution) at the cells of the image (Gull and Daniell, 1978). Our monkeys cannot throw their balls in accordance with the pdf of the actual image, because we do not know what the true image is. Instead, they throw their balls in accordance with some pre-assigned model m for the image. Quite often, our prior state of knowledge about the object is translation invariant among the cells i, so that m would have to be merely a constant, independent of i. For clarity of presentation, I shall at first adopt this simplification: the generalization to include m is straightforward.

Suppose that the monkeys throw N balls at the image. In any given realization of N throws, there will be a corresponding histogram n of the numbers of balls reaching the various cells. The number of ways of reaching a particular histogram is the corresponding classical degeneracy

$$\Omega(n) = N! \;/\; \prod_i n_i! \quad . \tag{9}$$

The "monkey" approach is to identify this with the prior probability required by Bayes' theorem,

$$\text{prior}(p) \propto \Omega\,(n) \quad , \quad p_i = n_i/N \tag{10}$$

which leads to the probabilistic conclusion

$$\text{post}(p) \propto L(p)\;\Omega\,(Np)\;. \tag{11}$$

I shall argue against this probabilistic approach.

For the simple case of hard constraints, where L(p) is either a constant (for "feasible" images which are consistent with the data) or zero, the probabilistic conclusion is

$$\text{post}(p) \propto \begin{cases} \Omega(Np) & \text{for feasible } p \\ 0 & \text{otherwise} \quad . \end{cases} \tag{12}$$

We are still required to select a single image from this distribution, but the choice is compelling. We must choose the maximum. The maximum is the only choice which is independent of the undefined number N. Moreover, as $N \to \infty$ to reduce the quantisation of the monkey-generated histogram to invisibility, the degeneracy becomes more and more sharply peaked about the maximum, so that the maximum is favoured more and more strongly. As $N \to \infty$, the degeneracy logarithm from (9) becomes

$$\log\Omega = N\,H \quad , \quad H = -\sum_i p_i \log p_i = \text{entropy} \quad , \tag{13}$$

so that we are required to select that single feasible image which has maximum entropy.

This prescription of maximizing entropy under hard constraints can be represented pictorially by drawing contours of constant entropy H through the set of feasible images. Sometimes the constraints are linear, as in the following diagram (Figure 3) for a three-cell image.

Figure 3. Maximum entropy with hard linear constraints.

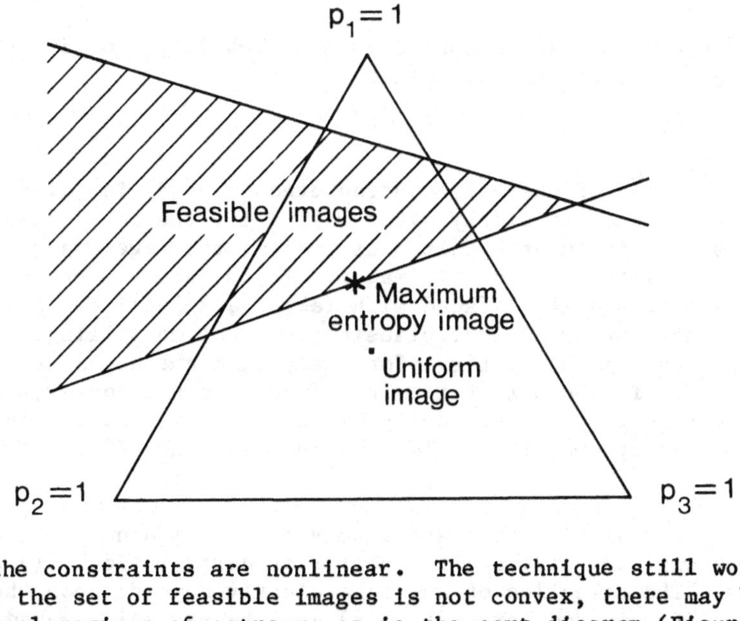

Sometimes the constraints are nonlinear. The technique still works although if the set of feasible images is not convex, there may be more than one local maximum of entropy, as in the next diagram (Figure 4).

Although it would presumably be wise in this case to display both the local entropy maxima A and B as alternative reconstructions, with image A of larger entropy preferred, this procedure presents a conceptual difficulty which foreshadows deeper difficulties with the parameter N. The difficulty is that of quantifying our preference for image A.

Figure 4. Maximum entropy with hard nonlinear constraint.

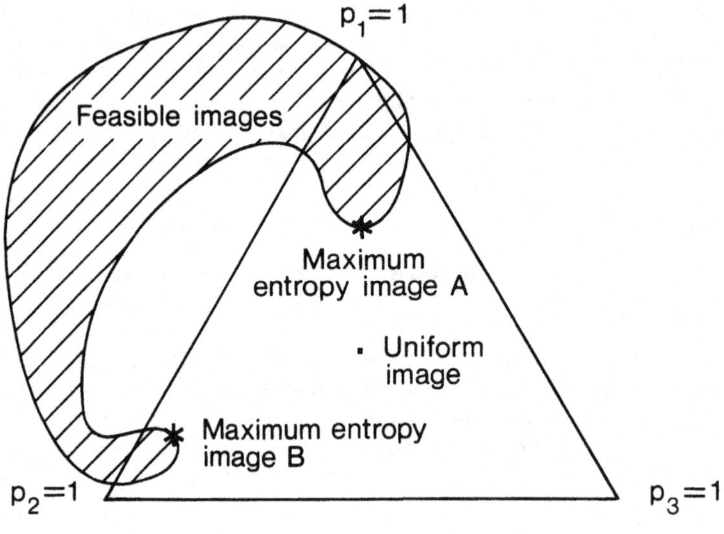

The monkey approach treats exp(NH) as a probability, but if we let N become large we reach the conclusion

$$\text{post}(B) \;/\; \text{post}(A) \to 0, \tag{14}$$

so that image B is effectively contradicted. Nobody is going to believe this. Although one can simply avoid saying it, the conclusion is clear nevertheless. Unfortunately, if N is finite, the degeneracy no longer acts as an absolute selector of images. It is merely a probability modulator and we are still faced with making a selection. Selecting the maximum becomes merely ad hoc, plausible but not compelling. Selecting the ensemble average of post(p), for example, might appear even more plausible. But if the set of feasible images is not convex, the ensemble average image might easily lie outside it, and it would be awkward to justify choosing an image which was contradicted by the data.

In practice, our difficulties are likely to be compounded because the constraints will be soft, and the posterior distribution (11) will be continuous. The selection of an image clearly depends upon the value we give to N. If N is too large, post(p) will be strongly peaked around the uniform image (Figure 5). Although the likelihood will not prohibit this absolutely and categorically, because the constraints are soft, the experimenter who provided the data would usually be most unhappy with such an uninformative result. Indeed, if he had useful data, he would wish to reject the uniform image at a high (though finite) significance level. If a theorist told him to accept an image which he wished to reject at the 99.9% significance level, he would justifiably reject the theorist instead.

Figure 5. Probability patch for N large
post(p) $\rightarrow \delta$ ($\underset{\sim}{p}$-uniform).

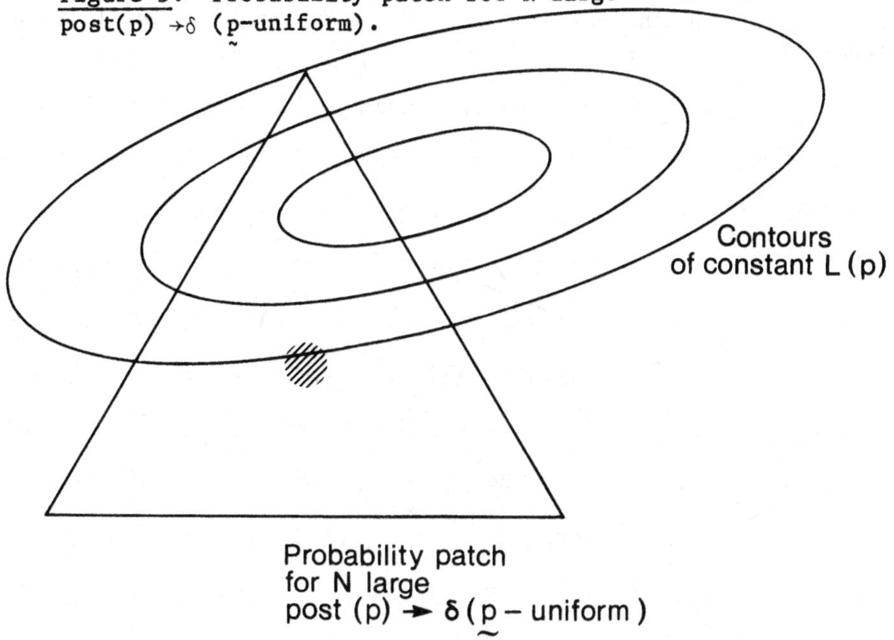

If, on the other hand, N is too small, post(p) reduces to L(p), and the experimenter has been told nothing fresh, apart from enforced positivity of p (Figure 6).

Figure 6. Probability patch for N small
post(p) $\rightarrow$ L(p).

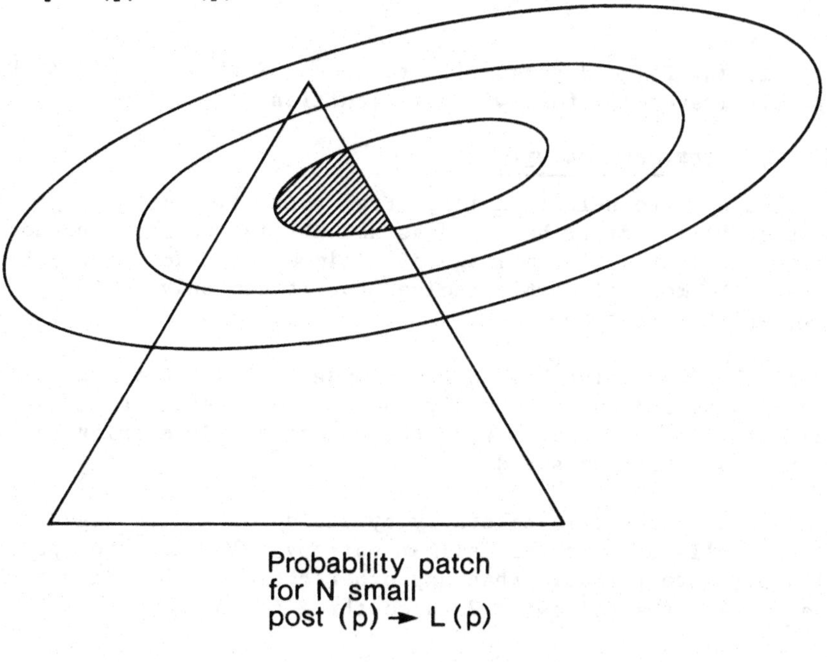

N must be given some intermediate value, which then gives a probability patch overlapping the uniform side of the tolerably feasible images (Figure 7).

Figure 7. Probability patch for N intermediate.

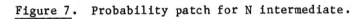

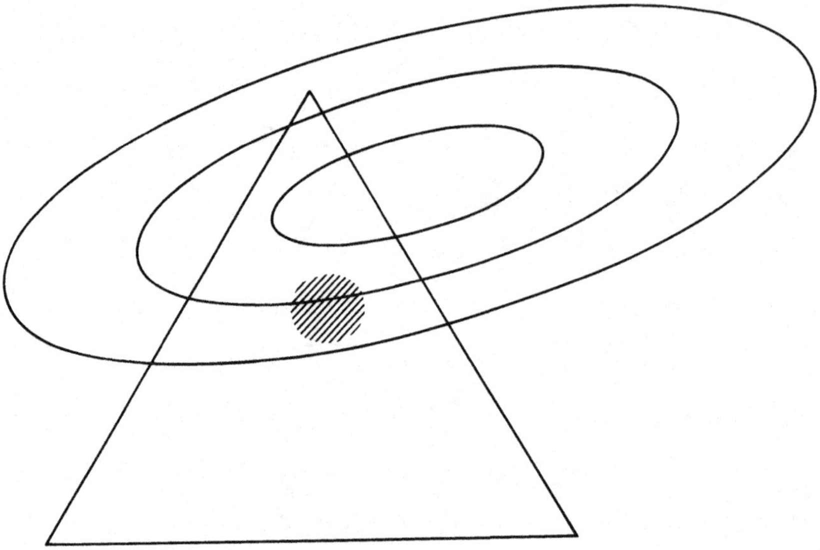

Probability patch
for N intermediate

Although it may be good pragmatism to choose N this way, it is bad theory. It leaves the following difficulties.

4 DIFFICULTIES WITH N

1) We STILL have to select an image from post(p). Altering the form of the probability distribution does not absolve us from the necessity of making a selection. The image of maximum posterior probability (11) is perhaps the most plausible choice, but it can only be fully justified if $N \to \infty$, which is not the case.

2) To fit the data tolerably, neither under-fitting nor over-fitting, N will have to be chosen differently for each dataset. It is difficult to reconcile this with the view that the degeneracy is a prior, assigned before the data were measured.

3) Having fitted the data tolerably by choice of N, the degeneracy factor now tells us that the maximum entropy selection from post(p) is quantifiably more probable than any other choice. This is unreasonable. Suppose we have the following data on three cells:

$$p_1 + p_2 + p_3 = 1 \quad \text{(hard constraint on normalization)}$$
$$p_1 = 2/3 \pm \sigma \quad \text{(soft Gaussian constraint on } p_1\text{),} \tag{15}$$

so that the likelihood is (Figure 8)

$$L(p) \propto \exp(- (p_1-2/3)^2/2\sigma^2)\ \delta(\ \Sigma p - 1\ ) \quad . \tag{16}$$

Figure 8. Soft data (15).

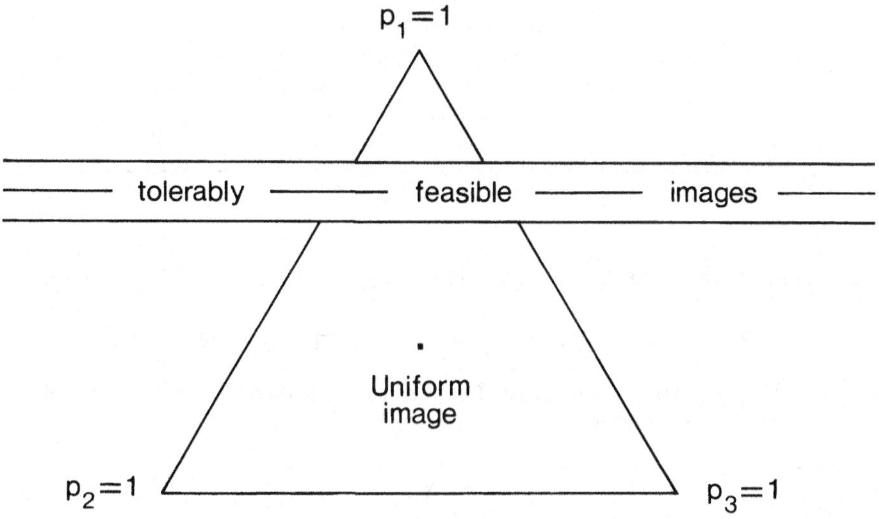

To fit the data tolerably, neither under-fitting nor over-fitting, we must take N of the order of $1/\sigma$. However, the degeneracy factor is then peaked in the horizontal (p_2-p_3) direction as well as in the vertical (p_1) direction, roughly with width $\sqrt{\sigma}$. This means that by measuring p_1 to one part in a million (say), we are also claiming to measure the difference (p_2-p_3) to about one part in a thousand. This claim is surely absurd, yet it is forced on us if we treat the degeneracy as a probability.

4) It may not be possible to find ANY value of N for which the maximum of post(p) is tolerably feasible, neither grossly under-fitting nor over-fitting. Maximizing post(p) at a given value of N is equivalent to maximizing its logarithm (NH + log L), which is achieved by the image which lies on the highest line of slope 1/N in the (H, -log L) plane (Figure 9).

Now consider the following data

$$\begin{aligned} p_1 + p_2 + p_3 &= 1 \\ p_1\ (p_2 - p_3) &= 0+ \\ p_2 - p_3 &= 1/2 \pm 1/2 \end{aligned} \tag{17}$$

Figure 9. Maximizing NH + log L for given N selects the image marked with a star.

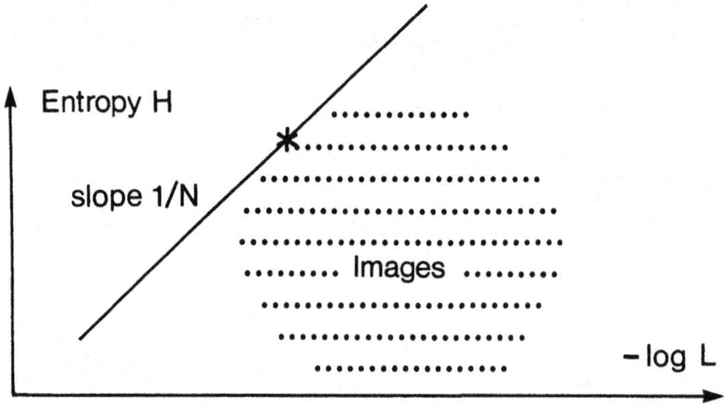

which are equivalent to the likelihod (Figure 10)

$$L(p) \propto \exp(-2(p_2-p_3-1/2)^2)\ \delta\ (p_1+p_2+p_3-1)\delta(p_1(p_2-p_3)) \quad (18)$$

Figure 10. Feasible images (17) lie along the dashed line in image space.

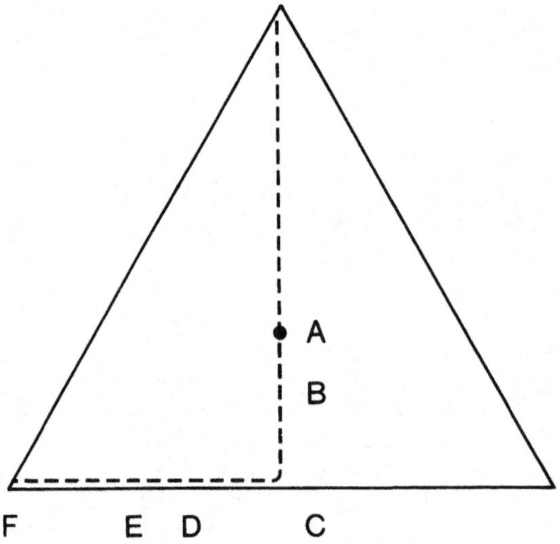

The feasible images lie along the following curve (Figure 11) in the (H, -log L) plane.

Values of N which are too large produce images between A and B which underfit the data. Values of N which are too small produce images

Figure 11. Entropy as a function of log-likelihood for feasible images (17).

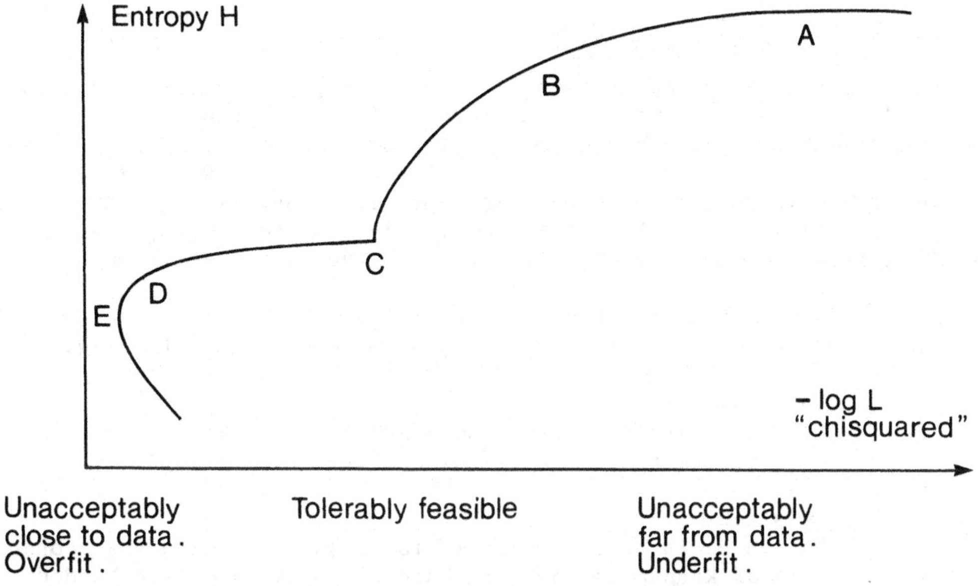

between D and E which overfit. No value of N gives the expected level of misfit near C. Although overfit is not a serious difficulty in such a small problem, it becomes very serious in practical problems with many degrees of freedom, because noise is then being interpreted as true signal, and the results of this can be disastrous (Gull and Daniell, 1978).

In a constrained maximization problem, a Lagrange multiplier (which is how N is being used here) may or may not lead to a solution with an acceptable value of the constraint. It depends on the problem.

5) Even if the practical difficulties could be surmounted, there is still the theoretical point that the degeneracy is not the true prior. The true prior may be quite different. A strong candidate (generalizing Jaynes, 1968) is

$$\text{prior}(p) = \Pi\ (1/p_i)\ .\ \delta(\ \Sigma p_i - 1\ ) \quad , \tag{19}$$

corresponding to ignorance of the scale of each component of p. With this prior, the proper Bayesian solution for post(p) is fully defined as L(p)prior(p). There is no place for an entropy or "monkey" argument.

We may conclude that although the monkey approach affords an interesting categorization of probability distributions (which are isomorphic to images), it does not help us to make a selection. Furthermore, it invites us to consider the degeneracy exp (NH) as a prior, when the true prior is actually quite different. We need to focus more clearly on the selection procedure. Probabilistic arguments have not helped us to select a single image.

5 THE KANGAROO APPROACH

The following argument is essentially due to Gull, and has appeared in the proceedings of a conference in Sydney (Gull and Skilling, 1984), hence its antipodean flavour. It was inspired by the far more careful and formal work of Shore and Johnson (1980, 1983), and I hope they will forgive this whimsical presentation. Tikochinsky, Tishby and Levine (1984) have given a closely related result, and a proof of intermediate formality is given by Livesey and Skilling (1985).

In the course of our Australian travels, we observed the following.

Information:
(1) One quarter of kangaroos have blue eyes.
(2) One quarter of kangaroos are left-handed.

Question:
On the basis of this information alone, estimate the proportion of kangaroos that are both blue-eyed and left-handed.

The joint proportions of left-handedness and incidence of blue eyes can be represented as a little 2 by 2 contingency table with one degree of freedom among the 4 entries. Acceptable solutions are positive throughout, and we show three examples, the independent case and those cases which display the maximum amount of positive and negative correlation.

Left-handed

Blue eyes		T	F
	T	1/16	3/16
	F	3/16	9/16

Uncorrelated

	T	F
T	1/4	0
F	0	3/4

Positive

	T	F
T	0	1/4
F	1/4	1/2

Negative

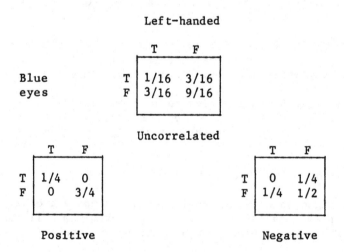

Suppose, though, that we must choose only ONE answer - which is the best? Clearly, the answer we select cannot be thought of as being any more likely than any other choice, because there may well be, and presumably is, some (small?) degree of genetic correlation between eye colour and handedness. However, it is nonsensical to select either positive or negative correlations without having any relevant data. Our prior knowledge should not lead us to expect any particular sign of correlation. Although the independent choice

$$\text{prob(blue and left)} = 1/16 \tag{20}$$

is not quantifiably more probable than any other, it is certainly to be preferred. Here is a selection problem for which we know the answer in advance. Let us use it to test our theory.

Suppose that we seek an optimal, or "best" solution, presumably by a variational principle, maximizing some functional

$$H(p) = \sum_{i=1}^{4} h(p_i) \tag{21}$$

which defines what we mean by "best". The table shows the results for a few currently recommended functionals (Nityananda and Narayan, 1982; Frieden, 1983).

Function		prob(blue eyes & left-handed)	Correlation
Shannon entropy	$-\sum p \log p$	1/16 = 0.0625	uncorrelated
Burg entropy	$\sum \log p$	0.09151	positive
Intermediate form	$\sum p^{1/2}$	0.07994	positive
Least squares	$-\sum p^2$	0	maximally negative

With one notable exception, these results are very peculiar: for example the least squares form predicts that no left-handed kangaroo has blue eyes.

This presentation in terms of contingency tables of probabilities is, of course, equally applicable to images. Consider the following re-statement of the problem.

Information (data):
(1) One quarter of the flux comes from the top half of the image.
(2) One quarter of the flux comes from the left half of the image.

Question:
What proportion of the flux comes from the top left quadrant?

Here again is a problem for which there is a single compelling candidate (answer = 1/16) for selection. Most selection procedures, though, give different answers. The least squares form, for instance, says that all the flux which comes from the top half of the image is concentrated on the right-hand side. Without further data, it is difficult to imagine any sensible basis for this general decision.

Every variational functional except $-\Sigma p \log p$ fails to be consistent on even this simplest non-trivial image problem. This result was already implicit in the work of Kullback (1959). Inconsistency will certainly not disappear just because practical data are more complicated: the only way to avoid it is to use maximum entropy. Four axioms are needed to prove this.

1) The form of H should not depend on the type of data being analyzed.
2) H should depend on the proportions, not the units, of the image.
3) Extra knowledge about the relative proportions within one area of the image should not affect the image elsewhere.
4) Independent data should give images which combine multiplicatively, as in the kangaroo example above.

This theorem is very important. It deals directly with the basic problem of selecting a single distribution p, given incomplete data. The proof uses a simple type of incomplete data for which there is one compelling selection. The only general way of obtaining this selection is to use maximum entropy.

The formal theorem explicitly requires hard constraints. Jaynes (1978) specifically warned against using maximum entropy with non-testable (soft) constraints, so that soft constraints must be made hard if the kangaroo approach is to be justified. To do this, we shall decide to accept any tolerably feasible distribution (say at 90% significance in some statistical test), and to use membership of this "feasible set" as a hard constraint C(p). By doing this, we will never be able to reach a reconstruction which is in serious disagreement with the experimental data. The data come first. They tell us which images are permitted, and which are not. They can do no more. Maximum entropy then tells us which image we should prefer.

There is an unexpected advantage to be gained from the apparently dubious step of forcing the soft likelihood function L(p) into a hard equality or inequality constraint C(p). Only the 2-norm (chisquared = total variance) of the residuals enters L(p), assuming Gaussian noise on the data. It may well be that only the total variance is important in the large space of all possible images. However, when a single image is selected, its residuals could be markedly non-Gaussian, even though the total variance is correct. If that is perceived as damaging, we could re-define C(p) to put a limit on the 1-norm or the ∞-norm of the residuals instead. Indeed we could apply any statistical test we liked to ensure that the residuals become "acceptably random". There is more information in the residuals than appears in the overall likelihood L(p) and the freedom to use it can be useful (Bryan and Skilling, 1980).

Yet more flexibility comes from invoking prior information. The simplified "kangaroo" derivation of entropy ignores any prior information one might have about the image. However, entropy is a relative quantity. It is always measured relative to some initial model or measure m, and the true generalized formula (Jaynes, 1968) is

$$H = -\sum_i p_i \log(p_i/m_i) \qquad (22)$$

as derived axiomatically by Shore and Johnson (1980, 1983). We shall discuss prior information in more detail later.

Summary of Kangaroo Approach

We are led to prefer the maximum entropy distribution on grounds of consistent reasoning. We can quantify this preference by evaluating differencees of entropy between different distributions, but we are not led to quantify this preference probabilistically. Returning to the earlier example (15) of a specific dataset $\{p_1+p_2+p_3=1, p_1=2/3 \pm \sigma$ with σ small$\}$, we are led to prefer the symmetric solution p = (2/3,1/6,1/6) over others such as (2/3,1/3,0). This preference can be quantified as H = 0.8675-0.6365 = (0.3333 bits) ln2, but we need not and do not claim that ($p_2=p_3$) has effectively been measured to any particular accuracy (such as $\sqrt{\sigma}$). The preference is well founded: the claim would be false.

Maximum entropy is seen to be very different (Figure 12) from the direct Bayesian approach. The Bayesian analysis is performed in the enormous space of all possible images p. In this space, results cannot be displayed and computations cannot be performed (except in specially simple cases). The maximum entropy analysis is performed with only a few images, in principle just m and p. The result can be displayed and the computations can be performed (except for pathologically difficult

Figure 12. Bayesian procedure contrasted with maximum entropy.

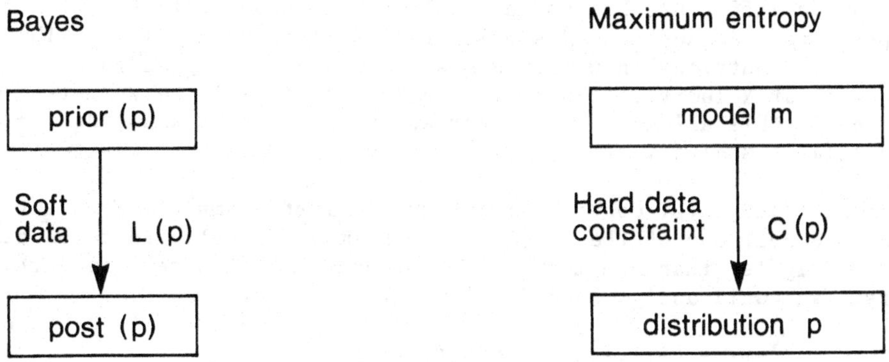

data constraints). However, the soft constraint L(p) has to be replaced by a hard constraint C(p) before maximum entropy can be applied.

There is a Bayesian approach that is connected to maximum entropy. It is the "monkey" argument, which ranks distributions according to their Bayesian degeneracy factors

$$\text{prob}(\text{histogram } n \mid \text{model } m) \propto \Omega = \exp(NH) \tag{23}$$

This is a probability distribution in the space of N samples from a model. It is not a distribution in the space of all images, and we must not confuse the two spaces.

6 INCORPORATING PRIOR KNOWLEDGE

The maximum entropy method is a good, general-purpose tool capable of reconstructing images from a wide variety of types of data. We gave a review (Skilling and Gull, 1981) at the first Laramie meeting, and a second review is published elsewhere (1984b). The method as developed above ignores many important types of prior knowledge. Of course, a maximum entropy image always uses an initial model m, but it appears at first that this can only incorporate simple prior knowledge. If, for example, we expected a particular part of the image to be bright, we could allow the model to take correspondingly large values there, and this would be reflected in the reconstruction (provided the data permitted).

However, although we might expect to find a bright object somewhere in the field of view, we might not know beforehand where it is to be found. This state of knowledge is translation invariant across the image, and it would seem that we must assign a uniform model m = constant in the entropy expression (22). I shall call this the austere model. It fails to encompass general prior knowledge.

We argued last year at Laramie that this difficulty can be resolved by using a larger space S^N of multiple-sample distributions. Maximum entropy can be applied to S^N just as well as to S. Only the interpretation is different, because a single sample from a distribution in S^N represents a sequence of N samples from a distribution in S. The model $m_{ijk...}$ (N suffices) now represents the measure assigned to a sequence of N individual samples i,j,k,...: it will be symmetric on the indices. This allows a far richer encoding of prior knowledge than is possible in S alone.

To illustrate this, consider the very simplest example. We are told that one cell out of the M cells in our image is believed to be six times brighter than the mean of the others. If this cell were the first, we would assign an initial model or measure

$$m_i = (6,1,1,1,..) = 5\ \delta_{i1} + 1 \tag{24}$$

on single samples. Likewise, we would assign a measure

$$m_{ij} = (5\ \delta_{i1} + 1)(5\ \delta_{j1} + 1) \tag{25}$$

on double samples, and so on, up to

$$m_{ijk...} = (5\ \delta_{i1} + 1)(5\ \delta_{j1} + 1)(5\ \delta_{k1} + 1)... = 6^{n_1} \tag{26}$$

for N samples, where n_1 is the number of "1" suffices in the sequence i,j,k,... However, we do not know that the cell in question is the first. It could be any of the cells, say the rth. This state of knowledge is encoded by adding the measures for the individual possibilities.

For single samples, we get

$$m_i = \sum_r (5\ \delta_{ir} + 1) = 5 + M\ , \tag{27}$$

which is independent of i, and hence uninformative as expected. For two samples, though, we get

$$m_{ij} = \sum_r (5\ \delta_{ir} + 1)\ (5\ \delta_{jr} + 1) = 25\ \delta_{ij} + 10 + M\ , \tag{28}$$

which is not constant, and hence is informative. The first sample i helps to indicate where the second sample j may be found. Continuing to N samples, the measure becomes

$$m_{ijk...} = \sum_{r-1}^{M} 6^{n_r}\ , \tag{29}$$

where n_r is the number of occurrences of "r" in the suffix list, defining the histogram of the samples.

Clearly, many other forms of prior knowledge can be coded likewise. N samples allow us to code correlations up to Nth order.

We are now in a position to use maximum entropy to find a distribution p_i in S, or p_{ij} in S^2, up to $p_{ijk...}$ in S^N, by maximizing

$$H^{(N)} = - \sum_{ijk..} p_{ijk..} \log(\ p_{ijk..}\ /m_{ijk..}) \tag{30}$$

subject to data constraints. We should remember that we are aiming to select one single image p_i in the single space S. In other words, we wish to present our "result" as a simple set of proportions p_i, isomorphic to a pdf from which successively independent samples i could be taken. Accordingly, we set

$$p_{ij} = p_i\ p_j \quad , \quad p_{ijk..} = p_i\ p_j\ p_k \cdots \tag{31}$$

in accordance with independent sampling, because this is our definition of an image in S^N. Note that the prior model m does not factorize, but our result p in S must.

Maximizing the austere entropy expression (22) in S using a model μ, over the constraint C(p), would have yielded the variational equation

$$- \log p_i + \log \mu_i = \lambda \, \partial C/\partial p_i + \alpha \tag{32}$$

with Lagrange multipliers λ for the constraint and α for normalization. Maximizing H(N) from (30) instead gives

$$- \log p_i + \sum_{jk..} p_j p_k .. \log m_{ijk..} = \lambda \, \partial C/\partial p_i + \alpha \, . \tag{33}$$

This is so similar to the austere equation (32) that we can think of it (and compute with it) in terms of a <u>result-dependent model</u>

$$\mu_i(p) = \exp\left(\sum_{jk..} p_j p_k .. \log m_{ijk..} \right) \tag{34}$$

in the single space S.

For the specific illustrative example above, the measure (29) becomes

$$m_{ijk...} = 6^g \tag{35}$$

as $N \to \infty$, where g is the greatest number of occurrences of any individual cell in the sequence ijk... The jk.. summation involved in the result-dependent model (34) is dominated by histogram values n_ℓ near Np_ℓ. Hence g, which includes a single "i" in its defining sequence, may be replaced by

$$g \cong \begin{cases} N\, p(\max) \quad , & \text{largest p not in cell i} \\ N\, p(\max) + 1 \quad , & \text{largest p in cell i} \ . \end{cases} \tag{36}$$

Finally, the result-dependent model (34) is

$$\mu_i = \begin{cases} \text{constant} \quad , & \text{largest p not in cell i} \\ 6 \ . \ \text{constant} \quad , & \text{largest p in cell i} \ , \end{cases} \tag{37}$$

which is an eminently reasonable way of encoding the prior knowledge.

The procedure can work in practice as well as in theory. The following graph (Figure 13) shows a noisy set of data for a one-dimensional 64-cell image, blurred by convolution with a 3-cell square wave.

Austere maximum entropy, with a uniform model, gives the reconstruction (Figure 14), containing a bright point on a relatively smooth background.

Figure 13. Blurred and noisy data.

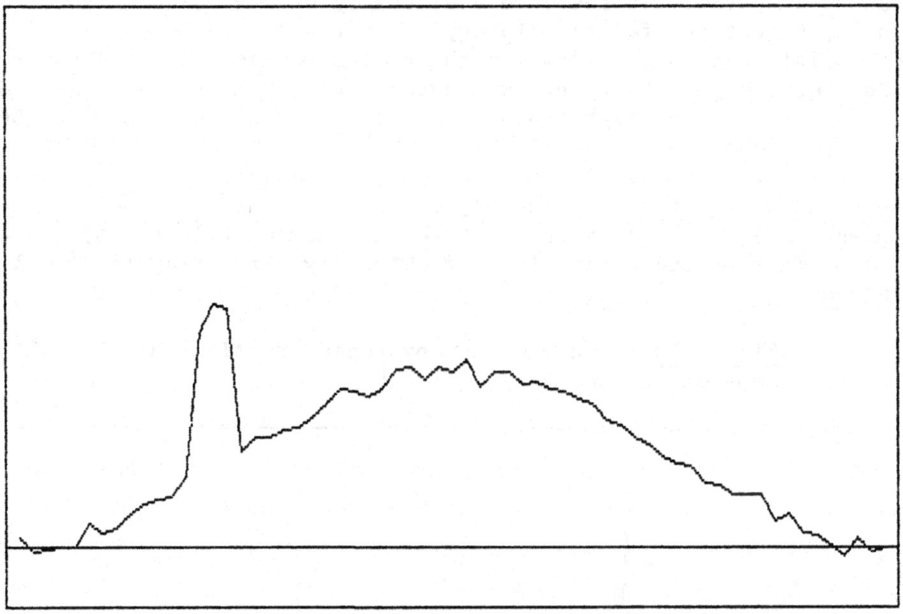

Figure 14. Austere maximum entropy reconstruction.

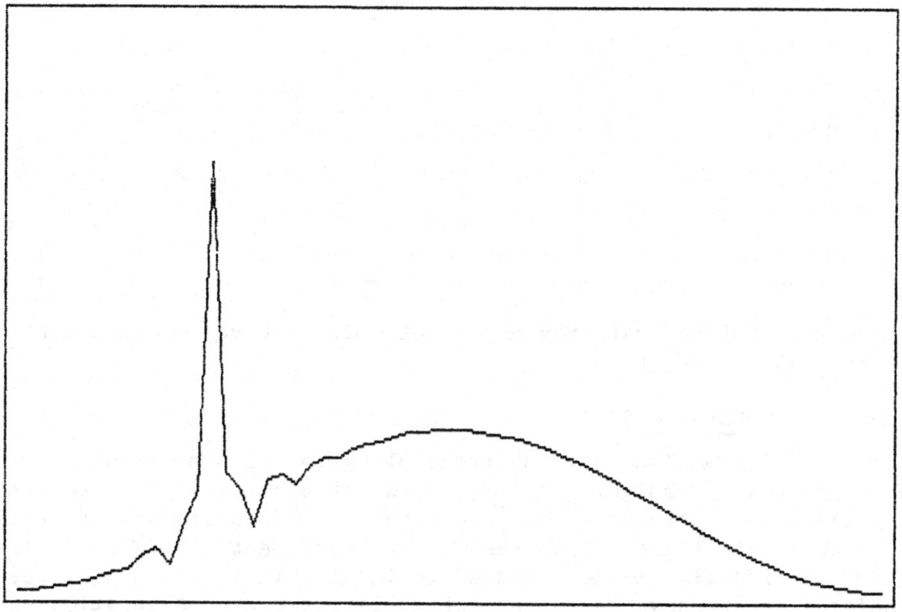

The neighbourhood of the bright point shows the oscillatory behaviour known as "ringing" in deconvolution problems. Ringing occurs because the austere entropy imposes a severe penalty on any point which is unusually bright (or faint). In trying to reduce the intensity of the bright point, the entropy damages the reconstruction of the local background. Of course, if our prior ignorance was such that we really did not expect to find a bright point, then this reconstruction would be perfectly defensible.

If we do know that such a point may exist, we should use the result-dependent model. It gives the next reconstruction (Figure 15), in which the point is reproduced cleanly, without badly corrupting the local background.

Figure 15. Maximum entropy reconstruction using result-dependent model (37).

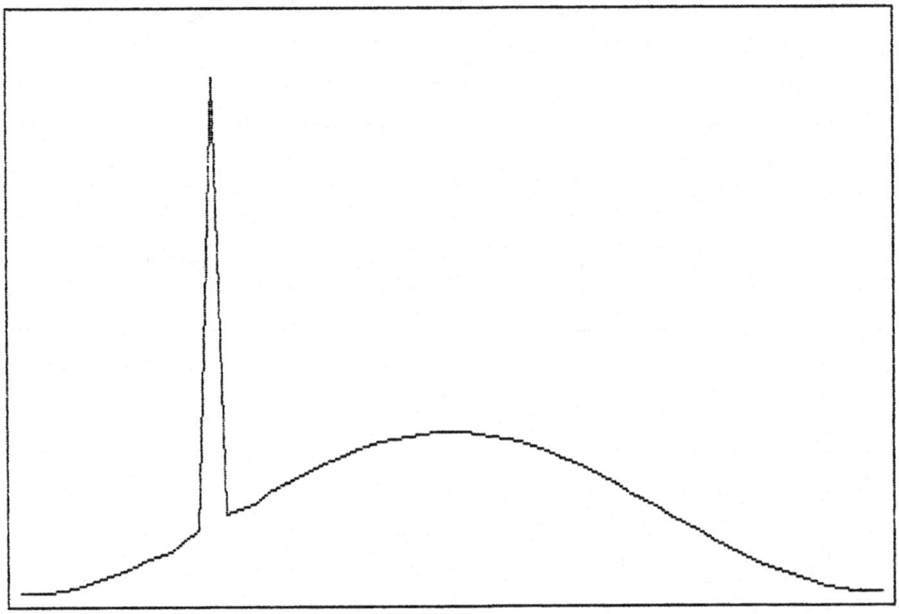

We hope that this almost trivial example will foreshadow much more powerful future results.

7 CONCLUSIONS

The maximum entropy principle gives us a powerful and general method of assigning distributions of probability or proportion (images) on the basis of incomplete data. An appealing way of deriving the entropy formula is to use the "monkey" argument in which N quanta are thrown randomly at the cells of an image. It is tempting to use the associated degeneracy $\Omega \propto \exp(NH)$ of a histogram n of occupation numbers as a prior probability for the corresponding distribution $p = n/N$ of proportions. However, this leads to increasingly severe difficulties over the role and meaning of N, and the probabilistic identification becomes untenable. The degeneracy is not the prior probability.

Fortunately, probabilistic reasoning can be bypassed entirely. There are some particularly simple problems, such as the "kangaroo" example, where the choice of distribution is compelling. The only generally applicable selection procedure which will give the correct answer for these simple problems is the maximum entropy method. Maximum entropy now appears as part of a hierarchical method of analysis. The data comes first. They are subsumed in a hard constraint which defines those images which are consistent with the data, and those which are not. (Incidentally, we are allowed a useful flexibility of choice of constraint function.) Only after applying the data constraint do we express our preference for images with larger entropy. Such images are not more probable than the others - they are merely to be preferred.

In the austere "kangaroo" approach, the parameter N effectively disappears from the theory. Most forms of prior knowledge are also absent from the theory, even though we know they can be useful in practice.

By retrieving N and working with N samples from our distribution instead of just one, it turns out that we can incorporate and use general prior knowledge in our reconstructions. N now measures the level of complexity (order of correlation) of prior knowledge which we are prepared to encode. We remain entitled to use maximum entropy. These more general calculations are equivalent to austere calculations with single samples, except that the "initial" model now depends on the reconstruction itself. Effectively, the computer programs which do the reconstructions can learn as they iterate towards the maximum entropy image, by recognizing and incorporating features expected on the basis of the prior knowledge.

REFERENCES

1 Bryan, R.K. and J. Skilling (1980). Deconvolution by Maximum Entropy as Illustrated by Application to the Jet of M87. Mon. Not. R. Astr. Soc., 191, 69-79.

2 Frieden, B.R. (1983). Unified Theory for Estimating Frequency-of-Occurrence Laws and Optical Objects. J. Opt. Soc. Am., 73, 927-938.

3 Gull, S.F. and G.J. Daniell (1978). Image Reconstruction from Incomplete and Noisy Data. Nature, 272, 686-690.

4 Gull, S.F. and J. Skilling (1984). The Maximum Entropy Method, in Indirect Imaging, J.A. Roberts (ed.), Cambridge Univ. Press.

5 Jaynes, E.T. (1968). Prior Probabilities. IEEE Trans., SCC-4, 227-241.

6 Jaynes, E.T. (1978). Where Do We Stand on Maximum Entropy? In The Maximum Entropy Formalism, R.D. Levine and M. Tribus (eds.), MIT Press; and, in Papers on Probability, Statistics and Statistical Physics, R. Rosenkrantz (ed.), Reidel.

7 Kikuchi, R. and B.H. Soffer (1977). Maximum Entropy Image Restoration. I: The entropy expression. J. Opt. Soc. Am., 67, 1656-1665.

8 Kullback, S. (1959). Information Theory and Statistics. Wiley.

9 Livesey, A.K. and J. Skilling (1985). Maximum Entropy Theory. Acta. Cryst. A41, 113-122.

10 Nityananda, R. and R. Narayan (1982). Maximum Entropy Image Reconstruction - A Practical Non-Information Theoretic Approach. J. Astrophys. Astron., 3, 419-450.

11 Shore, J.E. and R.W. Johnson (1980). Axiomatic Derivation of Maximum Entropy and the Principle of Minimum Cross-Entropy. IEEE Trans., IT-26, 26-37; and, (1983) IEEE Trans., IT-29, 942-943.

12 Skilling, J. and S.F. Gull (1981). Algorithms and Applications. Presented at first workshop on maximum entropy and Bayesian methods in Inverse problems, Laramie, Wyoming, C.R. Smith and W.T. Grandy (eds.) (Reidel).

13 Skilling, J. and S.F. Gull (1984a). The Entropy of an Image. SIAM Amer. Math. Soc. Proc., 14, 167-189.

14 Skilling, J. and S.F. Gull (1984b). The Maximum Entropy Method in Image Processing. Proc. IEE 131F, 646-659.

15 Tikochinsky, Y., N.Z. Tishby and R.D. Levine (1984). Consistent Inference of Probabilities for Reproducible Experiments. Phys. Rev. Lett., 52, 1357-1360.

THE CAMBRIDGE MAXIMUM ENTROPY ALGORITHM

John Skilling
Cambridge University
Department of Applied Mathematics and
Theoretical Physics
Silver Street
Cambridge CB3 9EW, England

ABSTRACT

The Cambridge algorithm is designed to compute high resolution maximum entropy images from large datasets. The underlying ideas are presented, together with recent developments which improve the algorithm's treatment of nonlinear data, for which there may be no unique maximum of entropy.

1 INTRODUCTION

This paper describes our work in Cambridge on practical methods for computing maximum entropy images. It divides into two main parts. The first part describes the technique of setting up an iterated subspace of search directions within which entropy is maximized. This multiple search direction method was originally reported at the first Laramie workshop (Skilling, 1981), and is now published as Skilling and Bryan (1984). The second part of the paper describes our recent ideas on how the image increment should be controlled within the subspace. For linear data (related to the object by a linear transformation), this new control algorithm effectively reduces to the method outlined in the 1981 and 1984 papers. For the more difficult problem of nonlinear data, such as the Fourier intensity data found in crystallography, the new control algorithm supersedes the old. It deals with the maximum entropy problem directly, instead of relying explicitly on Lagrange multipliers. These new ideas were first reported in Livesey and Skilling (1984).

The formal problem is to compute a maximum entropy image f_i ($i = 1,2,\ldots,N$) from a practical dataset D_k ($k=1,2,\ldots,M$). The data are related to the object by some transformation

$$F_k = F_k\,(f) \tag{1}$$

which defines the "mock" data F which would have been produced if the object were correctly represented by the image f. To use the data, we set up a constraint function C(f) such as chisquared (Gull and Daniell, 1978)

$$C(f) = \sum_k \,(F_k - D_k)^2/\sigma_k^2 \tag{2}$$

which measures the misfit between the mock data and the actual data.

The computational problem is then to find that image f which maximizes

$$S(f) = - \sum_i f_i \log(f_i /eA) \tag{3}$$

(where A is a constant defining the normalization of f) over a constraint

$$C(f) < C_o \tag{4}$$

(such as chisquared not exceeding the number M of observations). A severe restriction on programming is that the size of the problem precludes any explicit matrix operations in the mega-dimensional image or data space. We can, for example, Fourier transform between the two spaces in O(N log N) operations, but we must not attempt any $O(N^2)$ matrix operations. Since the entropy is nonlinear, the algorithm is necessarily iterative, and usually the computing time is dominated by the transforms.

2 SUBSPACE

Simple optimization algorithms such as steepest ascents or conjugate gradients (Fletcher and Reeves, 1964; Powell, 1977) normally maximize a single function Q(f) of the image by selecting a single direction e in which to increment the image:

$$f^i_{(new)} = f^i + x\, e^i \qquad (i=1,2,\ldots,N) \tag{5}$$

(note the contravariant indices, for consistency with later notation). The coefficient x which defines the new image is determined by maximizing Q along e, either using local gradient and curvature information near f or by an explicit line search. Conjugate gradient is a more powerful technique than steepest ascent because the search direction e is chosen more intelligently, but it relies on Q being a quadratic function, to at least a reasonable approximation. This means that the Hessian matrix $\partial^2 Q/\partial f^i \partial f^j$ must be effectively constant over several iterates of the algorithm. Unfortunately, if Q involves the entropy $-\Sigma f \log f$, the curvature involves terms like $1/f$. Such terms are actually very sensitive to changes in f, especially when components of f are small. Thus Q ceases to be reasonably quadratic precisely when the nonlinearity in the entropy should be most useful. We have found it impossible to construct a robust single-search-direction algorithm for maximum entropy.

We recommend always using several search directions e_μ ($\mu=1,2,\ldots,n$) and exploring images within the corresponding n-dimensional linear subspace

$$f^i_{(new)} = f^i + \sum_{\mu=1}^{n} x^\mu e^i_\mu \tag{6}$$

(Greek indices denote subspace quantities). With linear data, we usually use n=3.

However, it is not always possible to find a maximum entropy image on a given constraint surface by maximizing a single function Q=NS-C/2, whatever value of N is chosen. Accordingly, we model the entropy and the constraint by separate functions

$$S(f + \Sigma xe) \cong s(x) \tag{7}$$

$$C(f + \Sigma xe) \cong c(x) \tag{8}$$

within the subspace. For computational convenience, the particular functions we use are quadratic

$$s(x) = s(0) + \sum_{\mu} s_{\mu} x^{\mu} - \sum_{\mu\nu} g_{\mu\nu} x^{\mu} x^{\nu}/2 \tag{9}$$

$$c(x) = c(0) + \sum_{\mu} c_{\mu} x^{\mu} + \sum_{\mu\nu} h_{\mu\nu} x^{\mu} x^{\nu}/2 \tag{10}$$

corresponding to the first terms in the local Taylor series expansions of S(f) and C(f). Thus

$$s_{\mu} = \sum_{i} e^{i}_{\mu} \, \partial S/\partial f^{i} = \sum_{i} e^{i}_{\mu} (\log A - \log f^{i}) \tag{11}$$

$$c_{\mu} = \sum_{i} e^{i}_{\mu} \, \partial C/\partial f^{i} = \sum_{k} E^{k}_{\mu} \, \partial C/\partial F^{k} \tag{12}$$

$$g_{\mu\nu} = - \sum_{ij} e^{i}_{\mu} e^{j}_{\nu} \, \partial^{2}S/\partial f^{i} \partial f^{j} = \sum_{i} e^{i}_{\mu} e^{i}_{\nu}/f^{i} \tag{13}$$

$$h_{\mu\nu} = \sum_{ij} e^{i}_{\mu} e^{j}_{\nu} \, \partial^{2}S/\partial f^{i} \partial f^{j} = \sum_{kl} E^{k}_{\mu} E^{l}_{\nu} \, \partial^{2}S/\partial F^{k} \partial F^{l} \tag{14}$$

where we have written the transform of the search direction e into data space as

$$E^{k}_{\mu} = \sum_{i} e^{i}_{\mu} \, \partial F^{k} / \partial f^{i} . \tag{15}$$

Since the constraint function C is a sum of contributions from each datum separately, its gradients c_{μ} and curvatures $h_{\mu\nu}$ are more conveniently evaluated in data space: the major computing cost involved is then the transformation of each search direction e.

Distance limit

Quadratic models are only reliable in the vicinity of the current image f, where cubic and higher terms are negligible. Hence we should only use the models within a "trust region" of images within some distance r of the current image. This raises the question of choosing an appropriate metric on image space. The most natural measure of separation is the cross-entropy (Kullback, 1959)

$$\Delta S = \Sigma\ (f^i + \Delta f^i) \log\ (f^i + \Delta f^i)/f^i \tag{16}$$

which gives the entropy loss (or information gain) involved in changing one's knowledge of the image from f to f+Δf. For small Δf, this becomes (at fixed normalization)

$$\Delta S = \Sigma\ (\Delta f^i)^2 / \ 2f^i \tag{17}$$

which we now use in our definition of distance

$$\ell = (\ 2\Delta S\ )^{1/2} \tag{18}$$

The metric tensor corresponding to this definition is the entropy metric (Bryan, 1980)

$$g_{ij} = \begin{cases} 1/f^i & , \quad i=j \\ 0 & \text{otherwise} . \end{cases} \tag{19}$$

The same tensor appeared as the entropy curvature in (13), and we are led to define distances and angles in the subspace with $g_{\mu\nu}$ as the metric. Specifically, the length-squared of an image increment Σxe is defined to be

$$\ell^2 = \sum_{\mu\nu} g_{\mu\nu} x^\mu x^\nu \quad , \tag{20}$$

and the distance limit we impose is

$$\ell \leqq r \quad . \tag{21}$$

On dimensional grounds, r^2 is set to some fraction (typically 1/5) of the total image intensity Σf.

Search directions

The success of this algorithm is, of course, crucially dependent on choosing useful search directions. Obvious first choices are the gradients of the entropy (to increase entropy) and of the constraint function (to alter the constraint value). However, the differential operator $\partial/\partial f^i$ produces covariant vectors, which should not be used directly as increments to a contravariant image f^i. We use the entropy metric g and consider contravariant gradients

$$e_1^i = \sum_j g^{ij} \partial S/\partial f^j = f^i \partial S/\partial f^i \tag{22}$$

$$e_2^i = \sum_j g^{ij} \partial C/\partial f^j = f^i \partial C/\partial f^i \tag{23}$$

as search directions.

Quite apart from the aesthetic attraction, several practical advantages follow from using contravariant directions.

1) These are the directions which maximize the changes in entropy and constraint per unit distance, as defined by the entropy metric.

2) The metric-derived factors of f^i in the search directions discriminate in favour of high values (which presumably represent the most important parts of the image) and against low values, and this helps to keep all values positive from one iteration to the next.

3) Multiplication by f in image space corresponds to a nonlinear mixing in data space, so that the algorithm easily and naturally combines the information given by different data, incidentally allowing estimates of unmeasured data to be made.

Even for simple linear data, though, the two search directions (22) and (23) are in themselves barely adequate if the required image has high dynamic range f(max)/f(min). Algorithms such as quasi-Newton methods (Gill, Murray and Wright, 1981) and conjugate gradient obtain their power by using information from previous search directions, but we have not found it helpful to retain such memory. This is because the entropy curvature changes so much from iterate to iterate that old directions are useless: in fact if old directions are useful, we take this to mean that the distance limit r has been set too small.

However, we can look ahead to some extent by calculating from local information what the two search directions should become after the image has been incremented by $x^1 e_1 + x^2 e_2$. The entropy direction (22) would change by

$$x^1 f^i \sum_j e_1^j \, \partial^2 S/\partial f^i \partial f^j + x^2 f^i \sum_j e_2^j \, \partial^2 S/\partial f^i \partial f^j \quad . \tag{24}$$

Since $\partial^2 S/\partial f^i \partial f^j = g_{ij} = 1/f^i$, this is simply the original increment, so it gives us nothing new. The constraint direction (23) would change by

$$x^1 f^i \sum_j e_1^j \, \partial^2 C/\partial f^i \partial f^j + x^2 f^i \sum_j e_2^j \, \partial^2 C/\partial f^i \partial f^j \quad . \tag{25}$$

This does give us new directions, which we may select as

$$e_3^i = f^i \sum_j e_1^j \, \partial^2 C/\partial f^i \partial f^j \tag{26}$$

$$e_4^i = f^i \sum_j e_2^j \, \partial^2 C/\partial f^i \partial f^j \quad . \tag{27}$$

Each of these can be evaluated at a cost of two image/data transforms, and the resulting family of four search directions gives a very robust algorithm for linear data. One can recurse the lookahead procedure to generate more pairs of search directions, but the extra cost usually outweighs the benefits (one does not extend the radius of convergence of a series by taking more terms!).

Whatever search directions are used we always recommend checking that the final image one produces is indeed a maximum entropy image, and not merely an artefact of algorithmic inefficiency. This is done by calculating the angle

$$\theta = \cos^{-1} (\hat{S} \, . \, \hat{C}) \tag{28}$$

between the unit gradient vectors $\hat{S}$ and $\hat{C}$ of entropy and constraint (for consistency of approach, we use the entropy metric to define lengths and scalar products). Only if this angle is a reasonably small fraction of a radian is the image acceptable. It is, unfortunately, quite easy to write programs which purport to maximize entropy, but do not.

3 CONTROL

Within each iteration, we have now reduced the problem from an optimization in the full megadimensional image space to an optimization within a subspace of n search directions. Our standard programs for linear data use only three, but if the data are nonlinear functions of f, we may need more. Also, if the data are nonlinear, there need be no unique maximum of S for given constraint C. Especially in this case, we must control the increments of f with care.

To simplify the following algebra, we diagonalize g and h simultaneously: this is equivalent to choosing base vectors for the subspace with respect to which g is the identity and h is diagonal with eigenvalues h_μ. The reduced problem is then to maximize the quadratic entropy approximation

$$s(x) = s(0) + \sum_{\mu=1}^{n} (s_\mu x_\mu - x_\mu x_\mu/2) \tag{29}$$

over an appropriate limit on quadratically approximated constraint values

$$c(x) = c(0) + \sum_{\mu=1}^{n} (c_\mu x_\mu + h_\mu x_\mu x_\mu/2) \quad < \text{constant} \ , \tag{30}$$

within a trust region given by

$$\ell^2(x) = \sum_{\mu=1}^{n} x_\mu x_\mu \leqq r^2 \quad , \tag{31}$$

where r has a preassigned value. Even this smaller problem is tricky to analyze correctly.

Since s, c and ℓ^2 are all differentiable in x, any constrained maximum of s must lie at an extremum

$$\begin{aligned} \delta q &= 0 \quad , \\ q(x) &= as - c - (b-a)\ell^2/2 \end{aligned} \tag{32}$$

for Lagrange multipliers which we choose to write as a and (b-a)/2. All the extrema may be parameterized by just two variables a and b, but not all these extrema of q(x) correspond to maxima of s. We now proceed to retrict the possible values of a and b.

Clearly a>0, because reducing c at constant ℓ^2 must decrease s, if the entropy is to be truly maximized. Likewise we must have $a \leq b$ because reducing ℓ^2 at constant c must also decrease s: the special case a=b holds in the interior of the trust region, where the distance constraint is not operating. Thus

$$\begin{aligned} &0 < a < b \qquad \text{if } \ell = r \\ &0 < a = b \qquad \text{if } \ell < r \quad , \end{aligned} \tag{33}$$

which incidentally requires b to be positive.

The solution of (32) for x is

$$x_\mu = (c_\mu - a\, s_\mu)/\delta_\mu \quad , \quad \delta_\mu = -h_\mu - b \tag{34}$$

This is, of course, only relevant within the trust region

$$\ell^2 = \sum_\mu (c_\mu - a\, s_\mu)^2/\delta_\mu^2 \leq r^2 \tag{35}$$

and we may use this as a restriction on the permitted values of a.

If ℓ=r, a is restricted to either of the two roots

$$a_\pm = (\Sigma s_\mu c_\mu \delta_\mu^{-2} \pm D) \;/\; \Sigma s_\mu^2 \delta_\mu^{-2} \tag{36}$$

$$D^2 = r^2 \;\Sigma s_\mu^2 \;\delta_\mu^{-2} - \Sigma s_\mu^2 \delta_\mu^{-2} . \;\Sigma c_\mu^2 \delta_\mu^{-2} + (\Sigma s_\mu c_\mu \delta_\mu^{-2})^2 \quad . \tag{37}$$

On the other hand, if $\ell<r$ we must have a=b. All relevant extrema are now parameterized by one positive variable b, and it is feasible to instruct a computer to investigate a sufficiently dense set of its values directly.

If $D^2<0$, the trial value of b can be rejected immediately, since no relevant point can be inside the trust region. Otherwise, there are up to two choices for a:

$$\begin{aligned} a &= a_- \quad \text{is allowed if} \quad 0 < a_- < b \quad , \\ a &= a_+ \quad \text{is allowed if} \quad 0 < a_+ < b \quad , \\ a &= b \quad \text{is allowed if b is between } a_- \text{ and } a_+ \ . \end{aligned} \tag{38}$$

For each value of b and each choice of a, we must now investigate whether the corresponding extremum is a true maximum of s. The analysis is in two parts.

Distance limit not operating, $\ell < r$, $a = b > 0$.
Perturbing the stationary point (34) by $y_\mu = \Delta x_\mu$ yields changes

$$\begin{aligned} \Delta s &= \sum_\mu (t_\mu y_\mu - y_\mu y_\mu/2) \\ \Delta c &= \sum_\mu (a\, t_\mu y_\mu + h_\mu y_\mu y_\mu/2) \end{aligned} \tag{39}$$

where

$$t_\mu = s_\mu - x_\mu \tag{40}$$

is the local gradient of the entropy model. Since $\Delta c=0$, we may take Δs to be purely second order

$$\Delta s = \sum \delta_\mu y_\mu y_\mu / 2a \quad . \tag{41}$$

For the entropy to be a maximum, we require $\Delta s<0$ whenever $\Delta c=0$.

As y varies around a small perturbation sphere, the stationary points of Δs at fixed Δc occur when

$$\delta\left((1+a/p)\Delta s + \Delta c + a(1-q/p) \sum y_\mu^{\,2}/2 \right) = 0 \tag{42}$$

for suitable Lagrange multipliers related to p and q. The solution of (42) is

$$y_\mu = p t_\mu / (q - \delta_\mu) \quad . \tag{43}$$

The coefficient p is merely a scaling factor relating to the perturbation radius, but q is determined by the condition

$$\Sigma t_\mu y_\mu = 0 \tag{44}$$

that there be no first order change in the constraint. This condition reduces to

$$f(q) = \Sigma t_\mu^2/(q-\delta_\mu) = 0 . \tag{45}$$

Correspondingly, the change in s is

$$\Delta s = (q/2a) \, \Sigma \, y_\mu^2 \tag{46}$$

(after using (45)). It follows that s has the sign of q. Hence s will be a maximum if and only if all the roots of (45) for q are negative.

Fortunately, it is not necessary to solve (45) to determine whether or not its roots are all negative. Define the following functions of q.

$$\begin{aligned} \pi(q) &= \prod_\mu (q-\delta_\mu) \\ F(q) &= \pi(q)\, f(q) = \sum_\mu t_\mu^2 \prod_{\nu\neq\mu} (q-\delta_\nu) \end{aligned} \tag{47}$$

F (q) is clearly a polynomial of degree n-1, so that it has at most n-1 real roots. At $q=\delta_i$, it takes the following values

$$F(\delta_i) = t_i^2 \prod_{\nu\neq i} (\delta_i-\delta_\nu) \tag{48}$$

Assuming the eigenvalues h (and hence also δ) to be ordered with $h_1<h_2< \ldots <h_n$, we see that $F(\delta_i)$ has alternating sign in i ($i=1,2,\ldots,n$).

$$\text{sign}(F(\delta_i)) = (-1)^{i-1} \tag{49}$$

Hence the n-1 roots of F must all exist, and interleave the values δ_i. In particular, the rightmost (most positive) root of q lies between δ_2 and δ_1, and is the only root in that interval.

```
δ4           δ3           δ2           δ1      ———→ q
 .            .            .            .
 .            .            .            .
- - - - 0 + + + + + + 0 - - - - - - - 0 + + + + + + : Sign of F(q)
```

i) If, for our choice of b and a, we have $\delta_1 \leq 0$, then all roots q must be negative and s is maximized. This always happens for convex constraints because these have positive eigenvalues h and hence negative δ.

ii) Conversely, if $\delta_2 \geq 0$, then the rightmost root must be positive, and s is not maximized.

iii) The intermediate case $\delta_2 < 0 < \delta_1$ can be dealt with by inspecting the sign of F(0). If F(0) is positive, all the roots are negative and s is maximized: if F(0) is negative, s is not maximized.

This simple computational test tells us whether or not we have a true maximum of the entropy model.

Distance limit operating, $\ell = r$, $0 < a < b$.

Perturbing the extremum point (34) by y yields

$$\Delta s = \sum_\mu \left((s_\mu - x_\mu) y_\mu - y_\mu y_\mu / 2 \right) \tag{50}$$

as before, but now there is an additional restriction on

$$\Delta \ell^2 / 2 = \sum_\mu \left(x_\mu y_\mu + y_\mu y_\mu / 2 \right) . \tag{51}$$

The corresponding change in the constraint function near the stationary point x is

$$\Delta c = \sum_\mu \left(a(s_\mu - x_\mu) y_\mu - (b-a) x_\mu y_\mu + h_\mu y_\mu y_\mu / 2 \right) . \tag{52}$$

Since $\Delta\ell^2$ and Δc must both be zero, Δs is again of second order

$$\Delta s = \sum_\mu \delta_\mu y_\mu y_\mu / 2a . \tag{53}$$

The stationary values of this, subject to first order contraints

$$\Sigma x_\mu y_\mu = \Sigma s_\mu y_\mu = 0 \tag{54}$$

from $\Delta\ell^2$ and Δc, and to

$$\Sigma y_\mu^2 = \text{fixed} \tag{55}$$

from restriction to a small sphere around x, are at points of the form

$$y_\mu = (p s_\mu + r x_\mu) / (q - \delta_\mu) \tag{56}$$

where p, q, and r are related to the three Lagrange multipliers. Using (54), the corresponding entropy change is

$$\Delta s = (q/2a) \, \Sigma y_\mu^2 \tag{57}$$

which again has the sign of q. The allowed values of q are obtained from the condition

$$\phi(q) = \sum \frac{x_\mu x_\mu}{q - \delta_\mu} \cdot \sum \frac{s_\mu s_\mu}{q - \delta_\mu} - \left[\sum \frac{s_\mu x_\mu}{q - \delta_\mu} \right]^2 = 0 \qquad (58)$$

that (54) allows a solution with non-trivial p and r. A little algebra converts this to the form

$$\phi(q) = \frac{1}{2} \sum_{\mu \neq \nu} \frac{(s_\mu x_\nu - x_\mu s_\nu)^2}{(q - \delta_\mu)\,(q - \delta_\nu)} , \qquad (59)$$

from which we see that

$$\Phi(q) = \pi(q)\ \phi(q) \qquad (60)$$

is a polynomial of degree n-2. Accordingly there are at most n-2 real roots of $\phi(q)=0$. Now define functions

$$G(q) = \pi(q)\ \Sigma\, x_\mu x_\mu \,/(q-\delta_\mu)$$
$$H(q) = \pi(q)\ \Sigma\, s_\mu s_\mu \,/(q-\delta_\mu) \ . \qquad (61)$$

As above, G and H are polynomials of degree n-1, whose n-1 real roots interleave the values δ_i. Furthermore whenever G is zero, $\phi(q)$ is seen from (58) to be negative, so that the sign of $\Phi(q)$ alternates between successive zeros of G, and similarly for zeros of H. Consequently, the n-2 roots of Φ are all real, and interleave both the n-1 zeros of G and the n-1 zeros of H.

```
 δ4                δ3                δ2                δ1              q
 .                 .                 .                 .
- - - 0 + + + + + + + + + + + 0 - - - - - - 0 + + + + +   :  Sign of G(q)
 .  :              .           :  .           :   .
 .  :              .           :  .           :   .
- - - - - 0 + + + + + + 0 - - - - - - 0 + + + + + + + +   :  Sign of H(q)
    :   :              :  :          :  :
    :   :              :  :          :  :
+ + + + + + + + + + 0 - - - - - 0 + + + + + + + + + + +   :  Sign of Φ(q)
```

We can now test whether or not all the roots of ϕ (or Φ) are negative.

i) If $\delta_1 \leq 0$, all roots are certainly negative. This always happens for convex constraints.

ii) If $\delta_2 \leq 0 < \delta_1$, all roots are negative if and only if $\Phi(0)>0$.

iii) If $\delta_3 < 0 < \delta_2$, at least one root is positive if and only if $G(0) > 0$ or $H(0) > 0$ or ($G(0) < 0$ and $H(0) < 0$ and $\Phi(0) < 0$).

iv) If $\delta_3 \geq 0$, at least one root is certainly positive.

These straightforward tests tell us whether or not we have indeed maximized the entropy model.

4 FINAL SELECTION OF IMAGE INCREMENT

We can now instruct a computer to run through all positive values of b (using reasonably dense sampling with perhaps a hundred or so values), to calculate each possible value of a (a_+ or a_- or b), and to discard any pair which does not give a maximum of s. The surviving pairs form one or more sequences of permitted values, each with its own subspace increment x and entropy and constraint values s(x) and c(x).

Figure 1 shows a typical plot of s against c, for entropy and constraint models

$$s(x) = s(0) + 5x_2 + 5x_3 - (x_1^2 + x_2^2 + x_3^2)/2 \tag{62}$$

$$c(x) = c(0) - 2x_1 + 2x_3 + (x_1^2 + x_3^2)/2 \tag{63}$$

Figure 1. Entropy s as a function of constraint c for the model defined by equations 62 and 63, within the trust region. AB outlines the constrained entropy maxima.

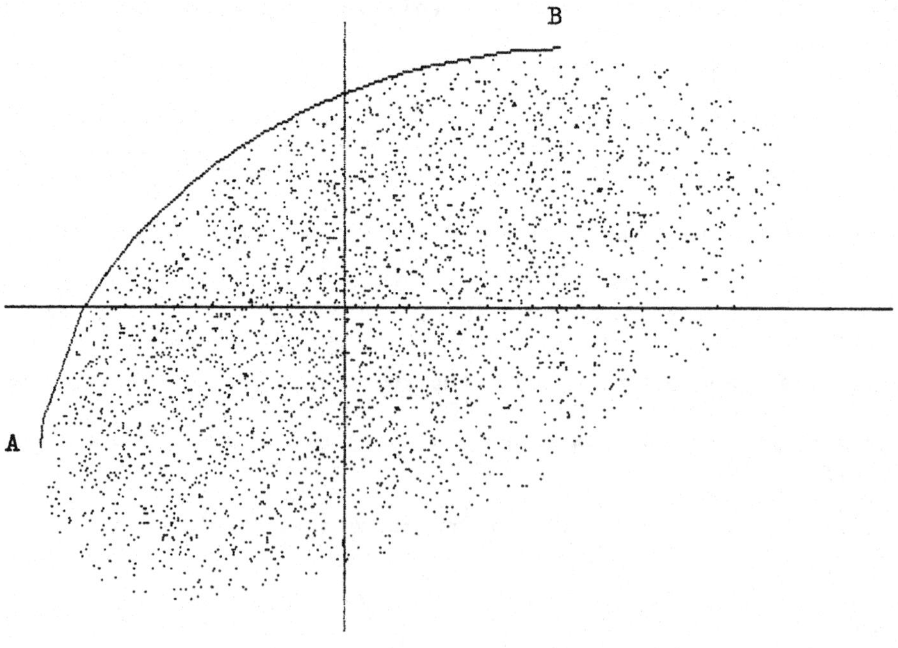

with the axes plotted through (c(0),s(0)). The curved line AB is the locus of points produced by the control algorithm, and the isolated points correspond to image increments x randomly scattered within the trust region

$$\ell^2(x) = x_1^2 + x_2^2 + x_3^2 \leq 1 . \tag{64}$$

AB accurately outlines the upper left boundary of the plotted points, which verifies that s(x) is indeed being maximized over c(x) ≤ constant within the relevant trust region. The algorithm rejected points on the upper right boundary beyond B because s could be further maximized (towards B) by allowing c to decrease. The particular model (63) had a convex constraint function c(x), for which the eigenvalues were all positive. Accordingly, there was a unique maximum of s for each value of c, and only one trajectory AB from the overall minimum of c to the overall maximum of s.

We still need to select a single image increment to be the output of the control algorithm. Although we usually wish to keep reducing the constraint value until the data are fitted correctly, our experience with such problems suggests that one should not try to reduce c more than about 4/5 of the way from the existing value c(0) to the overall minimum at A. Trying to reduce the constraint value faster is actually inefficient because the entropy s no longer has enough influence on the iterations. There seems to be no compelling logical reason favouring any particular compromise value: it is a matter for common sense and experience.

With nonlinear data, by contrast, there may be more than one local maximum of s at constant c. An example of this is the model

$$s(x) = s(0) + 5x_2 + 5x_3 - (x_1^2 + x_2^2 + x_3^2)/2 \tag{65}$$

$$c(x) = c(0) - x_1 + x_3 + (2x_1^2 - 5x_3^2)/2 \tag{66}$$

The corresponding plot of s against c, with axes plotted through (c(0),s(0)) is shown in Figure 2.

There are now two separate trajectories AB and PQ of local entropy maxima, which indicates a real ambiguity in the solution to be chosen. This may reflect a bifurcation in the locus of entropy maxima for given constraint values in the full image space, or it may not. In fact the multiplicity of solutions in the subspace bears no particular relation to the multiplicity in the full space. Separate maxima can develop in the full space, either in directions which are not spanned by the search directions or at distances greater than the trust limit. Equally, a bifurcation in the subspace can be merely an artefact of inadequate search directions. It is impossible to decide.

Returning of necessity to the subspace, an ideal computer program would investigate each branch separately. This may be impractical if there is

Figure 2. Entropy s as a function of constraint c for the model defined by equations 65 and 66. AB and PQ are two separate branches of constrained entropy maxima.

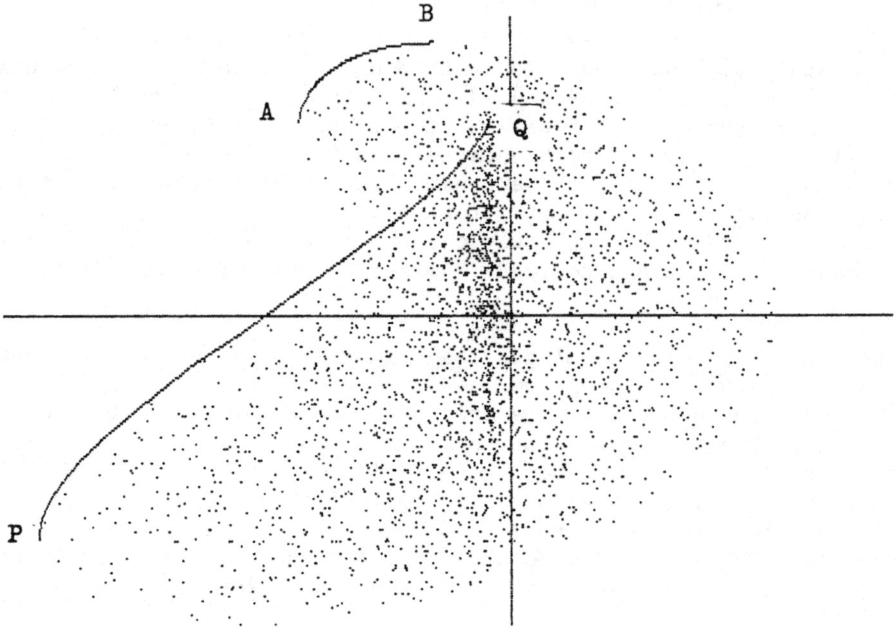

a high level of ambiguity. We may have to restrict our programs to the "most promising" branch, however that is to be defined, but allow the possibility of changing to alternative branches if the solution we obtain is physically unrealistic. This is not an easy problem. In Figure 2, for instance, should we select the upper branch AB because it has greater entropy, or the lower end of PQ because it fits the data better?

5 CONCLUSIONS

Programming maximum entropy is not trivial. This paper has outlined the "Cambridge algorithm", which deals reliably with linear data, and has discussed our latest ideas on controlling it for non-linear data. We see no rigorous way of treating all the ambiguities which might occur in the latter case, but we have realistic hopes of constructing maximum entropy programs which will be of value even for badly nonlinear data. The development of such programs is one of the frontiers of our current research.

ACKNOWLEDGEMENTS

Many colleagues deserve thanks for contributing to the Cambridge algorithm. In particular, Richard Bryan was closely involved with its initial development. Latterly, Gerard Bricogne (1984) catalysed considerable thought about nonlinear data because of his applications of maximum entropy to X-ray crystallography and the Fourier phase problem: Alastair Livesey has worked with us on this application.

REFERENCES

1 Bricogne, G. (1984). Acta Cryst, A40, 410.

2 Bryan, R.K. (1980). PhD thesis, University of Cambridge.

3 Fletcher, R., and C.M. Reeves (1964). Comp J, 7, 149.

4 Gill, P.E., W. Murray, and M.H. Wright (1981). Practical Optimization Theory. Academic Press.

5 Gull, S.F., and G.J. Daniell (1978). Nature, 272, 686.

6 Kullback, S. (1959). Information Theory and Statistics. Wiley, N.Y., and (1969) Dover, N.Y.

7 Livesey, A.K., and J. Skilling (1984). Paper presented at EMBO/LURE workshop on Maximum Entropy Methods in the X-Ray Phase Problem, Orsay, Paris. (CECAM report, G. Bricogne, ed., in press.)

8 Powell, M.J.D. (1977). Math Programming, 12, 141.

9 Skilling, J. (1981). Paper presented at first workshop on Maximum Entropy and Bayesian Methods in Inverse Problems, Laramie, Wyoming, C.R. Smith and W.T. Grandy (eds.), (Reidel).

10 Skilling, J., and R.K. Bryan (1984). Mon. Not. R. Astron. Soc. 211, 111.

MAXIMUM ENTROPY AND THE MOMENTS PROBLEM: SPECTROSCOPIC APPLICATIONS

C.G. Gray
Guelph Waterloo Program for Graduate Work in Physics
Physics Department
University of Guelph, Guelph, Ontario N1G 2W1

ABSTRACT

A classic inverse problem, arising in many contexts, is to determine a distribution function $I(\omega)$, given its moments $M_n = \int d\omega I(\omega)\omega^n$, $n = 0,1,2, \ldots$ If all the moments are known, the complete set $\{M_n\}$ uniquely determines $I(\omega)$, provided the M_n do not increase with n faster than n! In many applications, such as spectroscopy, we can calculate from statistical mechanics only a finite number of the moments. The problem then is: what is our best estimate of $I(\omega)$, given this finite amount of information about it. Standard maximum entropy arguments give us a simple solution to the problem, where 'best' is interpreted as 'least biased'. Examples to be discussed include infrared absorption in liquid and gaseous nitrogen. The N_2 molecule is nonpolar, so that isolated N_2 molecules do not absorb infrared radiation. In condensed phases, however, intermolecular forces induce dipoles in the molecules, thereby rendering the vibrational, rotational and translational motions of the molecules infrared-active. Theory and experiment agree in all cases qualitatively, and in many cases quantitatively.

1 INTRODUCTION

The moments problem is a classic inverse problem which has been studied for over a century.[1-20] We are given some moments M_n of a distribution function $I(\omega)$, where

$$M_n = \int_{-\infty}^{\infty} d\omega I(\omega)\omega^n \quad , \quad n = 0,1,2,\ldots \, , \tag{1}$$

and asked to find the distribution function. If the moments M_n do not increase with n faster than n!, then the complete set of $\{M_n\}$ for $n = 0,1,2,\ldots\infty$ determines $I(\omega)$ uniquely. Equivalently, $I(\omega)$ is determined uniquely by its moments if it decays at large ω no more slowly than $\exp(-\omega)$. Otherwise,[5,6] more than one distribution can correspond to the same set of moments. In the present context, that of trying to determine spectral line shapes from spectral moments, this theorem is of limited interest, since what we can calculate from statistical mechanics is a finite number of the moments, for $n = 0,1,2, \ldots n_{max}$. For gases, typically $n_{max} \sim 6$ or 8, and for liquids $n_{max} \sim 2$. The problem then is: What is our best estimate of $I(\omega)$ given the finite amount of information $\{M_0, M_1, \ldots, M_{n_{max}}\}$ about it? In what follows we review recent work in which maximum entropy/information theory methods[21] have been applied to this problem.

2 MOMENTS OF ABSORPTION SPECTRA

Absorption of radiation at angular frequency ω is measured by the absorption coefficient $A(\omega)$. An isolated molecule in the gas phase experiences only the electric field $\underline{E}_0$ of the exciting radiation, and in this case one can derive from statistical mechanics a rigorous expression for $A(\omega)$. In the classical ($\hbar \to 0$) limit, to which we restrict ourselves here, the standard Kubo-type linear response calculation yields[22]

$$A(\omega) = \frac{4\pi^2 \beta}{3cV} \omega^2 I(\omega) \tag{2}$$

where V is the system volume, c the speed of light, $\beta = (kT)^{-1}$ with T the temperature and

$$I(\omega) = \frac{1}{2\pi} \int_{-\infty}^{\infty} dt\, e^{-i\omega t} C(t) \tag{3}$$

where

$$C(t) = \langle \underline{\mu}(0) \cdot \underline{\mu}(t) \rangle \tag{4}$$

with $\underline{\mu}$ the total dipole moment of all the molecules. The brackets $\langle \ldots \rangle$ denote an equilibrium (canonical) ensemble average. For liquids (2) must be multiplied by a correction factor[16] $\kappa = \frac{1}{n}\left(\frac{n^2+2}{3}\right)^2$, where n is the optical refractive index, which takes into account (approximately) the fact that the local field $\underline{E}_\ell$ seen by a molecule differs from the externally applied field $\underline{E}_0$.

Eq. (3) is a Wiener-Khinchin type relation between the spectrum $I(\omega)$ and the time correlation function $C(t)$ for the system dipole moment $\underline{\mu}$. By expanding the exponential $\exp(i\omega t)$ in the inverse Fourier transform relation

$$C(t) = \int_{-\infty}^{\infty} d\omega\, e^{i\omega t} I(\omega) \tag{5}$$

we get the series expansion for $C(t)$,

$$C(t) = \sum_{n=0}^{\infty} M_n \frac{(it)^n}{n!} \tag{6}$$

where M_n are the moments defined by (1). The moments are thus related to $C(t)$ by

$$M_n = i^{-n} C^{(n)}(0) \tag{7}$$

where $C^{(n)}(0) \equiv (d^nC(t)/dt^n)_{t=0}$. From (7) and (4) we get

$$M_n = i^{-n} < \underline{\mu} \cdot \underline{\mu}^{(n)} > \tag{8}$$

In the classical limit $I(\omega)$ is an even function, so that the odd moments vanish. Using the standard relation[23] $< A\dot{B} > = - < \dot{A}B >$, where $\dot{A} \equiv dA/dt$, we rewrite the even moments M_{2n} as

$$M_{2n} = < \underline{\mu}^{(n)^2} > \tag{9}$$

In particular we have

$$M_0 = < \underline{\mu}^2 >, \quad M_2 = < \dot{\underline{\mu}}^2 >, \quad M_4 = < \ddot{\underline{\mu}}^2 > \tag{10}$$

Note the immense simplification in the expressions (10) for the moments, as compared to the expressions (4) and (3) for $C(t)$ and $I(\omega)$; (10) involves only static ($t=0$) equilibrium averages, whereas (4) involves the dynamical quantity $\underline{\mu}(t)$, the determination of which involves solving the equations of motion for the full many-body system.

The system dipole moment $\underline{\mu} = \underline{\mu}(\underline{r}^N \omega^N)$ depends on the positions $\underline{r}^N = \underline{r}_1 \dots \underline{r}_N$ and orientations $\omega^N = \omega_1 \dots \omega_N$ of the N molecules, and can be written

$$\underline{\mu} = \sum_i \underline{\mu}(i) + \sum_{i<j} \underline{\mu}(ij) + \sum_{i<j<k} \underline{\mu}(ijk) + \dots \tag{11}$$

The first term is a so-called one-body term, and involves a sum of all the permanent dipole moments $\underline{\mu}(i)$ of all the N molecules. This term is the dominant one for polar systems such as HCl and H_2O, but is absent for nonpolar species such as Ar, N_2 and CH_4 with which we shall be concerned. For a pair i,j of interacting nonpolar molecules, the pair moment $\underline{\mu}(ij) = \underline{\mu}(\underline{r}_{ij}\ \omega_i\ \omega_j)$ arises due to mutual distortion of the charge clouds of the two molecules. Specific distortion mechanisms are considered in Section 4. The three-body term $\underline{\mu}(ijk)$ in (11) is the additional dipole in a triplet i,j,k not already accounted for by the sum of the three pair terms, $\underline{\mu}(ij) + \underline{\mu}(ik) + \underline{\mu}(jk)$. The effect of this nonadditive term on the spectrum has been studied very little, and will be neglected, as will the higher terms in (11). For nonpolar molecules we therefore approximate (11) by

$$\underline{\mu} = \sum_{i<j} \underline{\mu}(ij) \quad , \tag{12}$$

where the sum $\sum_{i<j}$ is over all $N(N-1)/2$ pairs of molecules.

If we now substitute (12) in (10) we find, for example for M_0

$$M_0 = \underbrace{\Sigma <\underline{\mu}(12)^2>}_{\text{two-body}} + \underbrace{\Sigma <\underline{\mu}(12) \cdot \underline{\mu}(13)>}_{\text{three-body}} + \underbrace{\Sigma <\underline{\mu}(12) \cdot \underline{\mu}(34)>}_{\text{four-body}} \tag{13}$$

Similarly M_2, M_4 etc. will contain two-, three-, and four-body terms. Evaluation of (13) for liquids and dense gases requires the pair, triplet and quadruplet distribution functions for the liquid. Theoretical and computer simulation methods are available[23] for obtaining these distribution functions. For low density gases, where binary interactions predominate, only the first term in (13) contributes significantly, and the M_n become proportional to ρ^2, where $\rho = N/V$ is the fluid number density. The details of the calculations of the moments are given in refs. 12-19 and 24.

3 MAXIMUM ENTROPY RECONSTRUCTION OF A SPECTRUM FROM ITS MOMENTS

Given the moments (1) for $n = 0, 2, 4, \ldots n_{max}$, we wish to estimate $I(\omega)$. A least-biased estimate is obtained[21] if we maximize the information entropy S_I,

$$S_I = -\int_{-\infty}^{\infty} d\omega \, I(\omega) \, \ell n \, I(\omega) \quad , \tag{14}$$

subject to the constraints (1) for $n = 0, 2, 4, \ldots n_{max}$. Solution of this standard variational problem yields

$$I(\omega) = e^{-\left(\lambda_0 + \lambda_2\omega^2 + \lambda_4\omega^4 + \ldots + \lambda_{max}\,\omega^{n_{max}}\right)} \quad , \tag{15}$$

where the λ_n are Lagrange multipliers. These are determined from the constraint conditions. For $n_{max} \lesssim 6$, the solution is readily obtained[12-19] by standard numerical methods.

4 APPLICATIONS TO COLLISION-INDUCED ABSORPTION SPECTRA

Collision-induced absorption (CIA) is the standard name used for infrared absorption by nonpolar molecules. It is not entirely appropriate, since we are interested in absorption in liquids, as well as in gases, and even in gases there is absorption by pairs of molecules in bound states (dimers), as well as by pairs in collision. For general reviews, see refs. 24-26. We discuss in turn absorption by monatomic, diatomic and polyatomic species.

(a) Monatomics

When two nonidentical spherical atoms (e.g. Ar-Xe) interact, mutual distortion of the two charge clouds produces a pair dipole moment $\underline{\mu}(12) \equiv \underline{\mu}(\underline{r}_{12})$ which lies along the line of centres $\underline{r}_{12}$. The main effect occurs at short-range, when the two clouds overlap. Experimental and theoretical studies have shown that the short-range overlap dipole moment can be well represented by an exponential model

$$\mu(r) = \mu_0 \, e^{-r/\rho} \tag{16}$$

where μ_0 is the amplitude and ρ the range. Typically $\rho \sim 0.1\sigma$, where σ is the diameter of the atoms, i.e. the length parameter in the Lennard-Jones intermolecular potential model,

$$u(r) = 4\,\varepsilon\left[\left(\frac{\sigma}{r}\right)^{12} - \left(\frac{\sigma}{r}\right)^{6}\right], \tag{17}$$

where ε is the well depth. Also typically $\mu(r \sim \sigma) \sim 10^{-3}$ D; for comparison purposes the permanent dipole moment of a polar molecule such as HCl is of order 1D.

Figure 1. Maximum entropy/information theory and experimental[28] normalized line shapes $\hat{I}(\bar{\nu}) \equiv I(\bar{\nu})/\int d\bar{\nu}\, I(\bar{\nu})$ for Ar-Xe gas mixtures at T = 295 K.

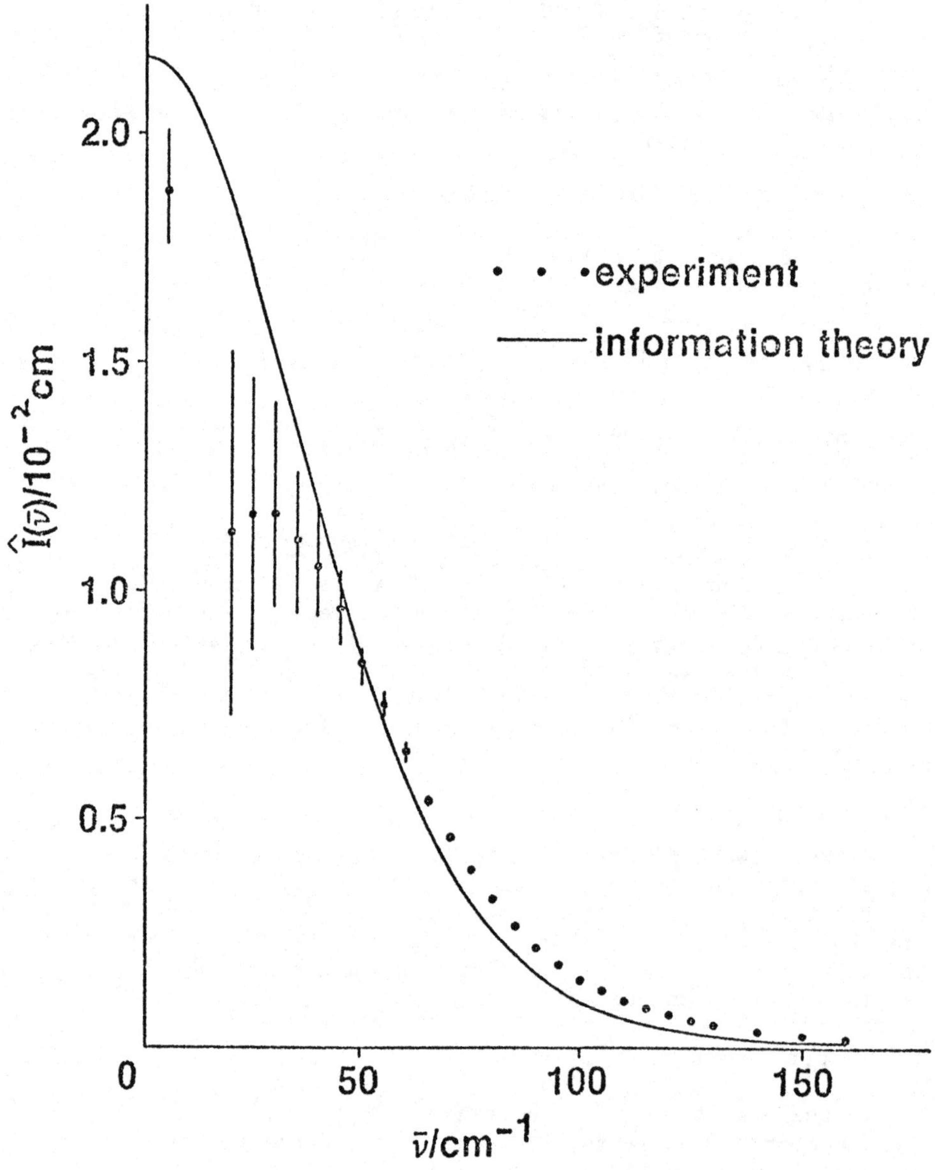

For gaseous Ar-Xe mixtures at T = 295 K moments M_n for the model (16) have been calculated[12,27] up to n_{max} = 8. The maximum entropy reconstruction of the spectrum using the set $\{M_0, M_2, M_4, M_6\}$ is shown in Fig. 1, along with the experimental data. Theory and experiment are in good qualitative agreement. The parameters in (16) and (17) were not adjusted to get good agreement; the parameters were simply taken from the literature. The small discrepancies in the range (50 - 150) cm^{-1} are probably due to small deficiencies in the models (16) and (17), or to the neglect of quantum effects, rather than to deficiencies in the maximum entropy method at the n_{max} = 6 level.

For rare gas liquid mixtures, the moments M_0, M_2 have been calculated[24], but have not yet been applied to spectrum reconstructions.

(b) Diatomics

When two homonuclear molecules (e.g. N_2) interact, the main contribution to the induced dipole moment of the pair arises from the quadrupolar fields surrounding the two molecules. The two molecules mutually polarize each other, with total pair moment $\underline{\mu}(12)$ given by

$$\underline{\mu}(12) = \frac{1}{3}\underline{\alpha}_1 \cdot \underline{T}^{(3)}(\underline{r}_{12}) : \underline{Q}_2 - \frac{1}{3}\underline{\alpha}_2 \cdot \underline{T}^{(3)}(\underline{r}_{12}) : \underline{Q}_1 \qquad (18)$$

The first term is the dipole moment induced in molecule 1 via its polarizability $\underline{\alpha}_1$ due to the electric field $\underline{E} = \frac{1}{3}\underline{T}^{(3)}(\underline{r}_{12}) \cdot \underline{Q}_2$ of the quadrupole moment $\underline{Q}_2$ of molecule 2 at position $\underline{r}_{12}$ relative to molecule 1. Here $\underline{T}^{(3)}(\underline{r}) = \underline{\nabla}\,\underline{\nabla}\,\underline{\nabla}(1/r)$ is a third-rank tensor and the dots denote tensor contractions[23]. Similarly the second term in (18) is the dipole induced in molecule 2 due to molecule 1. The dipole moment $\underline{\mu}(12)$ is modulated due to changes in the $\underline{Q}_i$ and $\underline{\alpha}_i$ (arising from molecular rotations) and in $\underline{T}^{(3)}(\underline{r}_{12})$ (due to molecular translational motions), giving rise to an induced rotational-translational spectrum in the far infrared. (There is also an induced vibrational spectrum, in the infrared, due to vibrational modulations of the $\underline{\alpha}_i$ and $\underline{Q}_i$. We shall not discuss the vibrational spectrum here.) In addition to the quadrupolar-induced moments, there are overlap-type terms[29], similar to those of the preceding section, but whose effect on the N_2 spectrum is smaller than that of the quadrupolar terms.

In calculating the correlation function (4), we shall assume, as a first approximation, that the molecules are freely rotating. There are in reality angle-dependent intermolecular forces[23] which couple the rotational and translational motions, but, for CIA spectra in N_2, we expect these effects could be included later as a smaller perturbation. We thus use the same model (17) for the intermolecular forces as was used for atoms. The correlation function now factorizes[14,15,16] into rotational and translational parts

$$C(t) = C_{rot}(t)\, C_{tr}(t) \qquad (19)$$

where $C_{rot}(t) \equiv \langle P_2(\cos\theta(t)\rangle$ is known exactly[15,16] for freely rotating molecules; here $\theta(t)$ is the angle through which a N_2 molecule rotates in time t, and P_2 is a Legendre polynomial. For gases $C_{tr}(t)$ is given by

$$C_{tr}(t) = \frac{1}{30}\alpha^2 Q^2 N(N-1) \langle \underline{T}^{(3)}(\underline{r}_{12}(0)) \vdots \underline{T}^{(3)}(\underline{r}_{12}(t))\rangle \tag{20}$$

where Q is the scalar quadrupole moment as normally defined[23], and α is the isotropic part[23] of $\underline{\alpha}$ (we have neglected the anisotropic part, which is small[23] for N_2). For liquids[16], $C_{tr}(t)$ contains three-body terms such as $\langle \underline{T}^{(3)}(\underline{r}_{12}(0)) \vdots \underline{T}^{(3)}(\underline{r}_{13}(t))\rangle$, as well as the two-body terms (20).

Corresponding to (19), the spectrum (3) can be written as a convolution of rotational and translational spectra,

$$I(\omega) = \int_{-\infty}^{\infty} d\omega' \, I_{rot}(\omega') \, I_{tr}(\omega - \omega') \tag{21}$$

where $I_{rot}(\omega)$ and $I_{tr}(\omega)$ are the Fourier transforms of $C_{rot}(t)$ and $C_{tr}(t)$ respectively.

Since $C_{rot}(t)$ and $I_{rot}(\omega)$ are known exactly, it is necessary to apply maximum entropy only to the estimation of $I_{tr}(\omega)$. Its moments M_n^{tr} have been calculated for n = 0, 2, 4, 6 for the gas[14] and for n = 0, 2 for the liquid[16]. $I_{tr}(\omega)$ is then given by (15), and the total spectrum by (21). Theory and experiment are compared in Figs. 2 and 3. The agreement is again impressive. In particular the qualitative difference between the gas and liquid spectra (note the shoulder in the gas spectrum) seen experimentally is explained by the theory. The shoulder arises in the gas spectrum because the component spectra $I_{tr}(\omega)$ and $I_{rot}(\omega)$ have different widths for the gas. For the liquid the two widths are about the same, and no shoulder arises. Similar results are found at other temperatures. As before, no adjustable parameters were used in the calculations; the parameters Q, α etc. were obtained from other sources.[23]

When the absolute absorption coefficients $A(\omega)$ (see (2)) are calculated, it is found that they are underestimated by the theory[14,16] by about 10% in the gas and 30% in the liquid. Possible reasons for this discrepancy are discussed in refs. 14 and 16, and include: (i) omission of shorter-range (e.g. overlap) induction mechanisms, (ii) omission of quantum effects, (iii) use of the simplified model (17) for the intermolecular potential.

(c) Polyatomics

For molecules with tetrahedral symmetry (e.g. methane CH_4), the first nonvanishing multipole moment is the third-rank octopole $\underline{\Omega}$ (i.e. $\underline{\mu} = 0$ and $\underline{Q} = 0$, see (23)). The pair moment analogous to (18), due to mutual octopole ($\ell = 3$) induction is

<u>Figure 2</u>. Normalized theoretical (_____) and experimental[30] (Φ) spectra $\hat{I}(\bar{\nu}) = I(\bar{\nu})/\int d\bar{\nu}\ I(\bar{\nu})$ for gaseous N_2.

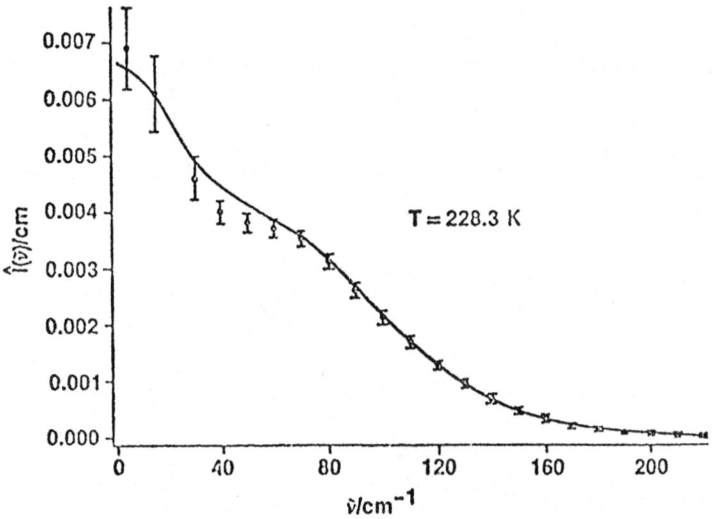

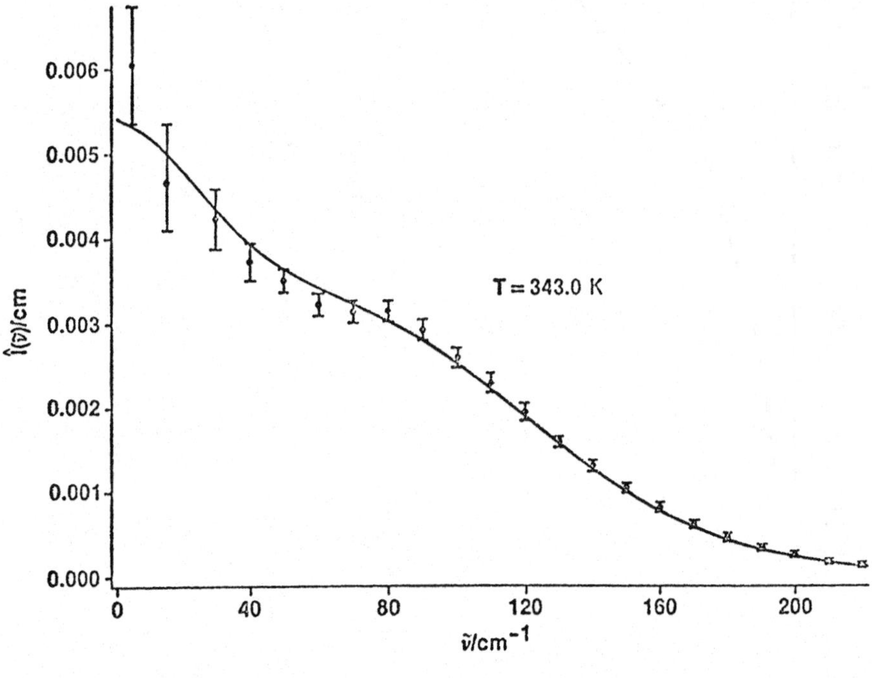

Figure 3. Normalized theoretical (_____) and experimental[31] (ϕ) spectra $\hat{I}(\bar{\nu}) = I(\bar{\nu})/\int d\bar{\nu}\, I(\bar{\nu})$ for liquid N_2.

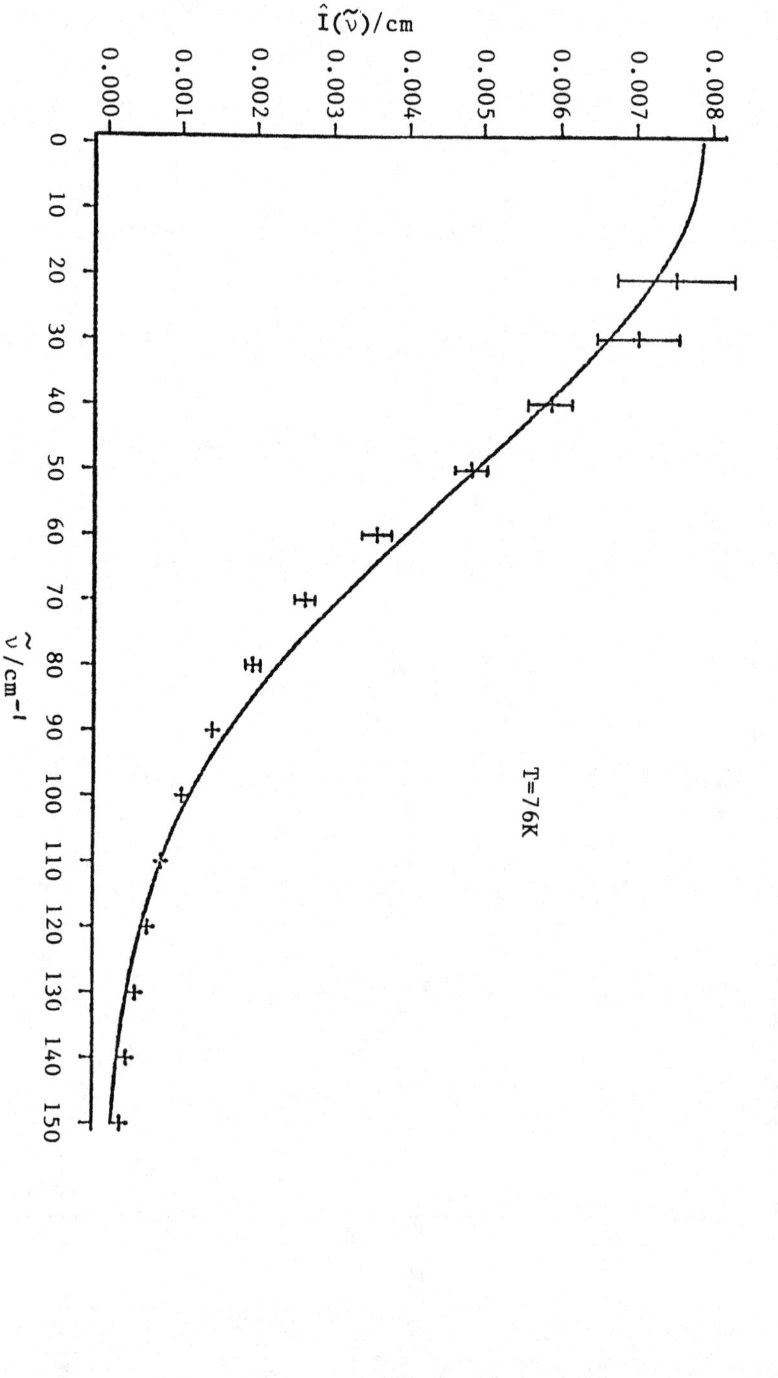

Figure 4. Normalized theoretical (_____) and experimental[32] (Φ) spectra $\hat{I}(\bar{\nu}) = I(\bar{\nu})/\int d\bar{\nu}\, I(\bar{\nu})$ for gaseous CH_4 at T = 296 K.

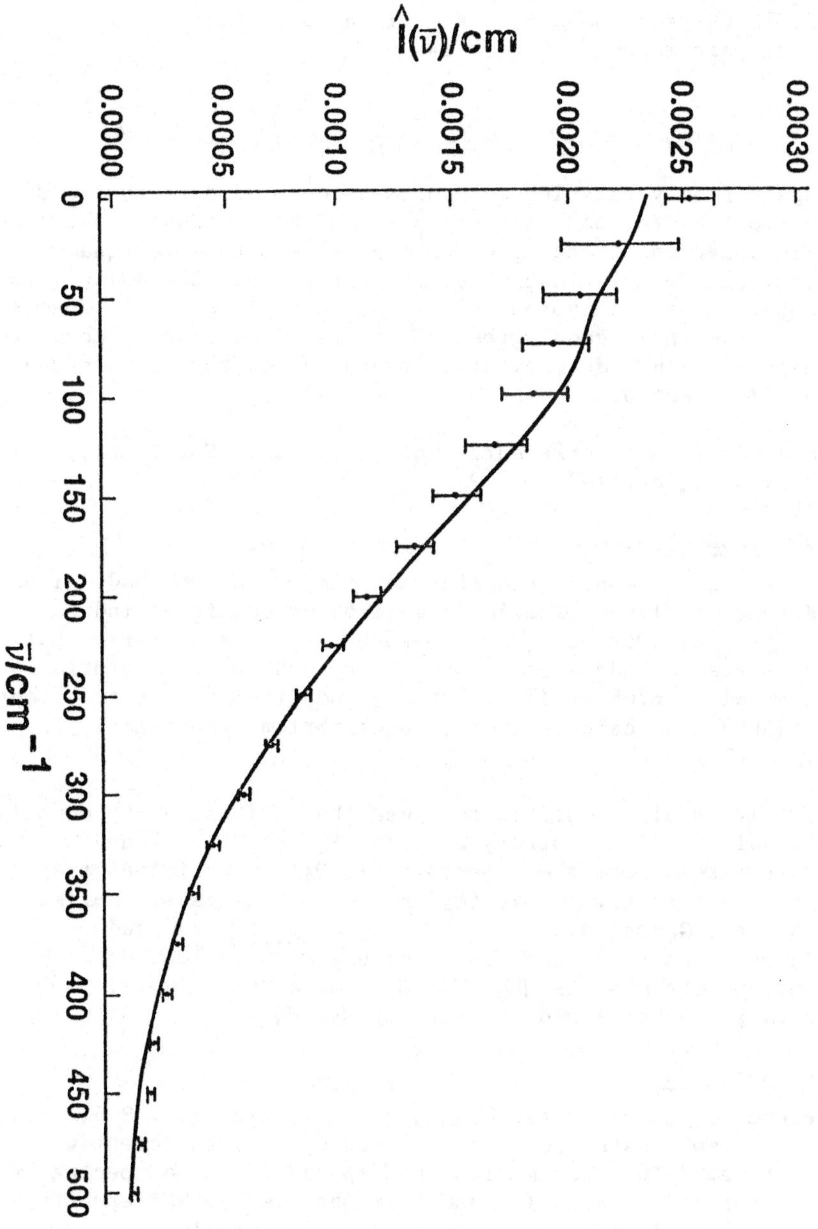

$$\underline{\mu}^{(3)}(12) = \frac{1}{15}\alpha\,\underline{T}^{(4)}(\underline{r}_{12}) \vdots \underline{\Omega}_2 + \frac{1}{15}\alpha\,\underline{T}^{(4)}(\underline{r}_{12}) \vdots \underline{\Omega}_1 \qquad (22)$$

where α is the polarizability (isotropic for tetrahedral molecules) and $\underline{T}^{(n)}(\underline{r}) = \underline{\nabla}^n(1/r)$. It turns out that mutual induction due to the next multipole (hexadecopole $\underline{\Phi}$, $\ell = 4$) is also important. The corresponding pair moment $\underline{\mu}^{(4)}(12)$ is

$$\underline{\mu}^{(4)}(12) = \frac{1}{105}\alpha\,\underline{T}^{(5)}(\underline{r}_{12}) \vdots\vdots \underline{\Phi}_2 - \frac{1}{105}\alpha\,\underline{T}^{(5)}(\underline{r}_{12}) \vdots\vdots \underline{\Phi}_1 \qquad (23)$$

Assuming again freely rotating molecules we can again factorize C(t) analogously to (19) for both the octopole and hexadecapole contributions to C(t). Calculation of the spectrum[17] parallels that for quadrupolar induction. Calculations were done for gaseous CH_4 and CF_4 for various temperatures. A typical result is shown in Fig. 4. Theory and experiment are again in good agreement for the line shape. Absolute intensities are again underestimated, probably for the same reasons as in the preceding section.

Calculations for liquid tetrahedral molecules (CH_4, CCl_4, CBr_4) are currently being carried out.[18]

5 CONCLUSIONS

We have demonstrated that maximum entropy methods give a simple and accurate line shape for a variety of collision-induced absorption spectra. Particularly noteworthy are the features that (a) the theory is free of adjustable parameters, and (b) the solution of the difficult dynamical problem is avoided by judicious use of the spectral moments, which can be calculated from equilibrium statistical mechanics.

We note that it is also possible to close the Mori hierarchy equations for C(t) by using maximum entropy closures.[13,15] This is useful in particular in cases where the moments ratio $\hat{M}_4/\hat{M}_2^2$ indicates that the line is more lorentzian-like ($\hat{M}_4/\hat{M}_2^2 \gg 1$) than gaussian-like ($\hat{M}_4/\hat{M}_2^2 = 3$ for a Gaussian). Here $\hat{M}_n = M_n/M_0$ is the reduced moment. In such cases one can sometimes use Mori/maximum entropy methods based on the moments $\{M_0, M_2, M_4\}$, or alternatively, pure maximum entropy methods based on $\{M_0, M_2, M_4, M_6\}$.

REFERENCES

1 For references to the classical literature, see refs. 2-6. For reviews, extensions and references to more recent work, see refs. 7-10. The maximum entropy approach to spectra was proposed in ref. 11, and there applied to NMR spectra. Applications to collision-induced absorption and light scattering spectra are described in refs. 12-19. The classical (polynomial and Pade) and maximum entropy approaches are compared in ref. 20.

2 Shohat, J.A. and J.D. Tamarkin (1963). The Problem of Moments. 2nd Ed., Amer. Math. Soc., Providence.

3 Vorobyev, Yu.V. (1965). Method of Moments in Applied Mathematics. Gordon and Breach, New York.

4 Akhiezer, N.I. (1965). The Classical Moment Problem. Oliver and Boyd, Edinburgh.

5 Kendall, M.G. and A. Stuart (1969). The Advanced Theory of Statistics. Vol. 1, 3rd Ed., p.109. Hufner, New York.

6 Feller, W. (1971). An Introduction to Probability Theory and its Applications. Vol. 2, 2nd Ed., p.227. Wiley, New York.

7 Nickel, B.G. (1974). J. Phys. C. 7. 1719.

8 Langhoff, P.W. (1980). In Theory and Applications of Moment Methods. B.J. Dalton et al. (eds.), p.191. Plenum, New York.

9 Reinhardt, W.P. (1982). Int. J. Quant. Chem. 21, 133.

10 Hadler, W. and K. Schulten (1983). Phys. Rev. Lett. 107, 249.

11 Powles, J.G. and B. Carazza (1970). In Magnetic Resonance. C.K. Coogan et al. (eds.), p.133. Plenum, New York. See also Berne, B.J. and Harp, G.D., Adv. Chem. Phys. (1970) 17, 63.

12 Gburski, Z., C.G. Gray and D.E. Sullivan (1983). Chem. Phys. Lett. 100, 383.

13 Gburski, Z., C.G. Gray and D.E. Sullivan (1984). Chem. Phys. Lett. 106, 55.

14 Joslin, C.G. and C.G. Gray (1984). Chem. Phys. Lett. 107, 249.

15 Joslin, C.G., C.G. Gray and Gburski, Z. (1984). Molec. Phys. 53, 203.

16 Joslin, C.G., S. Singh and C.G. Gray (1985). Can. J. Phys. 63, 76.

17 Joslin, C.G., C.G. Gray and S. Singh (1984). Molec. Phys. 54, 1469.

18 Joslin, C.G., C.G. Gray and S. Singh (1985). Molec. Phys. 55, 1075 and 1089, and Chem. Phys. Lett., in press.

19 Meinander, N. and G.C. Tabisz (1984). Chem. Phys. Lett., 110, 388.

20 Mead, L.R. and N. Papanicolaou (1984). J. Math. Phys. 25, 2404.

21 For a general review, see: Jaynes, E.T., 1979. In The Maximum Entropy Formalism. R.D. Levine and M. Tribus (eds.). M.I.T. Press, Cambridge, Ma. See also many of the other papers in this workshop volume.

22 MacQuarrie, D.A. (1976). Statistical Mechanics. Harper and Row, New York.

23 Gray, C.G. and K.E. Gubbins (1984). Theory of Molecular Fluids. Oxford University Press, Oxford.

24 Joslin, C.G. and C.G. Gray (1985). In Phenomena Induced by Intermolecular Interactions. G. Birnbaum (ed.), Plenum, New York.

25 Van Kranendonk, J. (ed.) (1980). Intermolecular Spectroscopy and the Dynamical Properties of Dense Systems. E. Fermi International School of Physics, Italian Physics Society, Vol. 75. North Holland, Amsterdam.

26 Birnbaum, G., B. Guillot and S. Brates (1982). Adv. Chem. Phys. 51, 49.

27 Recently B.G. Nickel (unpublished) has extended the calculations to $n_{max} = 20$.

28 Dagg, I.R., G.E. Reesor and M. Wong (1978). Can. J. Phys. 56, 1096; Quazza, J., P. Marteau, M. Vu and B. Vodar, 1976. J. Quant. Spec. Rad. Transf. 16, 491.

29 Poll, J.D. and J.L. Hunt (1981). Can. J. Phys. 59, 1448.

30 Stone, N.W.B., L.A.A. Read, A. Anderson and I.R. Dagg (1984). Can. J. Phys. 62, 338.

31 Buontempo, U., S. Cunsolo, P. Dore and P. Maselli (1979). Molec. Phys. 38, 2111.

32 Birnbaum, G., L. Frommhold, L. Nencini and H. Sutter (1983). Chem. Phys. Lett. 100, 292.

MAXIMUM-ENTROPY SPECTRUM FROM A NON-EXTENDABLE AUTOCORRELATION FUNCTION

Paul F. Fougere
AFGL/LIS
Hanscom AFB, MA. 01731

INTRODUCTION

In the first "published" reference on maximum entropy spectra, an enormously influential and seminal symposium reprint, Burg (1967) announced his new method based upon exactly known, error free autocorrelation samples. Burg showed that if the first n samples were indeed the beginning of a legitimate autocorrelation function (ACF) that the next sample (n+1) was restricted to lie in a very small range. If that (n+1) sample were chosen to be in the center of the allowed range, then the sample number (n+2) would have the greatest freedom to be chosen, again in a small range. In the same paper, Burg also showed that the extrapolation to the center point of the permissable range corresponds to a maximum entropy situation in which the available data were fully utilized, while no unwarranted assumptions were made about unavailable data. In fact unmeasured data were to be as random as possible subject to the constraint that the power spectral density produce ACF values in agreement with the known, exact ACF.

Some time later it was recognized that the realization of exactly known ACF values rarely if ever occurs in practice and that the ACF is usually estimated from a few samples of the time series or from some other experimental arrangement and is therefore subject to measurement error. Thus the concept of exact matching of given ACF values was weakened to approximate matching: up to the error variance. Since then there have been several extensions of the Maximum Entropy Method (MEM) to include error prone ACF estimates. Ables (1974) suggested an extension but gave no practical method or results. The earliest practical extension was due to Newman (1977) who showed that with a slightly generalized definition, one could obtain maximum entropy spectra given noisy ACF estimates.

The Newman method appeared to work well - at least for small problems. I am very much indebted to Bill Newman for sending me a copy of his FORTRAN program which I compiled to run on the AFGL CYBER - 750 computer. The problems that Newman had written-up as test cases worked very nicely. However, when I attempted to use the method on larger problems involving noisy ACF measurements, obtained from an interferometer (an interferogram), I had difficulties. For example, if we ignore errors in the ACF and simply solve the Yule-Walker equations using the simple program of Ulrych and Bishop (1975), at order 20 the

Yule-Walker equations failed, i.e., produced a reflection coefficient greater than one and a resultant negative error power.

Now applying Newman's program, I was able to extend the valid range to 30 or so but with difficulty. The program allows an initial adjustment of the zero-order correlation (ρ_0) upward until a solution is obtained and then tries to lower this value of ρ_0 as much as possible. In order to obtain convergence for Newman's iterative procedure, increasingly large ρ_0 values were required and even so convergence was painfully slow. I then decided that it is really improper to vary the ρ_0 lag, because the entropy will increase without bound as ρ_0 is increased and thus there will be no maximum entropy solution.

Other methods have appeared in the literature, notably Schot and McLellan (1984). This paper was written in such a general fashion, to allow for application to multichannel and/or multidimensional data, that I have been unable to utilize it properly for a test. Also the program was written in "C" programming language (not spoken by the CDC Cyber): an additional roadblock; even if the authors had made the program available it would have been a great labor to convert it to FORTRAN.

In addition the paper by Schot and McLellan mentioned the use of "mean lagged products as estimates of the ACF". The use of such estimates in any connection is decidedly "anti Maximum-Entropy" in a most fundamental way. Unmeasured data are assumed to be zero! This assumption is unjustified and completely contrary to the letter and the spirit of the Maximum Entropy Principle - which requires maximum use of known input data and maximum randomness of unmeasured data.

It was then decided to search for some viable and simple alternative to the above mentioned methods.

THE NEW METHOD

At the very heart of the maximum entropy method of Burg is the Fundamental Autocorrelation Theorem. This theorem, announced in the 1967 paper and proved by Burg in his Ph.D. thesis (1975), can be stated "the first n numbers $\rho_0, \rho_1, \cdots \rho_n$ constitute the beginning of an Autocorrelation Function (ACF) if and only if the Toeplitz autocorrelation matrix is non-negative definite. If we now write down the modern Yule-Walker equations as:

$$\begin{pmatrix} \rho_0 & \rho_1 & \rho_2 & \cdots & \rho_n \\ \rho_1 & \rho_0 & & & \\ \vdots & & & & \\ \rho_n & \cdots & & \cdots & \rho_0 \end{pmatrix} \begin{pmatrix} 1 \\ g_{n1} \\ \vdots \\ g_{nn} \end{pmatrix} = \begin{pmatrix} P_{n+1} \\ 0 \\ \vdots \\ 0 \end{pmatrix} \qquad (1)$$

the square matrix is the Toeplitz autocorrelation matrix where the prediction error filter is written 1, $g_{n1}, g_{n2}, \cdots g_{nn}$ and the error power at stage n is P_{n+1}. The numbers $g_{11}, g_{22}, \ldots g_{nn}$ are the so-called reflection coefficients (due to a geophysical analogy involving seismic reflection at interfaces). The fundamental ACF theorem is completely equivalent to the statement that the first (n+1) numbers, $\rho_0, \rho_1, \cdots \rho_n$, are the beginning of an ACF if and only if the reflection coefficients: (g_{jj}, j=1, n) are all of magnitude less than or equal to unity. That is: $|g_{jj}|<1$. This extremely powerful and simple condition was the basis for the successful and optimal non-linear methods of Fougere (1977, 1978), for real and complex data respectively, which solve the line splitting and line shifting problem associated with the Burg technique. We will now proceed to derive the new technique using a method which parallels the non-linear methods exactly. For if we have been given some numbers $\rho_0, \rho_1, \cdots \rho_n$ and we use the Yule-Walker equations (see for example the extremely simple program in Ulrych and Bishop (1975)) and find any $g_{jj}>1$, we know that there are errors in some or all of the ρ's. The given numbers simply cannot be the beginning of an ACF! But if we look at our Yule-Walker equations again we can see that using the bottom row, a very simple recursion for the ACF can be derived.

$$\rho_k = -\sum_{j=1}^{k} \rho_{k-j}\ g_{kj}\ ;\ k=1,2\ \ldots\ j-1 \tag{2}$$

For example $\rho_1 = -\rho_0\, g_{11}$; $\rho_2 = -(\rho_1 g_{21} + \rho_0 g_{22})$ etc. Given the prediction error filter (PEF) and using the simple Levinson Recursion:

$$g_{jk} = g_{j-1,k} - g_{jj} g_{j-1,j-k}\ ;\quad \begin{matrix} j=2,3,\ldots,n \\ k=1,2,\ldots\ j-1 \end{matrix} \tag{3}$$

we can obtain the PEF directly from the set of reflection coefficients.

Thus we now see that as long as we begin with reflection coefficients, all of which lie in the range $-1<g_{jj}<1$ we will always get an ACF.

This condition is trivial to enforce if we simply set $g_{jj} = U\sin\Phi_j$ where U is a constant, very slightly less than 1, and Φ_j is any real angle. This is the nub of the extremely simple new method. We start off by setting all $g_{jj} = U\sin\Phi_j$ and then find the ACF given by these numbers. We then minimize the distance, R, between our new acceptable ACF, ($\hat{\rho}_k$), and the given unacceptable ACF (ρ_k), where:

$$R^2 = \sum_{k=1}^{n} (\rho_k - \hat{\rho}_k)^2 \tag{4}$$

The result will always be a legitimate ACF and the extension (from n lags to ∞ lags) via the maximum entropy method will always produce a Maximum Entropy Spectrum.

It might now be argued that we have never written down an expression for entropy and maximized it. This is true but the extension of the allowable ACF will always be a "Maximum Entropy Extension". If our original given ACF were badly in error then our new ACF will fit the old ACF, but not very well. Nevertheless, the fit will always be as good as is allowed by the given ACF when the new method is allowed to converge.

The starting guess for the iterative solution is obtained by extrapolating the last acceptable ACF of order j up to the full length of the original given ACF, using equation (5) also to be found in Ulrych and Bishop (1975):

$$\hat{\rho}_k = \begin{cases} \rho_k & ; \quad k=1,2,\ldots,m \\ \sum_{j=1}^{m} \rho_{k-j}\, g_{mj} & ; \quad k=m+1,\ldots,n \end{cases} \tag{5}$$

The method, which has been programmed in FORTRAN, utilizes the IMSL subroutine ZXSSQ, which is a non-linear least squares routine. The required subroutine ZXSSQ can be obtained from IMSL; note that many computer libraries subscribe to IMSL and have copies of ZXSSQ already.

The program is available on 9 track tape on request by seriously interested scientists. Please do not send a blank tape but do include the required tape density, either 800 or 1600 BPI and the required code, either EBCDIC or ASCII.

ACKNOWLEDGEMENT

I am very grateful to Dr. William I. Neuman for letting me use his program and to Ms. Elizabeth Galligan and Mrs. Celeste Gannon for their excellent typing of this manuscript.

REFERENCES

1 Ables, J.G. (1974). Maximum Entropy Spectral Analysis. Astron. Astrophys. Suppl. Series 15, 383-393 (1972), Proc. Symp. on the Collection and Analyses of Astrophysical Data.

2 Burg, J.P. (1967). Maximum Entropy Spectral Analysis. Reprinted in Modern Spectrum Analysis (1978), D.C. Childers (ed.), IEEE Press, N.Y.

3 Burg, J.P. (1975). Maximum Entropy Spectral Analysis. Ph.D. Thesis, Stanford University, Palo Alto, CA, 123 pp.

4 Fougere, P.F. (1977). A Solution to the Problem of Spontaneous Line Splitting in Maximum Entropy Power Spectrum Analysis. J. Geophys. Res. 82, 1051, 1054.

5 Fougere, P.F. (1978). A Solution to the Problem of Spontaneous Line Splitting in Maximum Entropy Power Spectrum Analysis of Complex Signals. Proc. RADC Spectrum Estimation Workshop, Rome, N.Y.

6 Neuman, W.I. (1977). Extension to the Maximum Entropy Method. IEEE Trans. Inform. Theory, IT-23, 89-93.

7 Schott, J.P. and J.H. McClellan (1984). Maximum Entropy Power Spectrum Estimation with Uncertainty in Correlation Measurements. IEEE Trans. ASSP, 32, 410-418.

8 Ulrych, T.J. and T.N. Bishop (1975). Maximum Entropy Spectral Analysis and Autoregressive Decomposition. Rev. Geophys. and Space Phys. 13, 183-200.

MULTICHANNEL MAXIMUM ENTROPY SPECTRAL ANALYSIS USING LEAST SQUARES MODELLING

P.A. Tyraskis (*)
Dome Petroleum Limited
P.O. Box 200
Calgary, Alberta, Canada
T2P 2H8

* Presently at: Public Petroleum Corporation of Greece S.A.
199 Kifissias Av.
Maroussi, Greece 15124

ABSTRACT

Autoregressive data modelling using the least-squares linear prediction method is generalized for multichannel time series. A recursive algorithm is obtained for the formation of the system of multichannel normal equations which determine the least-squares solution of the multichannel linear prediction problem. Solution of these multichannel normal equations is accomplished by the Cholesky factorization method. The corresponding multichannel Maximum Entropy spectrum derived from these least-squares estimates of the autoregressive model parameters is compared to that obtained using parameters estimated by a multichannel generalization of Burg's algorithm. Numerical experiments have shown that the multichannel spectrum obtained by the least-squares method provides for more accurate frequency determination for truncated sinusoids in the presence of additive white noise.

1 INTRODUCTION

Multi-channel generalizations of Burg's[1-3] now-classical algorithm for the modelling of data as an auto-regressive sequence and therefore estimation of its equivalent maximum entropy spectrum have been obtained independently by several authors (Jones[4], Nuttal[5], Strand[6], Morf et al.[7], Tyraskis[8] and Tyraskis and Jensen[9]). For single-channel data, Ulrych and Clayton[11] have also introduced an alternative procedure which is commonly described as 'the exact-least-squares method' for the estimation of the autoregressive data model parameters from which a spectrum can be directly obtained. This method has been further developed and extended and efficient recursive computational algorithms have been provided by Barrodale and Errickson[16] and Marple[17]. The exact least-squares method has been demonstrated to allow much improved spectral resolution and accuracy when compared to Burg's algorithm for single-channel time series although Burg's algorithm requires somewhat less computational time and storage. In particular, spectral estimates based upon the Burg algorithm are now known[10-15] to be susceptible to significant frequency shifting and spectral line splitting in certain circumstances while the exact least-squares method is much less affected by these problems. In their original paper, Ulrych and Clayton[11] compared the performance of their exact least-squares method for autoregressive spectral estimation with respect to these effects with the performance of Burg - MEM spectra. Using a straightforward matrix inversion solution they showed that the exact least-squares method always leads to more stable frequency estimates. Barrodale and Errickson[16] have obtained a recursive algorithm for the formation of the normal equations obtained in the exact least-squares solution for the autoregressive parameters using forward and/or backward prediction. Marple[17] introduced a recursive algorithm for the solution of the normal equations in the autoregressive data-model parameters obtained via the exact-least-squares method using both forward and backward linear prediction.

Fougere[18] has studied several of the multichannel generalizations of the Burg algorithm[4-7] and has shown that they are also susceptible to serious frequency shifting and line splitting especially in the case of sinusoidal data in the presence of white additive noise.

With the intention of relieving the unsatisfactory performance of multichannel Burg-like algorithms for multichannel data, we, here, shall generalize the exact-least-squares linear prediction method for autoregressive data modelling to multichannel time series and examine the performance of this method in spectral estimation. We shall first develop the multichannel system of normal equations using the exact-least-squares modelling and then generalize the recursive algorithm of Barrodale and Errickson[16] for multichannel data. We shall solve the multichannel normal equations so-obtained using straightforward matrix inversion procedures based upon the Cholesky factorization. Later, we will compare multichannel maximum Entropy spectra derived from the solution of the multichannel normal equations obtained using the generalizations of Burg's algorithm and the generalization of the exact-least-squares method extended to multichannel data, presented here. The related problem of the selection of the order (or the length) of the autoregressive operator for multichannel time series is discussed; three different criteria are presented. Finally we shall present the results of a comparison between a generalization of Burg's algorithm to multichannel time series[9], and the multichannel least-squares linear prediction method, when they are used for the spectral estimation of multichannel sinusoidal data in the presence of additive Gaussian noise.

2 THEORY

Given a multichannel time series X_j, j=1, ..., M of the form $X_j=[x_1, \ldots, x_k]^T$, where the superscript T denotes the transpose operation and k is the number of channels, the one point prediction convolutional formula can be described as

$$\begin{bmatrix} X_N^T & X_{N-1}^T & \cdots & X_1^T \\ X_{N+1}^T & X_N^T & \cdots & X_2^T \\ & & \vdots & \\ X_{M-1}^T & X_{M-2}^T & \cdots & X_{M-N}^T \end{bmatrix} \cdot \begin{bmatrix} G_1 \\ G_2 \\ \vdots \\ G_N \end{bmatrix} = \begin{bmatrix} X_{N+1}^T \\ X_{N+2}^T \\ \vdots \\ X_M^T \end{bmatrix} \tag{1}$$

where G_j's are the k x k matrix-valued coefficients of the one-point forward prediction filter (henceforth, PF).

It is convenient for the following analysis to rewrite this system (1) in matrix form as

$$X \cdot G = Y \tag{2}$$

where X is the (M-N)x(N.k) matrix containing the input series, G is the (N.k) x k matrix of the one-point forward PF and Y is the (M-N) x k matrix of the input series as shown in the right hand side of the system (1).

If M-N>Nk (which is usually the case) the system is overdetermined and no G generally exists which satisfies equation (2), so it is useful to define a (M-N) x k matrix-valued residual vector

$$E = Y - X \cdot G$$

having the form

$$\begin{bmatrix} \varepsilon_{N+1} \\ \varepsilon_{N+2} \\ \vdots \\ \varepsilon_M \end{bmatrix}$$

A least squares (henceforth, LS) solution to equation (2) is then defined as any G which minimizes the sum of the squares of the residuals, equivalently the trace of

$$P = P(G) = E^T E$$

The LS solution to equation (2) is given from the solution of the (N.k)x(N.k) system of normal equations[19]

$$X^T X G = X^T Y \tag{3}$$

and it is unique if the matrix for the system is of full rank (equal to Nk for equation (2), assuming that M-N>Nk). We may adopt a more concise notation and refer to G as the solution to an (N.k)x(N.k) system of equations (usually called 'normal equations')

$$R \cdot G = S \tag{4}$$

where

$$R = [r_{i,j}] = X^T X \text{ and } S = [s_i] = X^T X \ .$$

The residual sum of squares is defined as

$$P\,(G) = E^T E = (Y-XG)^T (Y-XG) =$$
$$= Y^T Y - Y^T (XG) - (G^T X^T) Y + G^T (X^T X)\, G$$

Since the LS solution G satisfies the normal equations we obtain

$$P\,(G) = Y^T Y - Y^T XG \tag{5}$$

Similarly, the backward problem is described by the system of equations

$$\bar{X}\,\bar{G} = \bar{Y} \tag{6}$$

where $\bar{X} = \begin{bmatrix} x_2^T & x_3^T & \cdots & x_{N+1}^T \\ x_3^T & x_4^T & \cdots & x_{N+2}^T \\ & & \vdots & \\ x_{M-N+1}^T & x_{M-N+2}^T & \cdots & x_M^T \end{bmatrix}$, $\bar{G} = \begin{bmatrix} \bar{G}_1 \\ \bar{G}_2 \\ \vdots \\ \bar{G}_N \end{bmatrix}$ and $\bar{Y} = \begin{bmatrix} x_1^T \\ x_2^T \\ \vdots \\ x_{M-N}^T \end{bmatrix}$,

$\bar{G}$ being the backward PF.

The LS solution to equations (6) is characterized by the (N.k) x (N.k) system of normal equations

$$\bar{R} \cdot \bar{G} = \bar{S} \tag{7}$$

where $\bar{R} = (\bar{r}_{i,j}) = \bar{X}^T \bar{X}$ and $\bar{S} = (\bar{s}_i) = \bar{X}^T \bar{X}$

Also, the residual sum of squares using the backward PF is obtained as follows:

$$\bar{P}\,(\bar{G}) = \bar{Y}^T \bar{Y} - \bar{Y}^T \bar{X}\,\bar{G} \tag{8}$$

It is commonly known that the LS solution of an overdetermined system can lead to nontrivial difficulties since the matrices R and $\bar{R}$ of the normal equations (4, 7) can be difficult to invert accurately. The book by Lawson and Hanson[20] provides a very detailed coverage of this topic, complete with overall useful computer programs. Let us briefly summarize the current developments for the LS computations.

In general, for an overdetermined system of n equations in m unknown parameters, forming the normal equations requires $nm^2/2$ operations (an operation is a multiplication or division plus an addition or subtraction, and when comparing operation counts, only the highest powers of n and m are given) and solving them by the Cholesky method[20] requires $m^3/6$ operations. However, numerical analysts generally prefer to avoid this approach, since forming the normal equations considerably worsens the numerical condition of any LS problem for which the corresponding overdetermined system of equations is itself ill-conditioned.

Instead, it is usually advised that the LS solution be computed directly from the n x m overdetermined system by an orthogonalization method. Some popular algorithms of this type are the modified Gram-Schmidt process (nm^2 operations), Householder triangularization ($nm^2 - m^3/3$ operations), and singular value decomposition (at least $nm^2 + 5m^3/3$ operations)[20]. In a given precision floating-point arithmetic, these orthogonalization methods successfully process a wider class of LS problems than the normal equation algorithm.

For the applications in which we are primarily interested, it is frequently the case that $n>10m$ in which case the normal equations method involves less operations than any orthogonalization method. In addition, if the normal equations can be formed directly from the data rather than from an overdetermined system, then the normal equations algorithm requires only about $m(m+1)/2$ storage locations.

For the examples later in this paper, we use the normal equations method and we solved them using the Cholesky factorization (or decomposition) method which is known to possess remarkable numerical stability[21].

According to this method, for the inversion of any positive definite matrix, it is convenient to factor it into the product of a lower triangular matrix and its transpose, since it is a simpler matter to invert a triangular matrix. Use of the Cholesky factorization method can be made, since any normal equations matrix is positive definite provided the corresponding overdetermined system of equations is of full rank[20].

For the formation of the normal equations, the algorithm proposed by Barrodale and Erickson[16] can be used, provided that it is extended to account for multichannel time series.

Before summarizing the algorithm we first show how $\mathcal{R}_3$ and $\mathcal{S}_3$ (corresponding to N = 3) can be obtained from $\mathcal{R}_2$ and $\mathcal{S}_2$ (corresponding to N = 2). In the case when N = 2 we have

$$R_2 = X^TX = \begin{bmatrix} r^2_{1,1} & r^2_{1,2} \\ r^2_{2,1} & r^2_{2,2} \end{bmatrix} = \begin{bmatrix} \sum\limits_{j=1}^{M-2} X_{j+1}.X^T_{j+1} & \sum\limits_{j=1}^{M-2} X_{j+1}.X^T_{j} \\ \sum\limits_{j=1}^{M-2} X_{j} \,.\, X^T_{j+1} & \sum\limits_{j=1}^{M-2} X_{j} \,.\, X^T_{j} \end{bmatrix} = \begin{bmatrix} {}^{M-2}_{\ \ 2}R(0) & {}^{M-2}_{\ \ 1}R(-1) \\ {}^{M-2}_{\ \ 1}R(1) & {}^{M-2}_{\ \ 1}R(0) \end{bmatrix}$$

$$S_2 = X^TY = \begin{bmatrix} s^2_1 \\ s^2_2 \end{bmatrix} = \begin{bmatrix} \sum\limits_{j=1}^{M-2} X_{j+1}\, X^T_{j+2} \\ \sum\limits_{j=1}^{M-2} X_{j}\, X^T_{j+2} \end{bmatrix} = \begin{bmatrix} {}^{M-2}_{\ \ 2}R(1) \\ {}^{M-2}_{\ \ 1}R(2) \end{bmatrix}$$

where $${}^{L}_{K}R(t) = \sum_{j=1}^{L} X_{j+(K-1)}\, X^T_{j+(K-1)+t}$$

is an estimate of the multichannel autocorrelation coefficient at lag t, using a finite summation of L terms, with starting term involving the input vector series at time index K, and r and S are variables for referring to the elements of the R and S matrices.

In the case when N = 3, and suppressing the elements above the diagonal (since ${}^{L}_{K}R(-t) = {}^{L}_{K}R^T(t)$), we have

$$R_3 = \begin{bmatrix} r^3_{1,1} & . & . \\ r^3_{2,1} & r^3_{2,2} & . \\ r^3_{3,1} & r^3_{3,2} & r^3_{3,3} \end{bmatrix} = \begin{bmatrix} \sum\limits_{j=1}^{M-3} X_{j+2}\, X^T_{j+2} & . & . \\ \sum\limits_{j=1}^{M-3} X_{j+1}\, X^T_{j+2} & \sum\limits_{j=1}^{M-3} X_{j+1}\, X^T_{j+1} & . \\ \sum\limits_{j=1}^{M-3} X_j . X^T_{j+2} & \sum\limits_{j=1}^{M-3} X_j\, X^T_{j+1} & \sum\limits_{j=1}^{M-3} X_j . X^T_j \end{bmatrix}$$

$$= \begin{bmatrix} {}^{M-3}_{\ \ 3}R(0) & . & . \\ {}^{M-3}_{\ \ 2}R(1) & {}^{M-3}_{\ \ 2}R(0) & . \\ {}^{M-3}_{\ \ 1}R(2) & {}^{M-3}_{\ \ 1}R(1) & {}^{M-3}_{\ \ 1}R(0) \end{bmatrix}$$

Notice that the leading 2 x 2 submatrix R_3 can be obtained from R_2 by subtracting the first term in each summation which defines an element of R_2, i.e.,

$$r^3_{i,j} = r^2_{i,j} - X_{3-i} X^T_{3-j} \quad \text{for } 1 \leq i \leq 2$$

The last row of R_3, apart from its first element, can be obtained from the last row of R_2 as follows,

$$r^3_{3,j} = r_{2,j} - X_{M-2} X^T_{M+1-j} \quad \text{for } 2 \leq j \leq 3$$

and
$$r^3_{3,1} = s^2_2 - X_{M-2} X^T_M$$

Finally, S_3, apart from its last element, can be obtained from S_2 as follows

$$s^3_i = s^2_i - X_{3-i} X^T_3 \qquad \text{for } 1 \leq i \leq 2$$

and
$$s^3_3 = \sum_{j=1}^{M-3} X_j X^T_{j+3} .$$

Thus only one inner product S^3_3, has to be calculated again when we derive the normal equations for N = 3 from the normal equations for N = 2; each of the remaining elements of R_3 and S_3 is obtained at the cost of one operation.

This scheme generalizes so that R_{N+1} and S_{N+1} can be obtained from R_N and S_N as follows:

$$r^{N+1}_{i,j} = r^N_{i,j} - X_{N+1-i} X^T_{N+1-j} \quad \text{for } 1 \leq j \leq i \leq N,$$
$$r^{N+1}_{N+1,j} = r^N_{N,j-1} - X_{M-N} X^T_{M+1-j} \quad \text{for } 2 \leq j \leq N+1,$$
$$r^{N+1}_{N+i,1} = s^N_N - X_{M-N} X^T_M$$
$$s^{N+1}_i = s^N_i - X_{N+1-i} X_{N+1}^T \quad \text{for } 1 \leq i \leq N,$$

and
$$s^{N+1}_{N+1} = \sum_{j=1}^{M-N1} X_j X^T_{N+1+j} .$$

For the better understanding of the relation between multichannel linear prediction and the corresponding maximum entropy spectrum using exact least-squares modelling with the maximum entropy spectral analysis using generalized Burg algorithms, it is useful to extend the present analysis one step further.

In the case when N = 3 the residual sum of squares $P(G)$ can be easily shown to take the form

$$P(G) = Y^TY - Y^TXG = Y^TY - (X^TY)^TG$$

$$= {}^{M-3}_{4}R(0) - {}^{M-3}_{3}R(-1).G_1 - {}^{M-3}_{2}R(-2).G_2 - {}^{M-3}_{1}R(-3).G_3$$

The forward system of normal equations (4), using the analytical form of R and S, and augmented by the equation of the residual sum of squares can be written as follows

$$\begin{bmatrix} {}^{M-3}_{4}R(0) & {}^{M-3}_{3}R(-1) & {}^{M-3}_{2}R(-2) & {}^{M-3}_{1}R(-3) \\ {}^{M-3}_{3}R(1) & {}^{M-3}_{3}R(0) & {}^{M-3}_{2}R(-1) & {}^{M-3}_{1}R(-2) \\ {}^{M-3}_{2}R(2) & {}^{M-3}_{2}R(1) & {}^{M-3}_{2}R(0) & {}^{M-3}_{1}R(-1) \\ {}^{M-3}_{1}R(3) & {}^{M-3}_{1}R(2) & {}^{M-3}_{1}R(1) & {}^{M-3}_{1}R(0) \end{bmatrix} \cdot \begin{bmatrix} I \\ -G_1 \\ -G_2 \\ -G_3 \end{bmatrix} = \begin{bmatrix} P_4 \\ 0 \\ 0 \\ 0 \end{bmatrix}$$

In general for the forward system of order N we have,

$$\begin{bmatrix} {}^{M-N}_{N+1}R(0) & {}^{M-N}_{N}R(1)\ldots & {}^{M-N}_{2}R(-N+1) & {}^{M-N}_{1}R(-N) \\ {}^{M-N}_{N}R(1) & {}^{M-N}_{N}R(0)\ldots & {}^{M-N}_{2}R(-N+2) & {}^{M-N}_{1}R(-N+1) \\ & \vdots & & \\ {}^{M-N}_{2}R(N-1) & {}^{M-N}_{2}R(N-2)\ldots & {}^{M-N}_{2}R(0) & {}^{M-N}_{1}R(-1) \\ {}^{M-N}_{1}R(N) & {}^{M-N}_{1}R(N-1)\ldots & {}^{M-N}_{1}R(1) & {}^{M-N}_{1}R(0) \end{bmatrix} \cdot \begin{bmatrix} I \\ -G_1 \\ \vdots \\ -G_{N-1} \\ -G_N \end{bmatrix} = \begin{bmatrix} P_{N+1} \\ 0 \\ \vdots \\ 0 \\ 0 \end{bmatrix} \quad (9a)$$

where I, $-G_1$, ..., $-G_N$ is the (N+1)th order forward PEF.

Following the same procedure for the backward problem we obtain

$$\begin{bmatrix} {}^{M-N}_{N+1}R(0) & {}^{M-N}_{N}R(-1)\ldots & {}^{M-N}_{2}R(-N+1) & {}^{M-N}_{1}R(-N) \\ {}^{M-N}_{N}R(1) & {}^{M-N}_{N}R(0)\ldots & {}^{M-N}_{2}R(-N+1) & {}_{1}R(-N+1) \\ & \vdots & & \\ {}^{M-N}_{2}R(N-1) & {}^{M-N}_{2}R(N-2)\ldots & {}^{M-N}_{2}R(0) & {}^{M-N}_{1}R(-1) \\ {}^{M-N}_{1}R(N) & {}^{M-N}_{1}R(N-1)\ldots & {}^{M-N}_{1}R(1) & {}^{M-N}_{1}R(0) \end{bmatrix} \cdot \begin{bmatrix} -\bar{G}_N \\ -G_{N-1} \\ \vdots \\ -G_1 \\ I \end{bmatrix} = \begin{bmatrix} 0 \\ 0 \\ \vdots \\ 0 \\ \bar{P}_{N+1} \end{bmatrix} \quad (9b)$$

where $I, -G_1, \ldots -G_N$ is the (N+1)th order backward PEF. The matrix equations (9a,b) can be written together.

$$\begin{bmatrix} {}^{M-N}_{N+1}R(0) & {}^{M-N}_{N}R(-1) & \ldots & {}^{M-N}_{2}R(-N+1) & {}^{M-N}R(-N) \\ {}^{M-N}_{N}R(1) & {}^{M-N}_{N}R(0) & \ldots & {}^{M-N}_{2}R(-N+2) & {}^{M-N}_{1}R(-N+1) \\ & & \vdots & & \\ {}^{M-N}_{2}R(N-1) & {}^{M-N}_{2}R(N-2) & \ldots & {}^{M-N}_{2}R(0) & {}^{M-N}_{1}R(-1) \\ {}^{M-N}_{1}R(N) & {}^{M-N}_{1}R(N-1) & \ldots & {}^{M-N}_{1}R(1) & {}^{M-N}_{1}R(0) \end{bmatrix} \cdot \begin{bmatrix} I & -\bar{G}_N \\ -G_1 & -\bar{G}_{N-1} \\ \vdots & \vdots \\ -G_{N-1} & -\bar{G}_1 \\ -G_N & I \end{bmatrix} = \begin{bmatrix} P_{N+1} & 0 \\ 0 & 0 \\ \vdots & \vdots \\ 0 & 0 \\ 0 & P_{N+1} \end{bmatrix} \qquad (10)$$

This last system of equations (10) is similar to, but not identical with, the usual multichannel prediction-error system (see Appendix, A1), whose autocorrelation matrix has block-Toeplitz form. In this case the elements on each diagonal of the autocorrelation matrix are not equal resulting in a non-Toeplitz matrix. Each diagonal of the autocorrelation matrix contains estimates of the autocorrelation coefficients at the same lag, and although the order of the summation is the same, the starting data point in each summation is different resulting in a different estimate for every element of each diagonal. As the length of the input series M becomes much greater than the length of the PF, (N), and also much greater than the longest period of the input data, the contribution of the end points of the input series in the estimation of the autocorrelation matrix becomes less significant and the resultant spectra become practically identical to the spectra obtained from the solution of the usual multichannel prediction error system using any of the generalized Burg algorithms.

In the case of short-length data series the contribution of the end-points does become significant and where the Burg-like multichannel algorithms often produce appreciable line shifting effects the LS linear prediction method is much less sensitive. Line shifting effects using the LS method have only been observed in the spectral estimation of data for which the true mean of the signal is unknown and from which sample mean has been removed. Even in these cases the order of the frequency shift using the LS method is always much smaller than the frequency shift observed using multichannel generalizations of the Burg algorithm[22].

3 THE MULTICHANNEL MAXIMUM ENTROPY SPECTRUM

A principal use of the solution of the multichannel prediction error system using either generalized Burg algorithms or the LS modelling method, is for the maximum entropy spectral estimation.

Given a multichannel time series X_1, X_2, ..., X_M, the forward PEF and the corresponding prediction error power P_{N+1} (or residual sum of squares), we may obtain the spectral density matrix as (see, for example, (22)),

$$S(f) = G^{-1}(f) \cdot P_{N+1} \cdot (G^{\dagger}(f))^{-1} \tag{11a}$$

where G(f) is the Fourier transform of the forward PEF I, $-G_1$, ..., $-G_N$, and † denotes the complex conjugate transpose operation. Since the direction of time has no fundamental importance in spectral analysis of stationary time-series data, the maximum entropy spectrum can be derived in terms of the backward PEF and the backward prediction error power Q_{N+1} as

$$S(f) = \bar{G}^{-1}(f) \cdot Q_{N+1} \cdot (\bar{G}^{\dagger}(f))^{-1} \tag{11b}$$

where $\bar{G}(f)$ is the Fourier transform of the backward PEF ($-\bar{G}_N$, ..., $-\bar{G}_1$, I).

The estimation of the multichannel maximum entropy spectral density (11a,b) reduces to the determination of the appropriate length of the PEF, N or equivalently to the determination of the order of the corresponding autoregressive process that would best fit the given multichannel time series.

To select the order of the autoregressive model, Akaike[23] introduced the final prediction error criterion (FPE) and later generalized it for multichannel time series[24]. This criterion chooses the order of the autoregressive model so that the average error for a one-step prediction is minimized. Following our notation this criterion obtains minimization of the measure

$$(FPE)_N = |P_N| \cdot \left[\frac{M + 1 + kN}{M - 1 - kN}\right]^k$$

through selection of the length of the PEF, N. Here, M is the length of the input time series, k the number of channels of the multichannel time series, and P_N is the determinant of the prediction error power or equivalently the determinant of the residual sum of squares in the one-point prediction problem.

Akaike[25], obtained a second criterion based upon the maximization of the likelihood function. Given a sample of size M

$$x_1, x_2, \ldots, x_M$$

that have been drawn from a population characterized by AR parameters a_1, a_2, ..., a_N; N<M the likelihood function L (a_1, ..., a_N | x_1, ..., x_M) is equal to the joint probability density function for the x, P (x_1, ..., x_M | a_1, ..., a_N).

The identification procedure based on the maximization of the likelihood function is referred to as Akaike's information criterion (AIC) and has the form:

$$(AIC)_N = 2 \ln [\max(L(a_1, \ldots, a_N \mid x_1, \ldots, x_M))] + 2N.$$

Ulrych and Bishop[26] show that the likelihood function for a series that is generated by an autoregressive process of order N is a function of the residual sum of squares. Thus, for multichannel autoregressive time series with Gaussian errors the (AIC) criterion reduces to

$$(AIC)_N = M \ln |P_N| + 2k^2N.$$

For both (FPE) and (AIC) criteria, the order selected is the value of N for which they are minimized. The two criteria are asymptotically equivalent, and tests using both have shown that they almost invariably select the same order[27] for the autoregressive model of the data.

Parzen[28] introduced a third method known as the autoregressive transfer function criterion (CAT) and later also generalized if for multichannel time series[29]. According to this criterion the order of the autoregressive model is determined when the estimate of the difference of the mean square error between true, infinite length filter, which must exactly provide the true prediction-error power as residual and the estimated finite length filter is minimized. Parzen[29] showed that this difference can be estimated without explicitly knowing the exact infinite filter, rather replacing it with the current longest (N-length) filter available on

$$(CAT)_N = \mathrm{trace}\left[\frac{k}{M}\sum_{j=1}^{N}\frac{M-jk}{M}\,P_j^{-1} - \frac{M-jk}{M}\,P_N^{-1}\right]$$

The (AIC) and (CAT) criteria have been tested on synthetic Gaussian data sets and found to produce their minimum at the same value N[29].

Lander and Lacoss[30] have also studied the behavior of the three criteria (FPE, AIC and CAT) when applied to harmonic analysis of single-channel time series. They have reached the conclusion that all criteria give orders which produce acceptable spectra in the case of low levels of additive noise although they all underestimate the order for high noise levels.

4 NUMERICAL EXPERIMENTS USING SYNTHETIC DATA SETS

The Burg algorithm for the autoregressive spectral estimation of single-channel time series data is known to yield poor results for sinusoidal signals in presence of additive white noise under certain conditions.

Chen and Stegan[10], Ulrych and Clayton[11], Fougere et al.[12], Swingler[14], have noted: (a) a line splitting effect and (b) a frequency shifting effect of the spectral line. Line splitting effect is the occurrence of two or more closely-spaced peaks, in an autoregressive spectral estimate where only one spectral peak should be present. Frequency shifting effect is the bias in the positioning of a spectral peak with respect to the true frequency location of that peak.

As demonstrated by Chen and Stegan, for data consisting of 15 points of a sinusoid sampled 20 times per second with 10% additive white noise, the line shifting effect as a function of initial phase reached its maximum when the initial phase was an odd multiple of 45°.

Fougere[18] has studied the multichannel generalizations of the Burg algorithm (4) - (7) and has shown that they also produce line shifting for multichannel sinusoidal data in white additive noise.

Figure 1a. Frequency of the spectral peak as a function of the length of the data using Tyraskis algorithm with mean of the data sample removed, symbol (x), and with mean not removed, symbol (+). Initial phase in channels #1 and #2 is 45° and 135°.

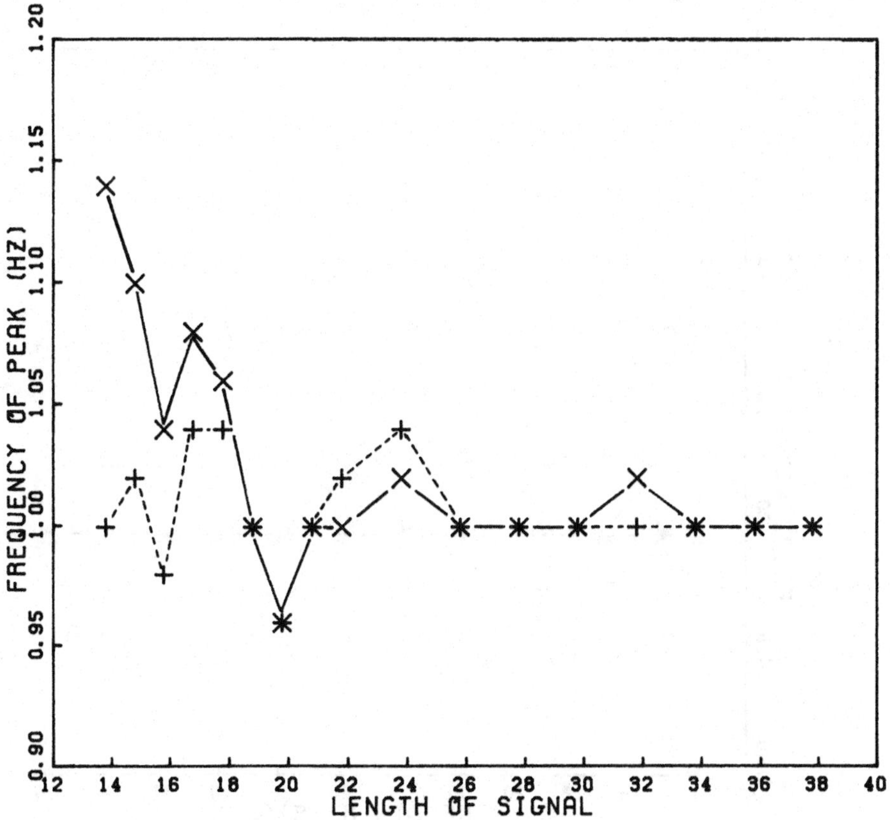

To compare the behavior of the multichannel generalization of the Burg algorithm with the LS method presented here, we shall use synthetic data similar to that employed by Fougere[18] in his analysis. For the multichannel generalization of the Burg algorithm, we use Tyraskis' (8) - (9) algorithm. An overview of this algorithm is presented in the Appendix.

The 3-channel input series consists of a single 1Hz sine wave sampled every 0.05 seconds to which has been added pseudo-gaussian white noise:

$$f_1(n) = \sin[(n-1)\pi/10 + \phi_1] + g_1(n)$$
$$f_2(n) = \sin[(n-1)\pi/10 + \phi_2] + g_2(n)$$
$$f_3(n) = g_3(n).$$

The ϕ_1 and ϕ_2 are the initial phase of each channel and $g_1(n)$, $g_2(n)$, $g_3(n)$ are independent realizations of a zero mean pseudo-gaussian process with standard deviation 0.01 units of amplitude.

Figure 1b. Frequency of the spectral peak as a function of the length of the data using the LS method with mean of the data sample removed, symbol (x), and with mean not removed (+). Initial phase in channels #1 and #2 is 45° and 135°.

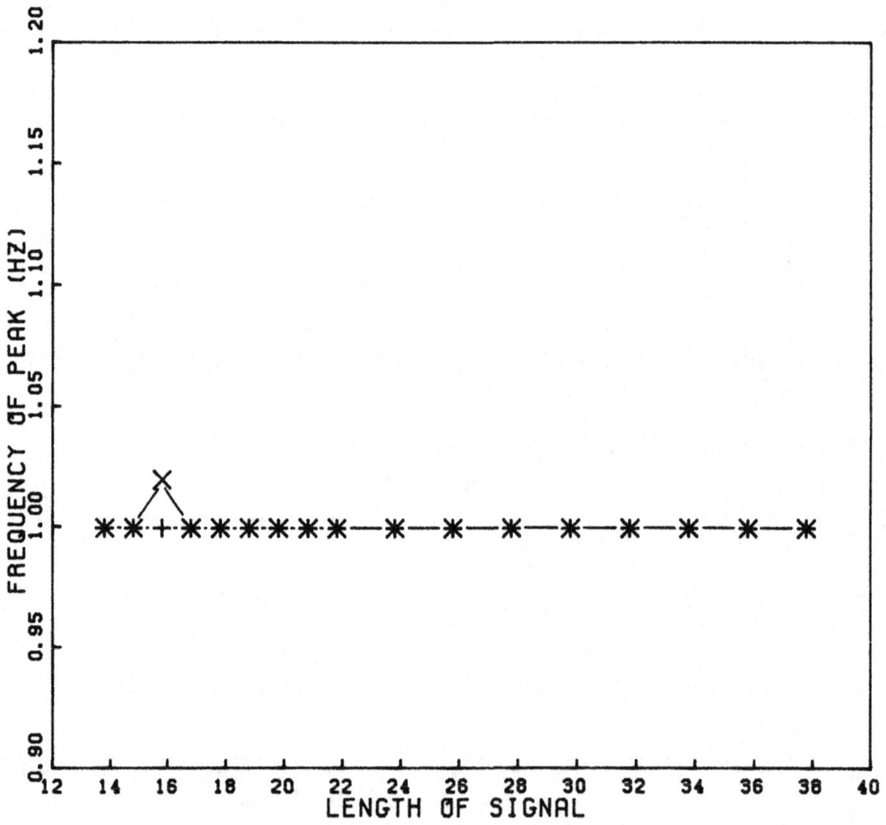

In the first case, we study the location of the spectral peak as a function of the length of the data with initial phases in channel #1 $\phi_1 = 45°$ and in channel #2, $\phi_2 = 135°$. Figure 1a shows the location of the spectral peak using the multichannel generalization of the Burg algorithm when the mean of the data sample has not been removed

(symbol, +). Figure 1b shows the location of the spectral peak using the LS method when the mean of the data sample has been removed (symbol, x) and when the mean of the data sample has not been removed (symbol, +). Note that both procedures place the spectral peaks at identical frequencies for channels #1 and #2 even though different noise realizations were used in the two channels.

Similarly, Figures 2a and 2b show the results of the two procedures when the initial phases in channel #1 is -45° and in channel #2 is -135°.

Figure 2a. Frequency of the spectral peak as a function of the length of the data using Tyraskis algorithm with mean of data sample removed, symbol (x), and with mean not removed, symbol (+). Initial phase in channels #1 and #2 is -45° and -135°.

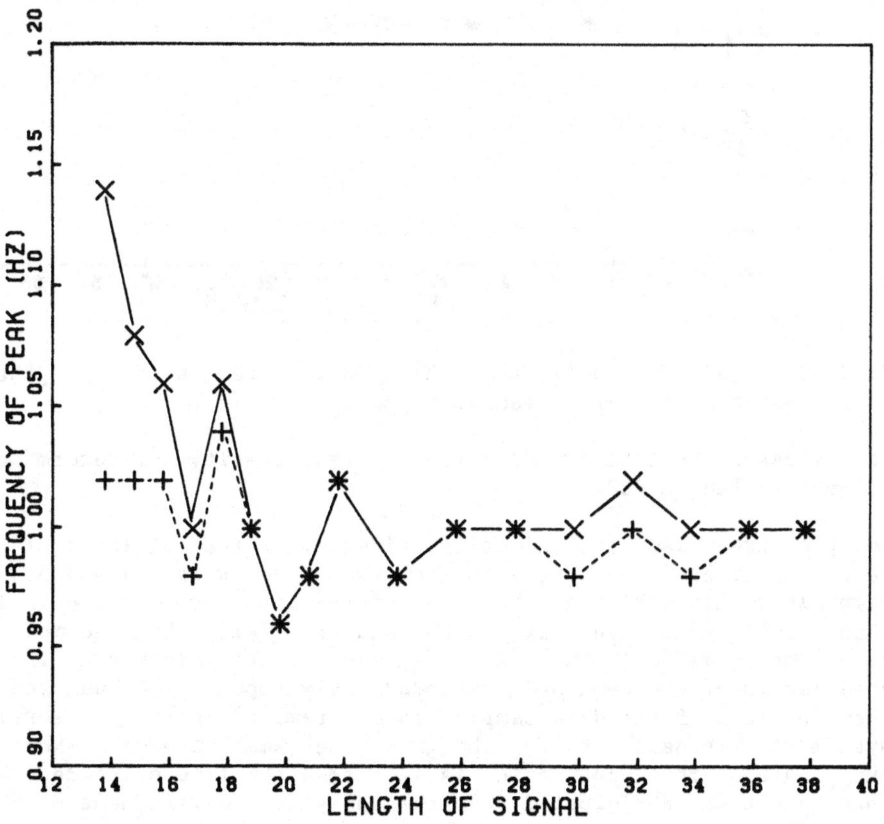

Figure 2b. Frequency of the spectral peak as a function of the length of the data using the LS method with mean of the data sample removed, symbol (x) and with mean not removed (+). Initial phase in channels #1 and #2 is -45° and -135°.

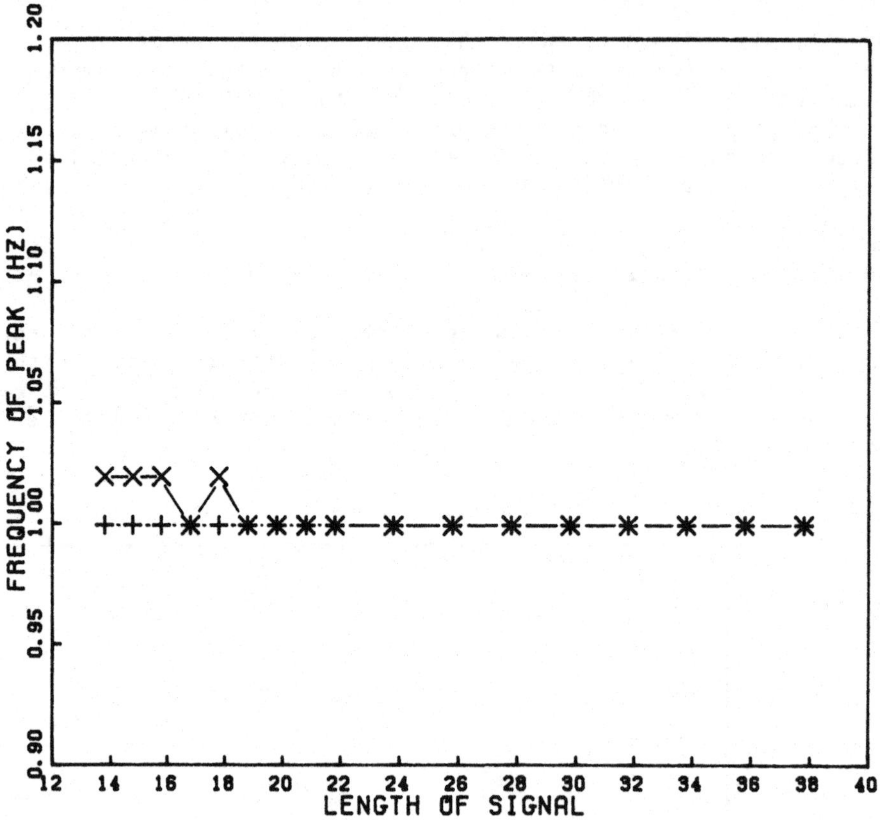

Finally, Figures 3a and 3b show the results when the initial phases are 0° and 90° for channel #1 and #2 respectively.

In all cases the estimation of the spectral density was determined using a 2-points length PEF.

For the three cases studied it is evident that the multichannel generalization of the Burg algorithm (see Figures 1a, 2a and 3a) produce significant line shifting when the length of the data sample is less than a full cycle (i.e., 21 points) and especially when the mean of the data sample has been removed. In this case, the frequency shift observed is of the same order as previously reported by Fougere[18]. When the mean of the data sample was not removed again we observe a systematic frequency shift although of much smaller order. Note that as the length of the input series is increased the line shift decreases; this result was anticipated from the analytical development of the two methods as presented previously. The location of the spectral peak using the LS method (see Figures 1b, 2b and 3b) was shown to be much less sensitive with respect to the length of the data sample. When the mean of the data sample was not removed the location of the spectral peak was always found at exactly 1Hz. When the mean of the data sample was removed the LS method sometimes produced slight line shifting when the length of the data was much less than a full cycle. Generally it was observed that the spectral line widths obtained using the LS method were extraordinarily narrower than those obtained using the generalized Burg algorithms.

Channel #3 contained only pseudo-gaussian white noise. However, the spectral density estimates always showed a small spectral peak at the same frequency as those peaks in channels #1 and #2 in all cases and with both methods. This is the effect of 'cross-talk' between the channels which has been previously observed by Fougere[18]. However, the amplitude of the 'cross-talk' peak in channel #3 is so much smaller

Figure 3a. Frequency of the spectral peak as a function of the length of the data using Tyraskis algorithm with mean of data sample removed, symbol (x), and with mean not removed (+). Initial phase in channels #1 and #2 is 0° and 90°.

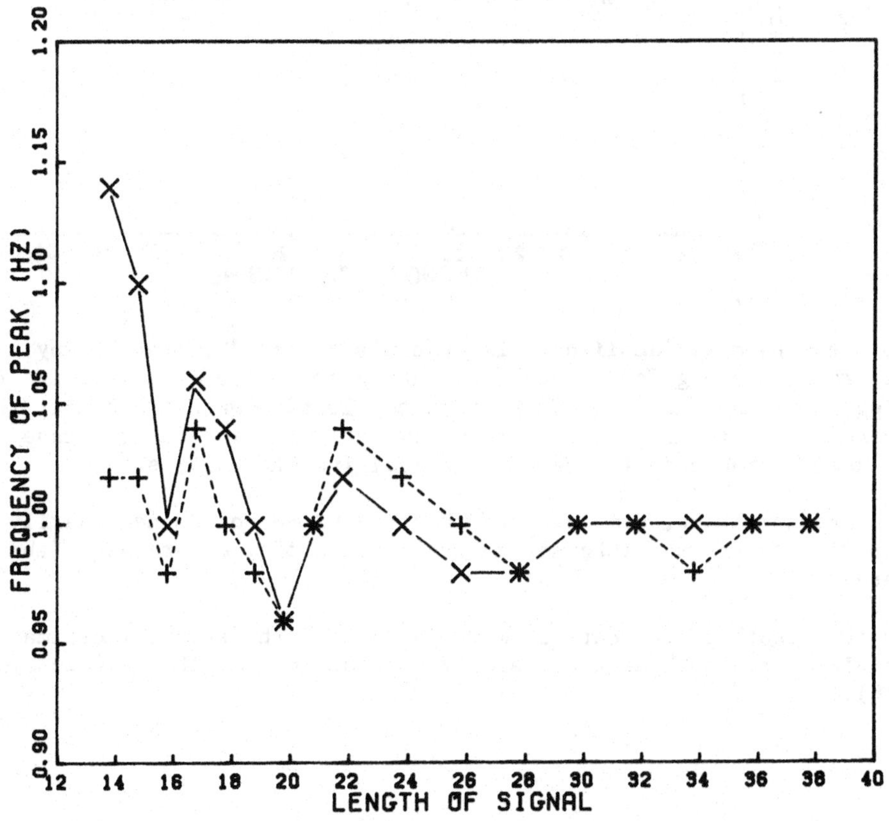

Figure 3b. Frequency of the spectral peak as a function of the length of the data using the LS method with mean of the data sample removed, symbol (x), and with mean not removed (+). Initial phase in channels #1 and #2 is 0° and 90°.

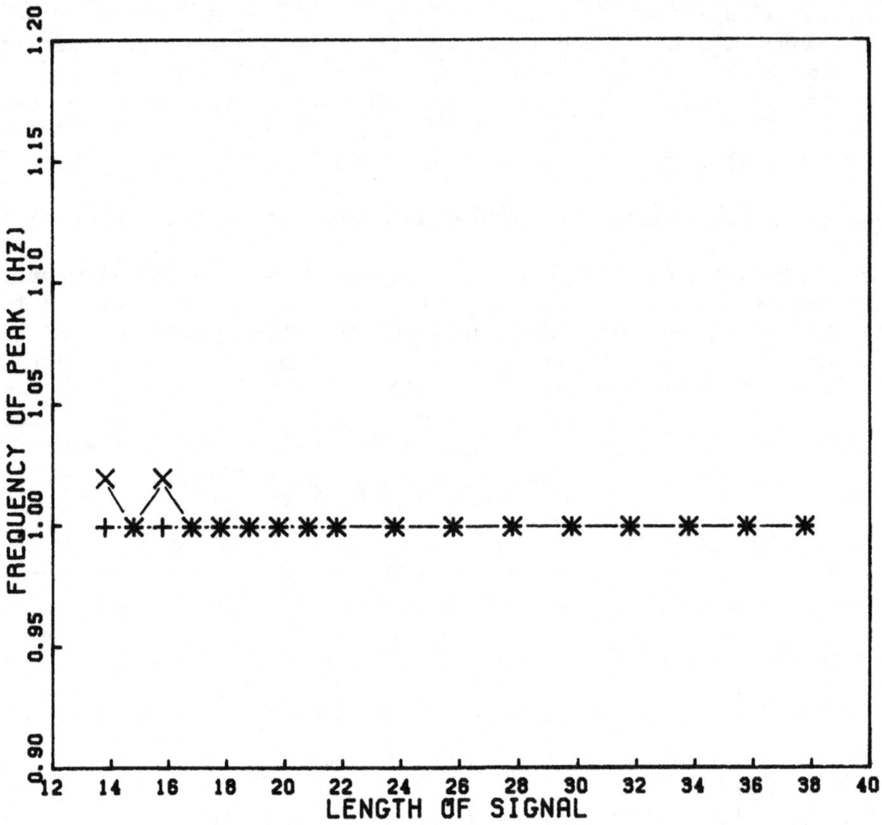

than the harmonic amplitudes in channels #1 and #2, that it may be regarded as a negligible problem. Using the above data, no line splitting has been observed. These results clearly demonstrate that the LS method leads to more stable frequency determination for the case of truncated sinusoids in presence of additive white noise.

We have not, here, demonstrated that the LS method is necessarily superior to the multichannel generalization of the Burg algorithm in all cases.

As the length of the data is much increased, the spectral estimates obtained from both methods usually approach one another. (See also: (16).)

5 CONCLUSIONS

We have presented a generalization of the 'exact least-squares' linear prediction method of Ulrych and Clayton[11] for multi-channel time series. We have further described a method of solution of the systems of normal equations derived in forward and backward prediction based upon the Cholesky factorization method. Employing synthetic data with additive white noise, we compared the multichannel Maximum Entropy spectra obtained by this method to those obtained using autoregressive parametric estimates as determined by Tyraskis'[8,9] generalization of Burg's[1,2,3] recursive algorithm. We showed that the spectra obtained via the new 'exact least-squares' formation were better than those derived from the Tyraskis estimates in two ways:

(1) The exact least-squares modelling provided much greater accuracy in the determination of the frequency of harmonic components in short data sets.

(2) Spectral estimates obtained by this new method were more stable; that is, the estimated power densities of harmonic components were much less dependent on the length of the data sequence.

Fougere[18], in his analysis of a multichannel data sets having similar properties to those employed in this analysis, showed that other multichannel generalizations of Burg's algorithm also lead to frequency-shift effects and unstable power density estimates. We, therefore, believe that this generalized exact least squares method offers a material advantage in the spectral and autoregressive parametric analysis of multichannel geophysical data. For short data sequences where this method shows its greater advantage, the additional computational costs involved in the 'exact least-squares' solution will not often prove to be significant.

REFERENCES

1 Burg, J.P. (1967). Maximum entropy spectral analysis, paper presented at the 37th Annual International Meeting, Soc. of Explor. Geophys., Oklahoma City, Oklahoma.

2 Burg, J.P. (1968). A new analysis technique for time series data, paper presented at Advanced Study Institute on Signal Processing, NATO, Netherlands.

3 Burg, J.P. (1975). Maximum entropy spectral analysis, Ph.D. dissertation, Stanford Univ., Stanford, California.

4 Jones, R.H. (1976). Multivariate Autoregression Estimation Using Residuals, Applied Time Series Analysis, D.F. Findley (ed.), Academic Press, N.Y.

5 Nuttal, A.H. (1976). Multivariate linear prediction spectral analysis employing weighted forward and backward averaging: A generalization of Burg's algorithm, Naval Underwater System Center, NUSC Tech. Doc. 5501, New London, CT.

6 Strand, O.N. (1977). Multichannel complex maximum (autoregressive) spectral analysis, IEEE Trans. Automat. Contr., Vol. AC-22, pp. 634-640.

7 Morf, M., A. Vieira, D.T.L. Lee, and T. Kailath (1978). Recursive multichannel maximum entropy spectral analysis estimation, IEEE Trans. on Geoscience Electronics, Vol. GE-16, No. 2, pp. 85-94.

8 Tyraskis, P.A. (1979). Multichannel Autoregressive Data Modelling in Geophysics, M.Sc. Thesis, McGill University, Montreal, Canada.

9 Tyraskis, P.A., and O.G. Jensen (1983). Multi-channel Autoregressive Data Models, IEEE Trans. Geosci. Remote Sensing, Vol. GE-21, No. 4, pp. 454-467.

10 Chen, W.Y. and G.R. Stegan (1974). Experiments with Maximum Entropy Power Spectra for Sinusoids, J. Geophys. Res., pp. 3019-3022.

11 Ulrych, T.J. and R.W. Clayton (1976). Time Series Modelling and Maximum Entropy, Phys. Earth Planet. Int., Vol. 12, pp. 188-200.

12 Fougere, P.F., E.J. Zawalick, and H.R. Radoski (1976). Spontaneous Line Splitting in Maximum Entropy Power Spectrum Analysis, Phys. Earth Planet. Int., Vol. 12, pp. 201-207.

13 Fougere, P.F. (1977). A Solution to the Problem of Spontaneous Line Splitting in Maximum Entropy Power Spectrum Analysis, J. Geophys. Res., Vol. 82, pp. 1052-1054.

14 Swingler, D.N. (1979). A comparison between Burg's maximum entropy method and a nonrecursive technique for the spectral analysis of deterministic signals, J. Geophys. Res., Vol. 84, pp. 679-685.

15 Kane, R.P. and N.B. Trivedi (1982). Comparison of maximum entropy spectral analysis (MESA) and least-squares linear prediction (LSLP) methods for some artificial samples, Geoph., Vol. 47, No. 12, pp. 1731-1736.

16 Barrodale, I., and R.E. Errickson (1980). Algorithms for least-squares linear prediction and maximum entropy spectral analysis - Part 1: Theory, Geoph., Vol. 45, No. 3, pp. 420-432.

17 Marple, S.L. (1980). A new autoregressive spectrum analysis algorithm, IEEE Trans. Acoust., Speech, Signal processing, Vol. ASSP-25, No. 5, pp. 423-428.

18 Fougere, P.F. (1981). Spontaneous line splitting in multichannel maximum entropy power spectra, Proceedings of the 1st ASSP workshop on spectral estimation, McMaster Univ., Hamilton, Ontario.

19 Claerbout, J.F. (1976). Fundamentals of Geophysical Data Processing, McGraw-Hill, New York.

20 Lawson, C.L., and R.J. Hanson (1974). Solving Least Squares Problems, Prentice-Hall, Inc., New Jersey.

21 Martin, R.S., G. Peters, and J.H. Wilkinson (1971). Symmetric decomposition of a positive definite matrix, In Handbook for automatic computation, Vol. II: J.H. Wilkinson and C. Reinsch (eds.), New York, Springer-Verlay, pp. 9-30.

22 Tyraskis, P.A. (1982). New Techniques in the analysis of geophysical data modelled as a multichannel autoregressive random process, Ph.D. Thesis, McGill University, Montreal, Canada.

23 Akaike, H. (1969). Fitting autoregressions for prediction, Ann. Inst. Statist. Math., Vol. 21, pp. 243-247.

24 Akaike, H. (1971). Autoregressive model fitting for control, Ann. Inst. Statist. Math., Vol. 23, pp. 163-180.

25 Akaike, H. (1974). A new look at the Statistical Model Identification, IEEE Trans. Autom. Control, Vol. AC-19, pp. 716-723.

26 Ulrych, T.J., and T.N. Bishop (1975). Maximum entropy spectral analysis and autoregressive decomposition, Rev. Geophys. Space Phys., Vol. 13, pp. 183-200.

27 Jones, R.H. (1974). Identification and Autoregressive Spectrum Estimation, IEEE Trans. Autom. Control, Vol. AC-19, pp. 894-897.

28 Parzen, E. (1974). Some Recent Advances in Time Series Modelling, IEEE Trans. Autom. Control, Vol. AC-19, pp. 723-730.

29 Parzen, E. (1977). Multiple Time Series: Determining the order of approximating autoregressive schemes in Multivariate Analysis - IV: P.R. Krishnaiah (ed.), pp. 283-295, New York, Academic Press.

30 Landers, T., and R.T. Lacoss (1977). Some geophysical applications of autoregressive spectral estimates, IEEE Trans. Geoscience Electronics, Vol. GE-15, pp. 26-32.

APPENDIX

Given the multichannel time series X_j, j = 1 ..., M as previously defined in Section II, the multichannel system of normal equations describing the forward and backward PEF's can be written (g) in the matrix form.

$$\begin{bmatrix} R(0) & R(1) & \cdots & R(-N) \\ R(1) & R(0) & \cdots & R(1-N) \\ & & \vdots & \\ R(N) & R(N-1) & \cdots & R(0) \end{bmatrix} \cdot \begin{bmatrix} I & B_N \\ F_1 & \cdot \\ \cdot & \cdot \\ \cdot & \cdot \\ \cdot & B_1 \\ F_N & I \end{bmatrix} = \begin{bmatrix} P_{N+1} & 0 \\ 0 & \cdot \\ \cdot & \cdot \\ \cdot & \cdot \\ \cdot & 0 \\ 0 & Q_{N+1} \end{bmatrix} \qquad \text{(A1)}$$

where F_i, B_i, i = 0,..., N are the forward and backward PEF's, P_{N+1}, Q_{N+1} are the forward and backward prediction error powers and R(t), the multichannel autocorrelation coefficient at lat t defined as:

$$R(t) = E\,[X_j X_{j+t}^{\dagger}] = R^{\dagger}(-t)$$

where the symbol E denotes the statistical expectation.

The solution of these normal equations when the autocorrelation matrix is known, is provided by the multichannel Levinson-Wiggins-Robinson (LWR) algorithm.

Briefly, this algorithm relates the solution of order N + 1 to the solution of order N according to the recursions:

$$U_{N+1} = \sum_{k=0}^{N} R\,(N+1-k) \cdot F_{k,N}$$

$$F_{N+1,N+1} = Q_{N+1}^{-1} \cdot U_{N+1}$$

$$B_{N+1,N+1} = -P_{N+1}^{-1} \cdot U_{N+1}^{\dagger}$$

$$P_{N+2} = P_{N+1}\,(I - B_{N+1,N+1}\,F_{N+1,N+1})$$

$$Q_{N+2} = Q_{N+1}\,(I - F_{N+1,N+1}\,B_{N+1,N+1})$$

$$F_{k,N+1} = F_{k,N} + B_{N+1-k,N}\,F_{N+1,N+1} \quad \text{for } 0 \le k \le N$$

$$B_{N+1-k,N+1} = B_{N+1-k,N} + F_{k,N}\,B_{N+1,N+1} \quad \text{for } 0 \le k \le N$$

with initial conditions

$$P_1 = Q_1 = R(0)$$

$$A_{o,k} = B_{o,k} = I \text{ for } 1 \leq k \leq N+1.$$

In the case where "a priori" estimates of the autocorrelation coefficients are not available, the last coefficient of the forward and backward PEF's, $F_{N+1,N+1}$ and $B_{N+1,N+1}$ must be otherwise calculated. Tyraskis[8], Tyraskis and Jensen[9] obtained an initial estimate of $F_{N+1,N+1}$, $B_{N+1,N+1}$ by separately minimizing the forward and backward prediction error powers:

$$\bar{F}_{N+1,N+1} = -(D_{N+1})^{-1} C^{\dagger}_{N+1}$$

and $$\bar{B}_{N+1,N+1} = -(G_{N+1})^{-1} C_{N+1}$$

where

$$C_{N+1} = \sum_{j=1}^{M-N-1} e_j^{N+1} (b_j^{N+1})^{\dagger}$$

$$D_{N+1} = \sum_{j=1}^{M-N-1} b_j^{N+1} (b_j^{N+1})^{\dagger}$$

$$G_{N+1} = \sum_{j=1}^{M-N-1} e_j^{N+1} (e_j^{N+1})^{\dagger}$$

and $$e_j^{N+1} = \sum_{k=0}^{N+1} F^{\dagger}_{N+1-k,N} X_{j+k}$$

$$b_j^{N+1} = \sum_{k=0}^{N+1} B^{\dagger}_{k,n} X_{j+k}$$

These estimates allow for the two estimates of the autocorrelation coefficient at log N+1

$$\bar{R}(N+1) = -\sum_{k=1}^{N+1} R(N+1-k) \cdot [F_{k,N} + B_{N+1-k,N} \bar{F}_{N+1,N+1}]$$

and $$\bar{R}(-N-1) = -\sum_{k=1}^{N+1} R(k-N-1) \cdot [B_{k,N} + F_{N+1-k,N} \bar{B}_{N+1,N+1}]$$

A unique autocorrelation coefficient at log N+1 is defined as the arithmetic mean of these two estimates:

$$R(N+1) = \frac{\bar{R}(N+1) + \bar{R}^{\dagger}(-N-1)}{2}$$

where, necessarily

$$R(-N-1) = R^{\dagger}(N+1)$$

The final estimates of the last coefficients of the forward and backward PEF's are those obtained as follows:

$$F_{N+1,N+1} = -\left[\sum_{k=1}^{N+1} R(N+1-k)\, B_{N+1-k,N}\right]^{-1} \cdot \sum_{k=0}^{N+1} R(N+1-k)\, F_{k,N}$$

and $$B_{N+1,N+1} = -\left[\sum_{k=1}^{N+1} R(k-N-1)\, F_{N+1-k,N}\right]^{-1} \cdot \sum_{k=0}^{N+1} R(k-N-1)\, B_{k,N}$$

The remaining coefficients of the forward and backward PEF's, and the forward and backward prediction error powers are calculated using the recursions of the multichannel LWR algorithm.

MULTICHANNEL RELATIVE-ENTROPY SPECTRUM ANALYSIS

Bruce R. Musicus
Research Laboratory of Electronics
Massachusetts Institute of Technology
Cambridge, Mass. 02139

Rodney W. Johnson
Computer Science and Systems Branch
Information Technology Division
Naval Research Laboratory
Washington, D.C. 20375

ABSTRACT

A new relative-entropy method is presented for estimating the power spectral density matrix for multichannel data, given correlation values for linear combinations of the channels, and given an <u>initial</u> estimate of the spectral density matrix. A derivation of the method from the relative-entropy principle is given. The basic approach is similar in spirit to the Multisignal Relative-Entropy Spectrum Analysis of Johnson and Shore, but the results differ significantly because the present method does not arbitrarily require the final distributions of the various channels to be independent. For the special case of separately estimating the spectra of a signal and noise, given the correlations of their sum, Multichannel Relative-Entropy Spectrum Analysis turns into a two stage procedure. First a smooth power spectrum model is fitted to the correlations of the signal plus noise. Then <u>final</u> estimates of the spectra and cross spectra are obtained through linear filtering. For the special case where p uniformly spaced correlations are known, and where the initial estimate of the signal plus noise spectrum is all-pole with order p or less, this method fits a standard Maximum Entropy autoregressive spectrum to the noisy correlations, then linearly filters to calculate the signal and noise spectra and cross spectra. An illustrative numerical example is given.

1 INTRODUCTION

We examine the problem of estimating power spectra and cross spectra for multiple signals, given selected correlations of various linear combinations of the signals, and given an <u>initial</u> estimate of the spectral density matrix. We present a method that produces final estimates that are consistent with the given correlation information and otherwise as similar as possible to the initial estimates in a precise information-theoretic sense. The method is an extension of the Relative-Entropy Spectrum Analysis (RESA) of Shore [1] and of the Maximum-Entropy Spectral Analysis (MESA) of Burg [2,3]. It reduces to RESA when there is a single signal and to MESA when the initial estimate is flat.

MESA starts with a set of p known data correlations. It then estimates a probability density for the signal that has as large an entropy as possible (is maximally "flat") but still satisfies the known correlations. Intuitively, the method seeks the most "conservative" density estimate that would explain the observed data. The resulting algorithm fits a smooth pth order autoregressive power-spectrum model to the known correlations. This technique gives good, high-resolution spectrum estimates, particularly if the signal either is sinusoidal or has been generated by an autoregressive process of order p or less.

RESA [1] is based on an information-theoretic derivation that is quite similar to that of MESA, except that it incorporates an initial spectrum estimate. This prior knowledge can often improve the spectrum estimates when a reliable estimate of the shape of the overall signal spectrum is available. In the case where the initial spectral density estimate is flat, RESA reduces to MESA.

In this paper we derive a multichannel RESA method that estimates the joint probability density of a set of signals given correlations of various linear combinations of the signal and given an initial estimate of the signal probability densities. The estimator was briefly presented in [4]. Our basic approach is similar in spirit to the multisignal spectrum-estimation procedure in [5,6], but the result differs significantly because that paper not only assumed that the initial probability-density estimates for the various signals were independent, but in effect imposed the same condition on the final estimates as well. We show that if this assumption is not made, the resulting final estimates are in fact not independent, but do take a form that is more intuitively satisfying. When applied to the case of estimating the power spectra and cross spectra of a signal and noise given selected correlations of their sum, our method first fits a smooth power spectrum model of the signal plus noise spectrum to the given correlations. It then uses a smoothing Wiener-Hopf filter to obtain the final estimates of the signal and the noise spectra. This Multichannel Relative-Entropy Spectrum Analysis method thus represents a bridge between the information theoretic methods and Bayesian methods for spectrum estimation from noisy data.

In certain filtering applications such as speech enhancement, relatively good estimates of a stationary noise background can be found during quiet periods when no signal is present. However, the signal spectrum may be changing relatively rapidly so that good initial estimates for this spectrum are not found as easily. Unfortunately, our technique, like the Bayesian methods, requires good initial estimates of both the signal and noise spectra. The simplest fix in the Bayesian estimation problem is to estimate the signal spectrum by spectral subtraction [7]. More sophisticated Bayesian methods estimate the signal model along with the signal and iterate between filtering steps and spectrum estimation steps [8,9]. With these methods in mind, we could consider several modifications to our Multichannel RESA method when a good initial signal estimate is not known. We could let the initial signal spectrum estimate be infinite or flat, we could try special subtraction, or we could try estimating the initial signal density along with the final joint signal and noise density. Unfortunately, none of these approaches gives a truly convincing solution to the problem, and so the issue remains open.

2 RELATIVE ENTROPY

The Relative-Entropy Principle [10] can be characterized in the following way. Let v be a random variable with values drawn from a set $v \in D$ with probability density $q^{\dagger}(v)$. We will assume that this "true" density is unknown, and that all we have available is an initial estimate p(v). Now suppose we obtain some information about the actual density that implies that $q^{\dagger}$, though unknown, must be an element of some convex set, Q, of densities. Suppose $p \notin Q$. Since Q may contain many (possibly infinitely many) different probability densities, which of these should be chosen as the best estimate q of $q^{\dagger}$? And how should the initial estimate be incorporated into this decision?

The Relative Entropy Principle states that we should choose this final density q(v) to be the one that minimizes the relative entropy:

$$H(q,p) = \int_D q(v) \log \frac{q(v)}{p(v)} \, dv \tag{1}$$

subject to the condition $q \in Q$. It has been shown [10] that minimizing any function other than H(q,p) to estimate q must either give the same answer as minimizing relative entropy or else must contradict one of four axioms that any "reasonable" estimation technique must satisfy. These axioms require, for example, that the estimation method must give the same answer regardless of the coordinate system chosen. The function H(q,p) has a number of useful properties: it is convex in q, it is convex in p, it is positive, and it is relatively convenient to work with computationally. If the convex set Q is closed and contains some q with $H(q,p)<\infty$, then there exists a $q \in Q$ that minimizes (1) [11]. This solution is unique up to a set of measure zero.

Relative-entropy minimization was introduced as a general method of statistical inference by Kullback [12] and has been advocated by a variety of authors [13,14,15] under a variety of names, including cross-entropy [16], expected weight of evidence [17,p.72], directed divergence [12,p.7], discrimination information [12,p.37], and relative entropy [18,p.19]. The principle of Maximum Entropy [19,20,21] is a special case of the Relative-Entropy principle [10,22] where the initial density is "flat" over the domain D.

One application in which we can explicitly state the form of the relative-entropy solution q is where we observe the expected values $\overline{g}_k$ of a finite set of known functions $g_k(v)$ given the actual density $q^\dagger(v)$. Then the set Q of possible densities is defined by the constraints:

$$\int_D g_k(v)q(v)dv = \overline{g_k} \qquad \text{for } k = 1,\dots,M \tag{2}$$

In addition, the density q(v) must be properly normalized:

$$\int_D q(v)dv = 1 \tag{3}$$

Because the constraints (2) and (3) are linear in q, the set Q of all probability densities satisfying these constraints must be convex. If the g_k are bounded functions, then Q is closed, and therefore there exists a density q that minimizes H(q,p) subject to the constraints (provided these are compatible with $H(q,p)<\infty$). In fact, even when the g_k are unbounded, the minimum-relative-entropy density q can be shown to exist under fairly general conditions; see [11] for a statement of such results.

Given the constraints (2) and (3), we wish to choose the final estimate q(v) of $q^\dagger(v)$ by minimizing the relative entropy (1) subject to (2) and (3). To do this, we introduce Lagrange multipliers λ_k, and construct the Lagrangian:

$$H(q,p) + (\lambda_o - 1)\left[\int_D q(v)dv - 1\right] + \sum_{k=1}^{M} \lambda_k \left[\int_D g_k(v)q(v)dv - \overline{g}_k\right] \tag{4}$$

and set the variation with respect to q to zero. We obtain:

$$q(v) = p(v)\exp\left[-\lambda_o - \sum_{k=1}^{M} \lambda_k g_k(v)\right] \tag{5}$$

It can be shown that if there is a solution q(v) to the constrained minimization problem, then it must have the form (5) with the possible exception of a set of points on which the constraints imply that q vanishes [12,p.38; 11]. Conversely, if there are multipliers λ_k such that q(v) in (5) satisfies the constraints (2) and (3), then q(v) must be the unique element of Q that minimizes the relative entropy subject to the constraints [11]. When the g_k are complex functions, (2) is equivalent to two real constraints for each k. We then write (5) with complex Lagrange multipliers, define complex conjugate quantities $g_{-r} = g_r^*$, $\lambda_{-k} = \lambda_k^*$, and let k in the sum range over negative as well as positive values. In general, it is difficult to find closed-form solutions for the λ_k in terms of the constraints $\overline{g}_k$. Computational methods using gradient search have been developed, however [23].

3 MULTICHANNEL RELATIVE-ENTROPY SPECTRUM ANALYSIS

Let us apply this theory to estimating the spectra and cross-spectra of a set of L signals, or "channels", $x_0(t)$, ..., $x_{L-1}(t)$, which we collect into a single vector-valued "multichannel" signal $x(t)=(x_0(t) \dots x_{L-1}(t))^T$. (In what follows, superscripts T and H denote the transpose and the Hermitian adjoint, and a star denotes the complex conjugate.)

We assume that x(t) is a bandlimited stationary random complex process. To simplify the mathematics, we will assume that x(t) is a finite sum of complex exponentials at frequencies w_n with random vector amplitudes c_n:

$$x(t) = \sum_{n=0}^{N-1} c_n e^{iw_n t} \tag{6}$$

This involves no essential loss of generality, since an arbitrary stationary complex random processes may be approximated by the form (6) with arbitrarily small mean square error on arbitrarily large finite intervals by choosing the number of frequencies large enough and their spacing close enough [24,p.36].

Let $q^{\dagger}(c_0, \dots, c_{N-1})$ be a joint probability density for the vector amplitudes c_n. We can express the correlation matrix of the signal as

$$\begin{aligned} R(\tau) &= E\left[\lim_{T\to\infty} \frac{1}{T}\int_0^T x(t)x^H(t-\tau)dt\right] \\ &= E\left[\sum_n c_n c_n^H e^{iw_n\tau}\right] \end{aligned}$$

(expectation with respect to $q^{\dagger}$). Fourier transformation gives the power spectral matrix

$$\begin{aligned} S(w_n) &= E\left[c_n c_n^H\right] \\ &= \int c_n c_n^H q^{\dagger}(c_0,\dots,c_{N-1})dc_0\dots dc_{N-1} \end{aligned} \tag{7}$$

Let us choose an initial probability density estimate p of $q^{\dagger}$ such that the c_n are independent Gaussian random variables with zero mean and covariance $P(w_n)$:

$$\begin{aligned} p(c_0,\dots,c_{N-1}) &= \prod_{n=0}^{N-1} p(c_n) \\ p(c_n) = N(0,P(w_n)) &= \frac{1}{\pi^L \det P(w_n)} \exp\left[-c_n^H P(w_n)^{-1} c_n\right] \end{aligned} \tag{8}$$

This choice of p corresponds to choosing the initial power spectrum estimate of $S(w_n)$ to be $P(w_n)$:

$$P(w_n) = \int c_n c_n^H p(c_o,\dots,c_{N-1})dc_o\dots dc_{N-1}$$

This Gaussian assumption is usually considered reasonable and is often implicit in spectrum-analysis approaches such as Blackman-Tukey periodograms [25] or estimation procedures such as Wiener-Hopf filters [26]. For further discussion of the assumed form, see [27].

Now suppose we learn correlations R_k at various lags τ_k of various pairs of linear combinations $\alpha_k^H x(t)$, $\beta_k^H x(t)$ of the vector signal components:

$$R_k = E\left[\lim_{T\to\infty} \frac{1}{T} \int_0^T \left(\alpha_k^H x(t)\right)\left(\beta_k^H x(t-\tau_k)\right)^* dt\right] \tag{9}$$

$$= \alpha_k^H R(\tau_k)\beta_k = \alpha_k^H \sum_n S(w_n) e^{iw_n\tau_k} \beta_k$$

This rather general form includes measurements of correlations of pairs of single signal components - individual matrix elements of $R(\tau_k)$. As another special case, treated in the next section, it includes measurements of autocorrelations of the sum of the signal components. With the help of (7), this gives constraints in the standard form of (2) as follows:

$$R_k = \int \left(\alpha_k^H \left(\sum_n c_n c_n^H e^{iw_n\tau_k}\right)\beta_k\right) q(c_o,\dots,c_{N-1})dc_o\dots dc_{N-1} \tag{10}$$

The Relative-Entropy final estimate of the probability density of the c_n coefficient given the initial estimate (8) and constraints (10) is then:

$$q(c_o,\dots,c_{N-1})=p(c_o,\dots,c_{N-1})\exp\left[-\lambda_o - \sum_k \lambda_k \sum_n c_n^H \beta_k \alpha_k^H c_n e^{iw_n\tau_k}\right] \tag{11}$$

for some set of Lagrange multipliers λ_k, which are chosen so that q(v) satisfies the constraints and is normalized (2). (Again we use the device of setting $\lambda_{-k} = \lambda_k^*$ and letting k in the sum run over negative values as well as positive. With the definitions $\tau_{-k} = -\tau_k$, $\alpha_{-k} = \beta_k$, $\beta_{-k} = \alpha_k$, this ensures a real result.)
Substituting the formula (8) for $p(c_o, \dots, c_{N-1})$ into (11) and simplifying puts the probability density estimate into the following elegant form:

$$q(c_o,\dots,c_{N-1}) = \prod_{n=o}^{N-1} q(c_n) \tag{12}$$

where

$$q(c_n) = N(0,Q(w_n))$$

$$Q(w_n) = \left(P(w_n)^{-1} + \sum_k \lambda_k \beta_k \alpha_k^H e^{iw_n \tau_k} \right)^{-1}$$

and where the unknowns λ_k must be determined from the constraints. Substituting this probability density into (10) and simplifying reduces the constraints to the form:

$$R_k = \alpha_k^H \left(\sum_n Q(w_n) e^{iw_n \tau_k} \right) \beta_k \tag{13}$$

Adjusting the λ_k until the latter equations are satisfied with $Q(w_n) \geq 0$ is a non-linear problem that must be solved, in general, by a non-linear gradient search technique.

The amplitudes c_n are a posteriori independent Gaussian random variables (i.e., have independent Gaussian final densities). Even if the channels of x(t) are a priori independent (i.e., have independent initial densities), so that the $P(w_n)$ matrices are all diagonal, the observation information concerns linear combinations of the channels, and as a result the covariance of the final density, $Q(w_n)$, will generally not be diagonal. Thus the final estimates of the various channels, unlike those in [5], will generally be correlated with each other.

4 SPECTRUM ESTIMATION FROM CORRELATIONS OF SIGNAL PLUS NOISE

A special case of great practical interest is that in which we observe autocorrelations only for the sum of the signal components

$$y(t) = \sum_{i=o}^{L-1} x_i(t) = e^T x(t) \tag{14}$$

where $e = (1\ 1\ \ldots\ 1)^T$. We then have:

$$R_k = e^T \sum_n Q(w_n) e\, e^{iw_n \tau_k} \tag{15}$$

These constraints are identical in form to those in (13) with $\alpha_k = \beta_k = e$ for all k. We may often take the signal components $x_i(t)$ to be a priori uncorrelated, so that the power spectral density matrix $P(w_n)$ is diagonal for all w_n. This restriction, however, is not necessary.

The Multichannel Relative-Entropy Spectrum Analysis estimate for x(t) from (12) is given by:

$$Q(w_n) = \left[P(w_n)^{-1} + e \left(\sum_k \lambda_k e^{iw_n \tau_k} \right) e^T \right]^{-1} \tag{16}$$

where the Lagrange multipliers $_k$ are chosen to satisfy (15). The structure of this estimate is quite similar to the single-channel Relative-Entropy Spectrum Analysis (RESA) estimate given by [1] except that the quantities involved are matrices. Namely, the second term inside the brackets is the product of a scalar Σ, the summation, with $e\ e^T$, a square matrix of all 1's. In the single-signal case, $P(w_n)$ and $Q(w_n)$ become scalars, and we can replace $e\ \Sigma e^T$ with Σ; the result is just the RESA estimate. On the other hand, there is also a close formal connection with the Multisignal RESA estimate given in [5]. That is equivalent to the result of replacing $e\ \Sigma e^T$ in (16) by ΣI, where I is an identity matrix.

The expression (16) can be put into another interesting form by using the Woodbury-Sherman formula $(A + BCD)^{-1} = A^{-1} - A^{-1}B(CA^{-1}B+D^{-1})^{-1}CA^{-1}$

$$Q(w_n) = P(w_n) - \frac{P(w_n)e\ e^T P(w_n)}{e^T P(w_n) e + \dfrac{1}{\sum_k \lambda_k e^{iw_n \tau_k}}} \tag{17}$$

Defining initial and final power-spectrum estimates for the summed signal y(t) by

$$P_{yy}(w_n) = e^T P(w_n) e$$

$$Q_{yy}(w_n) = e^T Q(w_n) e$$

we obtain from (17):

$$Q_{yy}(w_n) = P_{yy}(w_n) - \frac{P_{yy}(w_n)^2}{P_{yy}(w_n) + \dfrac{1}{\sum_k \lambda_k e^{iw_n \tau_k}}} \tag{18}$$

and thus

$$Q_{yy}(w_n) = \frac{1}{\dfrac{1}{P_{yy}(w_n)} + \sum_k \lambda_k e^{iw_n \tau_k}} \tag{19}$$

This is precisely the form of the single-signal RESA final estimate with initial estimate $P_{yy}(w_n)$. We can write (15) as

$$R_k = \sum_n Q_{yy}(w_n) e^{iw_n \tau_k} \tag{20}$$

The Lagrange multipliers in (19) must be chosen to make $Q_{yy}(w_n)$ satisfy the correlation constraints (20). We can thus determine λ_k in (16) by solving a single-channel problem. That provides everything necessary to determine the solution $Q(w_n)$ of the multichannel problem. We can in fact express $Q(w_n)$ directly in terms of $Q_{yy}(w_n)$ and $P(w_n)$; from (17) and (18) we obtain

$$Q(w_n) = P(w_n) + \frac{Q_{yy}(w_n)-P_{yy}(w_n)}{P_{yy}(w_n)^2} P(w_n) e\, e^T P(w_n) \tag{21}$$

These equations summarize the Multichannel Relative-Entropy Spectrum Analysis method for correlations of a sum of signals. The calculation of the final spectral matrix proceeds in two steps. First we must find Lagrange multipliers such that the final estimate of the power spectrum of the sum of signals, $Q_{yy}(w_n)$ in (19), has the observed correlation values (20). Computationally, this generally requires a nonlinear gradient search algorithm to locate the correct λ_k [23]. Next the final spectral density matrix, $Q(w_n)$ in (21), containing the cross-spectra as well as the power spectra of the individual signals, is formed by combining a linear multiple of the fitted power spectrum $Q_{yy}(w_n)$ with a constant term that depends only on the initial densities.

Frequently the multichannel signal $x(t)$ will comprise just two components, a signal $s(t)$ and an additive disturbance $d(t)$:

$$x(t) = \begin{pmatrix} s(t) \\ d(t) \end{pmatrix} = \sum_{n=0}^{N-1} \begin{pmatrix} \sigma_n \\ \delta_n \end{pmatrix} e^{iw_n t} \tag{22}$$

The initial estimate takes the form

$$P(w_n) = \begin{pmatrix} P_{ss}(w_n) & P_{sd}(w_n) \\ P_{ds}(w_n) & P_{dd}(w_n) \end{pmatrix} = E\left(\begin{pmatrix} \sigma_n \\ \delta_n \end{pmatrix} (\sigma_n^* \;\; \delta_n^*)\right)$$

The expression for $P_{yy}(w_n)$ specializes to:

$$P_{yy}(w_n) = \begin{pmatrix} 1 & 1 \end{pmatrix} P(w_n) \begin{pmatrix} 1 \\ 1 \end{pmatrix} = E\left(\left|\sigma_n + \delta_n\right|^2\right)$$

We also define the initial cross-power spectra of $s(t)$ and $d(t)$ with respect to $y(t)$ as follows:

$$\begin{pmatrix} P_{sy}(w_n) \\ P_{dy}(w_n) \end{pmatrix} = P(w_n) \begin{pmatrix} 1 \\ 1 \end{pmatrix}$$

$$\begin{pmatrix} P_{ys}(w_n) & P_{yd}(w_n) \end{pmatrix} = \begin{pmatrix} 1 & 1 \end{pmatrix} P(w_n)$$

We define the components $Q_{ss}(w_n)$, $Q_{sd}(w_n)$, $Q_{dd}(w_n)$, $Q_{yy}(w_n)$, $Q_{sy}(w_n)$, and $Q_{dy}(w_n)$ similarly. Then (21) becomes:

$$Q(w_n) = P(w_n) + \frac{Q_{yy}(w_n)-P_{yy}(w_n)}{P_{yy}(w_n)^2}\begin{pmatrix}P_{sy}(w_n)\\P_{dy}(w_n)\end{pmatrix}\begin{pmatrix}P_{ys}(w_n) & P_{yd}(w_n)\end{pmatrix} \tag{23}$$

An alternative formula for $Q(w_n)$ in terms of $Q_{yy}(w_n)$ is:

$$Q(w_n) = Q_{yy}(w_n)\begin{pmatrix}\frac{P_{sy}(w_n)}{P_{yy}(w_n)}\\ \frac{P_{dy}(w_n)}{P_{yy}(w_n)}\end{pmatrix}\begin{pmatrix}\frac{P_{ys}(w_n)}{P_{yy}(w_n)} & \frac{P_{yd}(w_n)}{P_{yy}(w_n)}\end{pmatrix} + \frac{\det P(w_n)}{P_{yy}(w_n)}\begin{pmatrix}1\\-1\end{pmatrix}\begin{pmatrix}1 & -1\end{pmatrix} \tag{24}$$

5 INTERPRETATION OF MULTICHANNEL RESA

The formulas defining Multichannel RESA have an interesting and profound structure that may not be obvious at first glance. First of all, as we show below, formula (21) makes it easy to state conditions under which the matrix estimate $Q(w_n)$ is positive definite. Next, the appearance of $Q_{yy}(w_n)$ in the constraint equation for the λ_k is actually something we should have expected by the property of subset aggregation that Relative-Entropy estimators satisfy [28]. Furthermore, formula (24), which builds the spectral density matrix estimate $Q(w_n)$ by linearly filtering the fitted model spectrum $Q_{yy}(w_n)$ is identical in form to the standard Bayesian formula for the final expected power and cross-power in two signals given the value of their sum. In particular, the first term in (24) applies the well-known Wiener-Hopf smoothing filter [26] to $Q_{yy}(w_n)$, while the second term can be interpreted as the expected final variance of σ_n and δ_n. Finally, as we will show, the relative entropy $H(q,p)$ has the same form as a generalized Itakura-Saito distortion measure [29]. Thus minimizing relative entropy in this problem is equivalent to finding the spectral matrix $Q(w_n)$ with minimum Itakura-Saito distortion.

A. Positive Definiteness

Assume that the initial spectral density matrices are positive definite, $P(w_n)>0$; then $P_{yy}(w_n)>0$ also for all w_n. This implies that $Q(w_n)$ in (21) is at least well-defined, provided we can find some $Q_{yy}(w_n)$ that satisfies the correlation constraints. Assume moreover that $Q_{yy}(w_n)$ is strictly positive, $Q_{yy}(w_n)>0$. Let u be any nonzero vector. Since $P(w_n)$ is positive definite, we can write $u = ae + v$ for some scalar a and some vector v such that $v^H P(w_n)e = 0$. Then (21) implies $u^H Q(w_n)u = |a|^2 Q_{yy}(w_n) + v^H P(w_n)v$, and at least one of the two terms on the right-hand side must be positive. Thus $u^H Q(w_n)u>0$ for every nonzero vector u; that is, $Q(w_n)$ is strictly positive definite, $Q(w_n)>0$.

B. Generalized Itakura-Saito Distortion Measure

If we substitute any zero-mean, Gaussian densities

$$q(c_n) = N(0,Q(w_n))$$

and

$$p(c_n) = N(0,P(w_n))$$

into the relative-entropy formula we get:

$$H(q,p) = \sum_n \mathrm{tr} \left\{ Q(w_n)P(w_n)^{-1} - I \right\} - \log \det(Qw_n)P(w_n)^{-1})$$

This is just a generalized version of the Itakura-Saito distortion measure [29]. We therefore could have derived the same spectrum estimate by minimizing the Itakura-Saito distortion measure over all possible spectral matrices $Q(w_n)$ subject to the constraints (13).

6 COMPUTATIONAL CONSIDERATIONS

The difficult step in the Multichannel RESA procedure is to solve for the Lagrange multipliers that will give $Q_{yy}(w_n)$ the appropriate correlations in (20). Gradient search algorithms for computing this in general are given by Johnson [23]. Once $Q_{yy}(w_n)$ is known, the components of $Q(w_n)$ may be easily found by filtering $Q_{yy}(w_n)$ and adding in the final covariance estimate, an amount of computation that is linear in the number of frequency samples.

Once special case is particularly easy to solve. This is when correlations of $y(t)$ are given for uniformly spaced lags $\tau_k = -p,-p+1, \ldots, p$ and when the initial spectral density of the signal plus noise, $P_{yy}(w_n) = (1\ 1)P(w_n)(1\ 1)^T$ is autoregressive of order at most p. Let us take the limiting form of our equations for equispaced frequencies as the spacing becomes extremely small, so that we can treat the spectral densities as continuous functions of w. Then because $P_{yy}(w_n)$ is autoregressive (all-pole), the term $1/P_{yy}(w_n)$ in the denominator of

$$Q_{yy}(w) = \frac{1}{\frac{1}{P_{yy}(w)} + \sum_{k=-p}^{p} \lambda_k e^{iwk}}$$

has the same form as the sum over k. We can therefore combine coefficients in the two sums and write

$$Q_{yy}(w) = \frac{1}{\sum_{k=-p}^{p} \beta_k e^{iwk}}$$

where $\beta_{-p}, \ldots, \beta_p$ are to be determined so that:

$$R_k = \int_{-\pi}^{\pi} Q_{yy}(w) e^{iwk} \frac{dw}{2\pi}$$

This, however, is the standard Maximum Entropy Spectrum Analysis problem (MESA), and can be solved with $O(p^2)$ calculation by Levinson recursion [30,31,32]. Thus in this special case, the power spectrum of the signal plus noise $Q_{yy}(w)$ will be set to the MESA estimate. The initial estimate $P_{yy}(w)$ will be completely ignored in this step. (This is an example of "prior washout", as discussed by Shore and Johnson [28].) The spectral density matrix estimate $Q(w)$ for $s(t)$ and $d(t)$ will then be formed as the appropriately filtered function of this MESA spectrum. In this case, therefore, the prior information is used solely to control how the estimates for the signal and the disturbance are to be obtained from the MESA estimate. Note that although $P_{yy}(w)$ and $Q_{yy}(w)$ will both be autoregressive spectra, the individual components of the spectral density matrices $P(w)$ and $Q(w)$ will generally not be autoregressive.

7 EXAMPLE

We define a pair of spectra, S_B and S_S, which we think of as a known "background" and an unknown "signal" component of a total spectrum. Both are symmetric and defined in the frequency band from $-\pi$ to π, though we plot only their positive-frequency parts. (The abscissas in the figures are the frequency in Hz, $w/2\pi$, ranging from 0 to 0.5.) S_B is the sum of white noice with total power 5 and a peak at frequency $0.215 \times 2\pi$ corresponding to a single sinusoid with total power 2. S_S consists of a peak at frequency $0.165 \times 2\pi$ corresponding to a sinusoid of total power 2. Figure 1 shows a discrete-frequency approximation to the sum $S_S + S_B$, using 100 equispaced frequencies. From the sum, six autocorrelations were computed exactly. S_B itself was used as the initial estimate P_B of S_B - i.e., P_B was Figure 1 without the left-hand peak. For P_S we used a uniform (flat) spectrum with the same total power as P_B. Figures 2 and 3 show multisignal RESA final estimates Q_B and Q_S by the method of [5] - independence was assumed for the final joint probability densities of the two signals. Figures 4 and 5 show final spectrum estimates obtained by the present method from the same autocorrelation data and initial spectrum estimate. The initial cross-spectrum estimates were taken to be zero (P was diagonal). No such assumption was made for the final estimate, of course, and indeed the final cross-spectrum estimates (not shown) are non-zero.

In the results for both methods, the signal peak shows up primarily in Q_S, but some evidence of it is in Q_B as well. Comparison of Figures 4 and 5 with Figures 2 and 3 shows that both final spectrum estimates by the present method are closer to the respective initial estimates than are the final estimates by the method of [5].

In view of the fact that the present method has the logically more satisfying derivation and is computationally cheaper, the comparison of Figure 5 with Figure 3 is somewhat disappointing; the signal peak shows up less strongly in Figure 5. It must be pointed out, however, that in this example the signal and noise are truly uncorrelated. Our technique does not use this information, and in fact estimates a non-zero

Figure 1. Assumed total original spectrum for example.

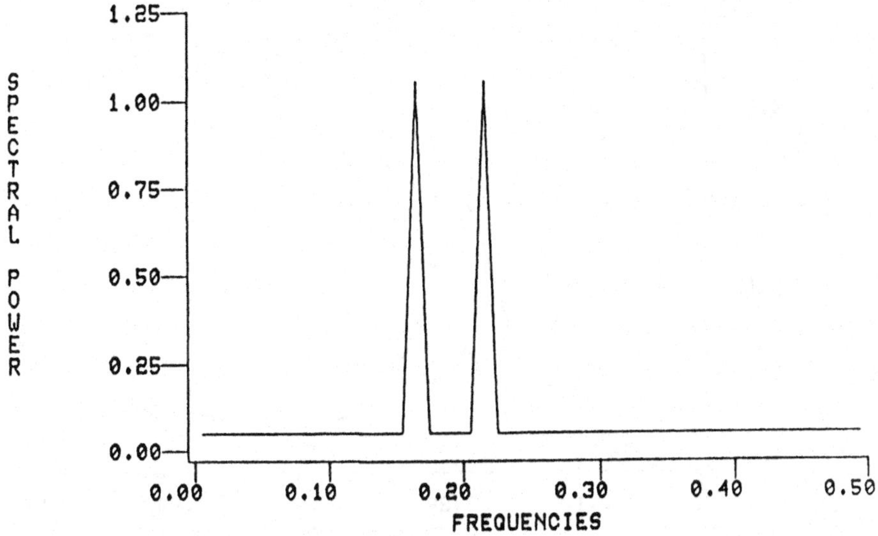

Figure 2. Final spectrum estimate Q_B by the method of [5].

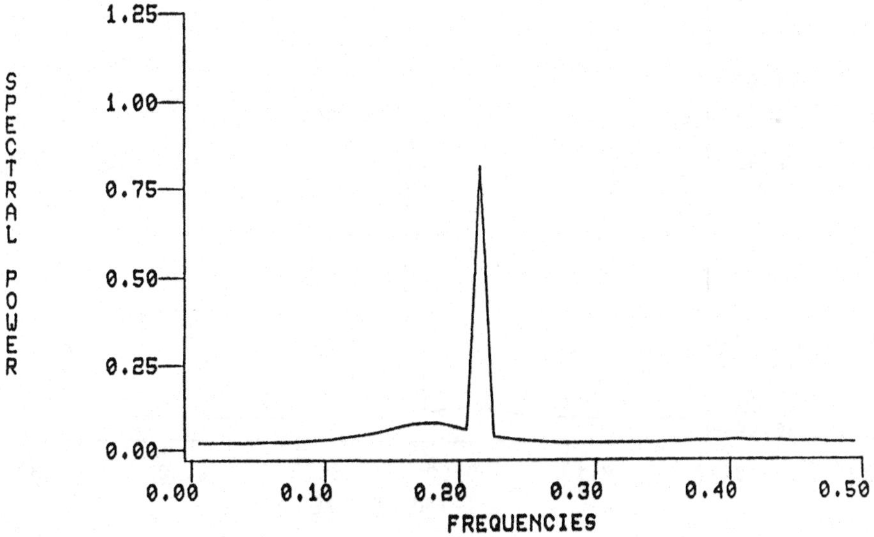

Figure 3. Final spectrum estimate Q_S by the method of [5].

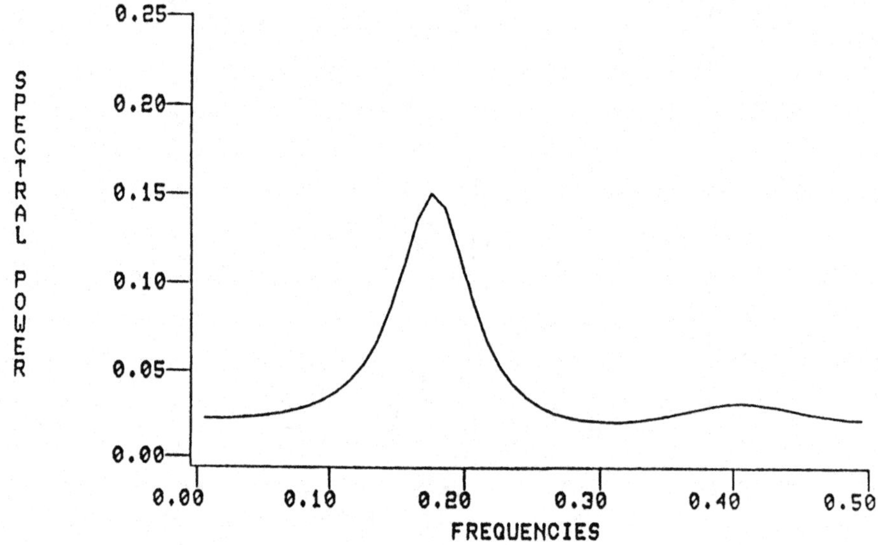

Figure 4. Final spectrum estimate Q_B by the method of this paper.

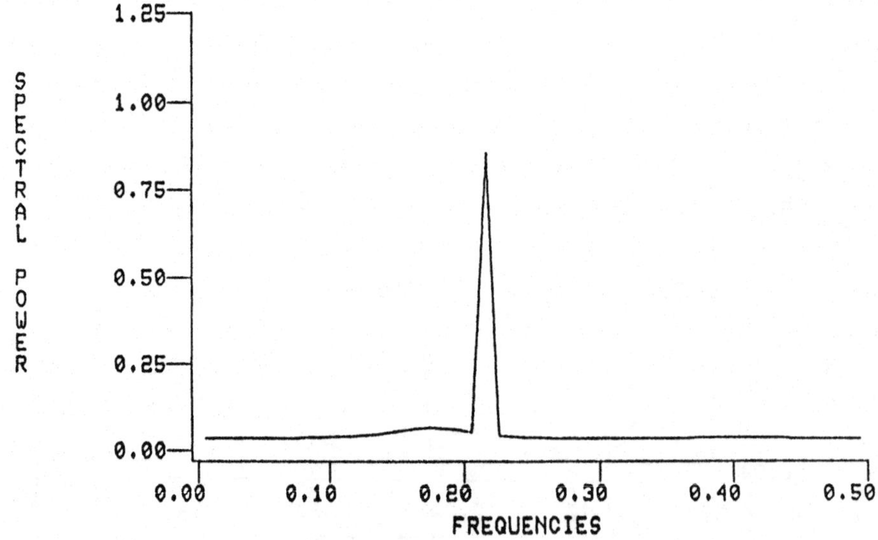

Figure 5. Final spectrum estimate Q_S by the method of this paper.

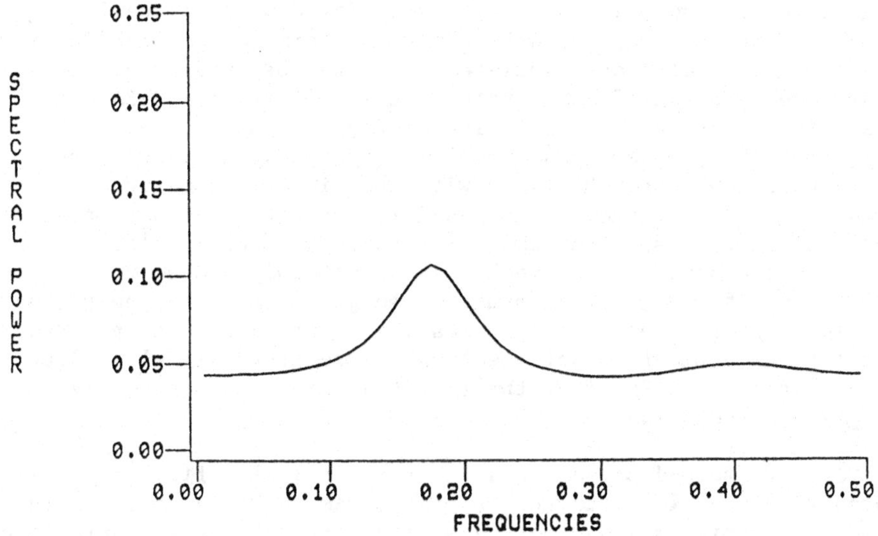

cross-correlation between the signal and noise. The method in [5], however, uses this additional knowledge and therefore, in this case, is able to produce better estimates than our technique.

8 DISCUSSION

In this paper we have derived a Multichannel Relative-Entropy Spectrum Analysis method that estimates the power spectra and cross-spectra of several signals, given an initial estimate of the spectral density matrix and given new information in the form of correlation values for linear combinations of the channels. Both this method and the multisignal method of [5] will estimate the power spectra of a signal and noise when prior information is available in the form of an initial estimate of each spectrum and given selected correlations of the signal plus noise. The present method can accept more general forms of correlation data and also produces cross-spectrum estimates, which are implicitly assumed to be zero in [5]. Even when the only correlation data are for the signal plus noise, and cross-spectrum estimates are not desired, there is a persuasive argument for preferring the present method to that of [5] - if the discrepancy between the given correlation values and those computed from the initial estimates can be accounted for in part by correlations between the signal and noise, then the correlation data should be regarded as evidence for such correlations, and correlated final estimates should be produced.

Estimates by the present method are considerably more economical to compute than estimates by the method of [5]. The algorithm first fits a smooth model power spectrum to the noisy signal using the given correlations. The available prior information is then used to linearly filter this spectrum estimate in order to obtain separate estimates for the signal and the noise. This allocation formula is virtually identical to that used by the usual Bayesian formula in which the signal and noise power spectra are estimated from the observed signal plus noise spectrum. The difference between the Multichannel RESA and Bayesian methods is that the relative-entropy technique starts by fitting a smooth power spectrum model to the observed correlations, while the Bayesian approach starts with the directly observed power spectrum. This Multichannel Relative-Entropy technique thus provides a smooth model fitting spectrum analysis procedure that is closely analogous to the Bayesian approach. If p uniformly spaced correlations are given, and if the prior information suggests that the power spectrum of the signal plus noise is autoregressive of order at most p, then the step of fitting a smooth model spectrum to the noisy signal is identical to using a standard MESA algorithm to fit a smooth autoregressive model to the given correlations.

In general, the method presented in this paper yields final spectral estimates that are closer to the initial estimates than those of [5]. This is not surprising. Our method starts with an initial estimate of the signal and noise spectra, and uses correlations of the signal plus noise to get better power spectra estimates. The method in [5] uses the same information, but also assumes that the signal and noise are uncorrelated. This additional knowledge further restricts the constraint space Q in which the probability density is known to lie, effectively leaving less unknown aspects of the density to estimate, and thus improving the final spectra. In general, the resulting spectral estimate will have higher relative entropy than the solution from our method, and will thus be "farther" from the initial density p than the solution from our method.

Our estimate of $Q_{yy}(w_n)$ can be improved by observing more and more correlations of the signal plus noise. Regardless of how much data is gathered, however, our method relies exclusively on the initial estimate of the signal and noise spectra and cross spectra to allocate $Q_{yy}(w_n)$ between the signal, noise, and cross terms. The fundamental difficulty is that observing correlations of the signal plus noise gives no insight into how this observation energy should be partitioned between the signal and the noise. Achieving accurate estimation of the signal and noise spectra separately requires a different type of observation data. Learning that the signal and noise are uncorrelated as in [5], for example, will improve our spectral estimates. The best solution, of course, would be to use an accurate model of the signal and noise processes, or to directly observe the signal and/or noise correlations.

ACKNOWLEDGEMENT

The work of Bruce Musicus has been supported in part by the Advanced Research Projects Agency monitored by ONR under Contract N00014-81-K-0742 NR-049-506 and in part by the National Science Foundation under Grant ECS80-07102.

REFERENCES

1 Shore, J.E. (1981). Minimum Cross-Entropy Spectral Analysis. IEEE Trans. Acoust., Speech, Signal Processing ASSP-29, 230-237.

2 Burg, J.P. (1967). Maximum Entropy Spectral Analysis. Presented at the 37th Annual Meeting Soc. of Exploration Geophysicists, Oklahoma City, Okla.

3 Burg, J.P. (1975). Maximum Entropy Spectral Analysis. Ph.D. Dissertation, Stanford University, Stanford, California. University Microfilms No. 75-25,499.

4 Johnson, R.W. and B.R. Musicus (1984). Multichannel Relative-Entropy Spectrum Analysis. In Proc. ICASSP84, San Diego, California. IEEE 84CH1945-5.

5 Johnson, R.W. and J.E. Shore (1983). Minimum-Cross-Entropy Spectral Analysis of Multiple Signals. IEEE Trans. Acoust. Speech, Signal Processing ASSP-31, 574-582.

6 Johnson, R.W., J.E. Shore and J.P. Burg (1984). Multisignal Minimum-Cross-Entropy Spectrum Analysis with Weighted Initial Estimates. IEEE Trans. Acoustics, Speech, Signal Processing ASSP-32.

7 Lim, Jae S. (1978). Evaluation of a Correlator-Subtractor Method for Enhancing Speech Degraded by Additive White Noise. IEEE Trans. Acoust. Speech and Signal Processing ASSP-26, 471-472.

8 Musicus, Bruce R. (1979). An Iterative Technique for Maximum Likelihood Parameter Estimation on Noisy Data. S.M. Thesis, M.I.T., Cambridge, Mass.

9 Musicus, Bruce R. and Jae S. Lim (1979). Maximum Likelihood Parameter Estimation of Noisy Data. In 1979 IEEE Int. Conf. on Acoust., Speech, Signal Proc., Washington, D.C.

10 Shore, J.E. and R.W. Johnson (1980). Axiomatic Derivation of the Principle of Maximum Entropy and the Principle of Minimum Cross-Entropy. IEEE Trans. Inform. Theory IT-26, 26-37. (See also comments and corrections in IEEE Trans. Inform. Theory IT-29, Nov. 1983).

11 Csiszár, I. (1975). I-Divergence Geometry of Probability Distributions and Minimization Problems. Ann. Prob. 3, 146-158.

12 Kullback, S. (1968). Information Theory and Statistics. Dover, New York. Wiley, New York, 1959.

13 Johnson, R.W. (1979). Axiomatic Characterization of the Directed Divergences and their Linear Combinations. IEEE Trans. Inf. Theory IT-25, 709-716.

14 Jaynes, E.T. (1968). Prior Probabilities. IEEE Trans. Systems Science and Cybernetics SSC-4, 227-241.

15 Hobson, A. and B.K. Cheng (1973). A comparison of the Shannon and Kullback Information Measures. J. Stat. Phys. 7, 301-310.

16 Good, I.J. (1963). Maximum Entropy for Hypothesis Formulation, Especially for Multidimensional Contingency Tables. Ann. Math. Stat. 34, 911-934.

17 Good, I.J. (1950). Probability and the Weighing of Evidence. Griffin, London.

18 Pinsker, M.S. (1964). Information and Information Stability of Random Variables and Processes. Holden-Day, San Francisco.

19 Elsasser, W.M. (1937). On Quantum Measurements and the Role of the Uncertainty Relations in Statistical Mechanics. Phys. Rev. 52, 987-999.

20 Jaynes, E.T. (1957). Information Theory and Statistical Mechanics I. Phys. Rev. 106, 620-630.

21 Jaynes, E.T. (1957). Information Theory and Statistical Mechanics II. Phys. Rev. 108, 171-190.

22 Gray, R.M., A.H. Gray, Jr., G. Rebolledo and J.E. Shore (1981). Rate-Distortion Speech Coding with a Minimum Discrimination Information Distortion Measure. IEEE Trans. Inform. Theory IT-27, 708-721.

23 Johnson, R.W. (1983). Algorithms for Single-Signal and Multisignal Minimum-Cross-Entropy Spectral Analysis. NRL Report 8667, Naval Research Laboratory, Washington, D.C.

24 Yaglom, A.M. (1962). An Introduction to the Theory of Stationary Random Functions. Prentice-Hall, Englewood Cliffs, N.J.

25 Blackman, R.B. and J.W. Tukey (1959). The Measurement of Power Spectra. Dover, New York.

26 Van Trees, H.L. (1968). Detection, Estimation and Modulation Theory. Wiley, New York.

27 Shore, J.E. (1979). Minimum Cross-Entropy Spectral Analysis. NRL Memorandum Report 3921, Naval Research Laboratory, Washington, D.C. 20375.

28 Shore, J.E. and R.W. Johnson (1981). Properties of Cross-Entropy Minimization. IEEE Trans. Inform. Theory IT-27, 472-482.

29 Itakura, F. and S. Saito (1968). Analysis Synthesis Telephony Based on the Maximum Likelihood Method. Reports of the 6th Int. Cong. Acoustics.

30 Makhoul, J. (1975). Linear Prediction: A Tutorial Review. Proc. IEEE 63(4), 561-580. Reprinted in Modern Spectrum Analysis, Donald G. Childers (ed.), IEEE Press, New York, 1978.

31 Levinson, N. (1947). The Wiener RMS (root mean square) Error Criterion in Filter Design and Prediction. J. Math Phys. 25, 261-278.

32 Durbin, J. (1960). The Fitting of Time Series Models. Rev. Inst. Int. de Stat. 28, 233-244.

MAXIMUM ENTROPY AND THE EARTH'S DENSITY

E. Rietsch
Texaco Houston Research Center
P.O. Box 770070
Houston, Texas 77215-0070

ABSTRACT

The maximum entropy approach to inversion appears at its best when conclusions are to be drawn from very limited information. An example is the estimation of the density profile of the earth (assumed to be spherically symmetric) on the basis of only its mean density and its relative moment of inertia. With conventional methods giving rather unsatisfactory results, the maximum entropy method provides a density profile which agrees surprisingly well with the one presently considered to be the best.

1 INTRODUCTION

Inverse problems in geophysics frequently confront us with one of two extreme situations. While we generally have a large number of unknowns to estimate, we may also have a tremendous amount of data - actually, more data than we can reasonably process - and more or less elaborate data reduction schemes are employed to reduce the wealth of data to a more manageable size. Of course, this reduction is performed in a way which improves the quality of the retained data in some sense (e.g., increases the signal-to-noise ratio).

At other times we may still have large numbers of unknowns to contend with but very few data. In fact, the data may be so inadequate that any attempt at estimating the unknowns appears bound to fail. It is this situation which I now want to address by means of an example.

We are given the radius, the mass, and the moment of inertia of the earth and are asked to determine its density as a function of depth. The assumption, of course, is that the earth can be considered to be spherically symmetric. From radius, mass, and moment of inertia we can derive two relevant data, the mean density $\bar{\rho}$ and the relative moment of inertia y.

$$\bar{\rho} = 5.517 \text{ g/cm}^3, \quad y = 0.84. \tag{1}$$

The latter is the ratio of the actual moment of inertia to the moment of inertia of a homogeneous sphere with the same size and mass. Slightly different values of $\bar{\rho}$ and y are given by Romanowicz and Lambeck (1977).

The fact that y is less than 1, i.e., that the moment of inertia of the earth is lower than that of a homogeneous sphere of the same size and mass indicates that the density of the earth tends to increase with depth; but, of course, we would like to make more quantitative statements.

2 THE BACKUS-GILBERT TECHNIQUE

Most geophysicists can be expected to be familiar with the Backus-Gilbert technique (Backus and Gilbert, 1967, 1968, 1970). It is, therefore, sufficient to say that this technique produces weighted averages of the unknown function - in this case the density as a function of radius. We would like these averaging functions (called resolving kernels) to be non-negative and highly localized. However, with the information we have, this turns out to be impossible (a brief outline of the application of the Backus-Gilbert technique to this problem is given in Appendix A). Figure 1 compares what is believed to be a good estimate of the density as a function of depth (Bullen, 1975) with the averages obtained by this technique. To illustrate the reason for this poor match, this figure also shows the resolving kernel for a depth of 2000 km. While indeed having a maximum at about 2000 km, the kernel is by no means localized, and the average does, therefore, include also the high densities in the core. It is thus significantly higher than the actual density at this depth. The density estimate for the center of the earth, on the other hand, is much lower than what we believe to be correct. The Backus-Gilbert technique fails to give us reasonably accurate answers; we are not supplying enough information for the construction of sufficiently localized resolving kernels which would allow one to equate the averaged density with its value at the location of the "peak" of the kernel.

3 EXTREME MODELS

In 1972, Parker suggested that bounds on parameters of interest were the appropriate information that should be extracted from so little information. For this particular example he showed that no earth model can have a density which is everywhere lower than ρ_0, where

$$\rho_0 = \bar{\rho}/y^{3/2} = 7.166 \text{ g/cm}^3. \tag{2}$$

It is interesting to note that the highest density obtained with the Backus-Gilbert technique is 7.20 g/cm^3, and thus barely higher than ρ_0.

Parker's equation can be generalized to allow one to specify a minimum density ρ_1 (Rietsch, 1978).

$$\rho_0 = \rho_1 + (\bar{\rho} - \rho_1)^{5/2}/(y\bar{\rho} - \rho_1)^{3/2}. \tag{3}$$

If we assume that the density in the earth is at least 1 g/cm^3, then ρ_0, the lower bound on the largest density, turns out to be 7.26

g/cm^3; it is thus higher than the highest average density predicted by the Backus-Gilbert technique.

While Parker's approach provides us with rigorous constraints that have to be satisfied by the density in the earth, these constraints are less restrictive than we would like them to be.

4 PARAMETERIZATION

A very popular approach to inversion involves parameterization. There are many different ways to parameterize this problem, and I will just concentrate on one.

Let us subdivide the earth into a number of concentric shells and assume that the density is contant in each shell. If N is the number of shells, then there are N unknown densities ρ_n and N-1 unknown radii r_n. Mean density and relative moment of inertia are connected with these parameters by

$$\sum \rho_n V_n = \bar{\rho} \tag{4}$$

$$\sum \rho_n I_n = y\bar{\rho} \; . \tag{5}$$

Here and in the following, summations are performed from n=1 to N. The factors V_n are proportional to the volume of the n-th shell

$$V_n = (r_n^3 - r_{n-1}^3)/R^3 \tag{6}$$

and the I_n are proportional to the contribution to the moment of inertia of the n-th shell

$$I_n = (r_n^5 - r_{n-1}^5)/R^5 \tag{7}$$

Of course, $r_N = R$ is the radius of the earth and $r_0 = 0$.

For N=2, the smallest meaningful number of shells, we have a total of three unknowns (the two densities and the outer radius of the inner shell) and thus more unknowns than data.

But actually, for N=2, the model - while overparameterized with respect to the data - is definitely underparameterized with regard to the density distribution in the earth. We need many layers, and this suggests that we resort to the either generalized inverses or to probabilistic methods.

5 THE GENERALIZED INVERSE

In the generalized inverse method the desired parameters - in this case the function $\rho(r)$ - are represented in terms of eigenvectors (or eigenfunctions) in the so-called model space (Jackson, 1972; Parker, 1975). This approach is sketched in Appendix B. For N = 100, the density distribution turns out to have the form

$$\rho_n = 10.15 - 4.63\,\{[n^{5/3} - (n-1)^{5/3}]/N^{2/3}\} \tag{8a}$$

in units of g/cm^3. For $N \to \infty$ we get

$$\rho(r) = 10.15 - 7.72(r/R)^2 \quad [g/cm^3] \tag{8b}$$

Both, the discrete and the continuous density distribution, are shown in Figure 2. While it most certainly represents an improvement over the results obtained with the Backus-Gilbert technique, the agreement with Bullen's density is far from satisfactory.

6 THE MAXIMUM ENTROPY APPROACH

In this approach we look for a probability density function $p(\underline{\rho}) = p(\rho_1, \rho_2, \cdots, \rho_N)$ (in the following called probability distribution to avoid using the term "density" for two different things). While satisfying certain conditions, this probability distribution should be as non-committal, as unspecific as possible; in other words, it should have higher entropy than all other probability distributions that satisfy the same constraints. One of these constraints is, of course, the normalization condition

$$\int p(\underline{\rho})d\rho = 1 \tag{9}$$

where $d\rho$ denotes the N-dimensional volume element. Here and in the following, the integration is performed over all positive values of the ρ_n.

The easiest way to include the information about mean density and relative moment of inertia is to request that the expectation values of these quantities agree with the corresponding measurements. Thus equations (4) and (5) are replaced by

$$\int (\sum \rho_n V_n) p(\underline{\rho}) d\rho = \bar{\rho} \tag{10}$$

and

$$\int (\sum \rho_n I_n) p(\underline{\rho}) d\rho = y\bar{\rho} \, , \tag{11}$$

respectively.

Maximizing the entropy

$$H = \int p(\underline{\rho}) \log[p(\underline{\rho})/w(\underline{\rho})] d\rho \, , \tag{12}$$

subject to the above three constraints leads to the desired probability distribution.

Before we can do this, we have to find the prior distribution $w(\underline{\rho})$ which appears in the expression for the entropy. This prior distribution is actually an "invariant measure" function, and, as discussed previously (Rietsch, 1977), there are reasons to assume that it is constant provided all the shells have the same volume. Thus the radii of the shells are given by

$$r_n = (n/N)^{1/3}R. \tag{13}$$

The probability distribution which maximizes the entropy subject to conditions (9)-(11) with constant $w(\underline{\rho})$ turns out to have the form

$$p(\underline{\rho}) = \prod_n (1/\bar{\rho}_n)\exp(-\rho_n/\bar{\rho}_n) \tag{14}$$

where $\bar{\rho}_n$ is the expectation value of the density in the n-th shell (Rietsch, 1977)

$$\bar{\rho}_n = 1/(\lambda_1 + I_n \lambda_2) \ . \tag{15}$$

$$I_n = N(r_n^5 - r_{n-1}^5)/R^5 = [n^{5/3} - (n-1)^{5/3}]/N^{2/3} \tag{16}$$

The Lagrange multipliers λ_1 and λ_2 are determined in such a way that the $\bar{\rho}_n$ satisfy conditions (10) and (11). For N=100 they turn out to be (Rietsch, 1977)

$$\lambda_1 = 0.0667, \quad \lambda_2 = 0.13640 \tag{17}$$

in units of cm^3/g. Passing to the limit $N \to \infty$, we get

$$\bar{\rho}(r) = 1/[\lambda_1 + \frac{5}{3}\lambda_2(r/R)^2] \tag{18}$$

with

$$\lambda_1 = 0.06674, \quad \lambda_2 = 0.13633 \tag{19}$$

Both, the discrete and the continuous density distribution, are shown in Figure 3. This Figure illustrates the surprisingly good match between $\bar{\rho}(r)$ and our best estimate of the density in the outer part of the mantle. The agreement deteriorates for greater depths. This is, of course, due to the fact that the influence of the density on the moment of inertia increases with increasing distance from the center of rotation.

It is worth pointing out that the generalized inverse solution given in Eq. (8) is related to the density determined by means of the maximum entropy method (Eq. (18)) in exactly the same way the power spectrum based on the autocorrelation function is related to the maximum entropy spectrum.

7 RELATED RESULTS

In maximum entropy power series analysis, it is customary to represent the entropy in terms of the expectation value of the power spectrum

$$H \propto \int \log[\bar{P}(\omega)]d\omega$$

An analogous formula can be derived from the probability distribution for the density. As shown in Appendix C, for practical purposes

$$H \propto \int \log[\bar{\rho}(r)]r^2dr \tag{20}$$

Maximizing this expression subject to the conditions

$$\frac{3}{R^3}\int \bar{\rho}(r)r^2dr = \bar{\rho} \tag{21}$$

$$\frac{5}{R^5}\int \bar{\rho}(r)r^4dr = y\bar{\rho}, \tag{22}$$

which are the continuous analogs of Eqs. (4) and (5), leads to $\bar{\rho}(r)$ as defined in Eq. (18). This is less cumbersome than going through the probability distribution formalism. It is, however, not clear whether (20) would still be equivalent to (12) if different constraints (constraints non-linear in the ρ_n, for example) were present.

It has become very popular to regard an unknown, non-negative parameter as a probability. This parameter is then used in the definition of the entropy which in this case would have the form

$$H = -\int \rho(r)\ \ln[\rho(r)]r^2dr \tag{23}$$

For some parameters this may be justifiable, for others it appears quite artificial. In any case, the above definition has been used by Graber (1977) to derive the density distribution within the earth based on the same information. Since he realized that the entropy in this formulation depends on the scale or the units of measurements, he introduced a scale parameter k. Maximizing the modified entropy

$$H = -\int \rho(r)\ \ln[\rho(r)/k]r^2dr \tag{24}$$

subject to the conditions

$$\int\rho(r)f_1(r)r^2dr = \bar{\rho} \qquad f_1(r) = 3/R^3 \tag{25}$$

$$\int\rho(r)f_2(r)r^2dr = y\bar{\rho} \qquad f_2(r) = 5r^2/R^5\ , \tag{26}$$

which are essentially identical with Eqs. (21) and (22), leads to the representation

$$\rho(r) = \frac{1}{k}\exp[-k - \frac{\lambda_1}{k}f_1(r) - \frac{\lambda_2}{k}f_2(r)]. \tag{27}$$

And now there is a little problem - at least in principle. There are three unknown parameters, namely the Lagrange multipliers λ_1,λ_2 and the scale factor k, but only the two equations (25) and (26) for their computation. However, since $f_1(r)$ does not depend on r, λ_1 and k are not independent, and (27) can be written as

$$\rho(r) = A \exp[-\lambda(r/R)^2] \tag{28}$$

and so there are only two independent parameters.

The two new parameters A and λ can be determined from Eqs. (25) and (26) and turn out to be

$$A = 12.14 \text{ g/cm}^3, \quad \lambda = 1.435. \tag{29}$$

The resulting density distribution is shown in Figure 4. It agrees quite well with Bullen's density but appears to be somewhat poorer than (18) near the surface of the earth.

It should, however, be borne in mind that the parameters in (27) could only be determined because $f_1(r)$ happened to be constant. If no non-trivial linear combination of $f_1(r)$ and $f_2(r)$ were constant (i.e., independent of r), then the parameters in (27) could not have been determined without additional information.

8 EPILOG

It is worth mentioning that Graber has done more than I have shown here. In addition to radius, mass and moment of inertia he also used three zero-node torsional normal modes. His results are quite interesting but outside the scope of what I intended to discuss here. I wanted to demonstrate the remarkable ability of the maximum entropy method to extract information from very few data, a feat that becomes particularly remarkable when compared with the results of more conventional methods. I also wanted to show one more example that is not related to either time series analysis or image enhancement, and finally I wanted to use this opportunity to humbly suggest not automatically regarding unknown physical parameters as probabilities simply because they happen to be non-negative.

9 ACKNOWLEDGEMENT

I am indebted to Texaco, Inc. for the permission to publish this paper.

REFERENCES

1 Backus, G. and F. Gilbert (1967). Numerical Applications of a Formalism for Geophysical Inverse Problems. Geophys. J.R. astr. Soc. 13, 247-276.

2 Backus, G. and F. Gilbert (1968). The Resolving Power of Gross Earth Data. Geophys. J. 16, 169-205.

3 Backus G. and F. Gilbert (1970). Uniqueness in the Inversion of Inaccurate Gross Earth Data. Phil. Trans. Roy. Soc. London, Ser. A 266, 123-192.

4 Bullen, K.E. (1975). The Earth's Density. London, Chapman and Hall.

5 Graber, M.A. (1977). An Information Theory Approach to the Density of the Earth. Technical Memorandum 78034, NASA Goddard Space Flight Center, Greenbelt, MD 20771.

6 Jackson, D.D. (1972). Interpretation of Inaccurate, Insufficient and Inconsistent Data. Geophys. J.R. Astr. Soc. 28, 97-109

7 Lanczos, C. (1961). Linear Differential Operators. D. Van Nostrand Co., London

8 Parker, R.L. (1972). Inverse Theory with Grossly Inadequate Data. Geophys. J.R. astr. Soc. 29, 123-138.

9 Parker, R.L. (1977). Understanding Inverse Theory. Rev. Earth Planet. Sci. 5, 35-64.

11 Rietsch, E. (1977). The Maximum Entropy Approach to Inverse Problems. J. Geophysics 42, 489-506.

12 Rietsch, E. (1978). Extreme Models from the Maximum Entropy Formulation of Inverse Problems. J. Geophysics 44, 273-275.

13 Romanowicz, B. and K. Lambeck (1977). The Mass and Moment of Inertia of the Earth. Phys. Earth Planet. Inter. 15, P1-P4.

APPENDIX A

The Backus-Gilbert Approach

The information about the earth's density can be written in the form

$$\bar{\rho} = \frac{3}{R^3} \int r^2 \rho(r)dr \qquad \text{(A-1)}$$

$$y\bar{\rho} = \frac{5}{R^5} \int r^4 \rho(r)dr \qquad \text{(A-2)}$$

A linear combination of (A-1) and (A-2) has the form

$$(\mu_1 + \mu_2 y)\bar{\rho} = \int K(r)\rho(r)dr \qquad \text{(A-3)}$$

where

$$K(r) = [3\mu_1 (\frac{r}{R})^2 + 5\mu_2 (\frac{r}{R})^4]/R \qquad \text{(A-4)}$$

and μ_1, μ_2 are yet undetermined parameters.

The right hand side of (A-3) can be regarded as a weighted average of the density with weight function K(r). The parameters μ_1 and μ_2 are now chosen in such a way that K(r), the so-called "resolving kernel", is unimodular and resembles the Dirac δ-distribution δ(r-r') for some radius r'. With r' included in the argument list, the first condition reads

$$\int K(r,r')dr = 1. \tag{A-5}$$

The requirement that K(r,r') be peaked at radius r' is usually formulated as

$$\int (r - r')^2 K^2(r,r')dr = \text{minimum}. \tag{A-6}$$

The functional in (A-6) attains its minimum subject to condition (A-5) for

$$\mu_1(r') = [\frac{20}{11} - \frac{15}{4} x + \frac{40}{21} x^2]/(3D) \tag{A-7}$$

$$\mu_2(r') = -[\frac{40}{21} - \frac{15}{4} x + \frac{12}{7} x^2]/(5D) \tag{A-8}$$

where

$$D = \frac{52}{231} - \frac{1}{2} x + \frac{92}{315} x^2 \tag{A-9}$$

$$x = r'/R \tag{A-10}$$

These values of μ_1 and μ_2 have been used to compute the density estimate

$$\rho(r') = \bar{\rho}[\mu_1(r') + \mu_2(r')y] \tag{A-11}$$

shown in Figure 1. The graph of the resolving kernel for r' = 4371 km (x = 0.686 illustrates its resolving power - or rather the lack of it.

APPENDIX B

The Generalized Inverse Method

Let $\underline{A}$ denote a matrix with M rows and N columns. Such a matrix can be represented in the form (Lanczos, 1961)

$$\underline{A} = \underline{U}\,\underline{D}\,\underline{\tilde{V}} \tag{B-1}$$

where $\underline{U}$ is an M x p matrix whose columns are the eigenvectors $\underline{u}_i$ of the "data space". These eigenvectors are defined by

$$\underline{A}\,\underline{\tilde{A}}\,\underline{u}_i = \mu_i^2 \underline{u}_i \qquad i = 1,\ldots,p \tag{B-2}$$

where p denotes the number of non-zero eigenvalues of $\underline{A}\underline{\tilde{A}}$ and $\underline{\tilde{A}}\underline{A}$. Similarly, $\underline{V}$ is an N x p matrix whose columns are the eigenvectors $\underline{v}_i$ of the "model space". They are defined by

$$\underline{\tilde{A}}\,\underline{A}\,\underline{v}_i = \mu_i^2\,\underline{v}_i \qquad i = 1,\ldots,p, \tag{B-3}$$

and $\underline{D}$ denotes the p x p diagonal matrix of non-zero eigenvalues μ_i. The tilde denotes transposition.

The generalized inverse $\underline{A}^I$ of the matrix $\underline{A}$ can then be expressed in the form

$$\underline{A}^I = \underline{V}\,\underline{D}^{-1}\,\underline{\tilde{U}}\,, \tag{B-4}$$

and a solution of

$$\underline{A}\,\underline{x} = \underline{b} \tag{B-5}$$

is

$$\underline{x} = \underline{V}\,\underline{D}^{-1}\,\underline{\tilde{U}}\,\underline{b}\,. \tag{B-6}$$

Let $\underline{f}_j$ denote the j-th row vector of the matrix $\underline{A}$. Then

$$\underline{\tilde{A}}\,\underline{A} = \sum_j \underline{f}_j\underline{\tilde{f}}_j \tag{B-7}$$

Similarly

$$\underline{A}\,\underline{\tilde{A}} = \begin{pmatrix} \underline{\tilde{f}}_1\underline{f}_1 & \underline{\tilde{f}}_1\underline{f}_2 & \cdots & \underline{\tilde{f}}_1\underline{f}_M \\ & \cdots & & \\ \underline{\tilde{f}}_M\underline{f}_1 & \underline{\tilde{f}}_M\underline{f}_2 & \cdots & \underline{\tilde{f}}_M\underline{f}_M \end{pmatrix} \tag{B-8}$$

According to (B-7) the matrix $\underline{\tilde{A}A}$ is the sum of the M matrices of rank 1, and its eigenvectors $\underline{v}_i$ can be represented as a linear combination of the M vectors $\underline{f}_i$.

$$\underline{v}_i = \sum_k c_{ik}\underline{f}_k \tag{B-9}$$

Substituting (B-7) and (B-9) in (B-3), we get

$$\sum_k \underline{\tilde{f}}_j\underline{f}_k c_{ik} = \mu_i^2\, c_{ij} \tag{B-10}$$

which shows that the $\underline{c}_i$, the vectors whose components are c_{ik}, are proportional to the eigenvectors $\underline{u}_i$. From the normalization condition $\underline{v_i v_i} = 1$ follows that the factors of proportionality are $1/\mu_i$. Thus

$$\underline{c}_i = \frac{1}{\mu_i} \underline{u}_i \, , \tag{B-11}$$

and $\underline{V}$ can be written as

$$\underline{V} = \underline{\tilde{A}} \; \underline{C} \tag{B-12}$$

where $\underline{C}$ denotes the matrix whose columns are the vectors $\underline{c}_i$. Because of (B-11)

$$\underline{C} = \underline{U} \; \underline{D}^{-1} \qquad \underline{V} = \underline{\tilde{A}} \; \underline{U} \; \underline{D}^{-1} \, , \tag{B-13}$$

and

$$\underline{A}^I = \underline{V} \; \underline{D}^{-1} \underline{\tilde{U}} = \underline{\tilde{A}} \; \underline{U} \; \underline{D}^{-2} \underline{\tilde{U}} \tag{B-14}$$

Substituting (B-14) in (B-6), we find that the desired solution $\underline{x}$ of (B-5) can be represented as a linear combination of the row vectors $\underline{f}_i$ of $\underline{A}$

$$\underline{x} = \sum_i \lambda_i \underline{f}_i \tag{B-15}$$

The λ_i are the components of a vector $\underline{\lambda}$ given by

$$\underline{\lambda} = \underline{U} \; \underline{D}^{-2} \; \underline{\tilde{U}} \; \underline{b}. \tag{B-16}$$

Equations (4) and (5) represent a linear system of two equations (M=2) of the form (B-5). The vectors $\underline{f}_1$ and $\underline{f}_2$ have components V_n and I_n, respectively.

As a specific example, let us assume that the shells have all the same volume. Then r_n is given by Eq. (13), and

$$f_{1n} = 1/N, \quad f_{2n} = [n^{5/3} - (n-1)^{5/3}]/N^{5/3} \tag{B-17}$$

$$\underline{A} \; \underline{\tilde{A}} = \frac{1}{N} \begin{pmatrix} 1 & 1 \\ 1 & \sum\limits_n [n^{5/3} - (n-1)^{5/3}]^2/N^{2/3} \end{pmatrix} \tag{B-18}$$

For N = 100, the eigenvalues of $\underline{A\tilde{A}}$ turn out to be

$$\mu_1^2 = 2.09975/N, \quad \mu_2^2 = 0.09070/N \, . \tag{B-19}$$

Since they are non-zero, the two vectors $\underline{f}_1$ and $\underline{f}_2$ are linearly independent (which in this case was, of course, obvious), and

$$\underline{U} = \begin{pmatrix} 0.67276 & 0.73986 \\ 0.73986 & -0.67276 \end{pmatrix} \tag{B-20}$$

From (B-16) follows then with $b_1 = \bar{\rho}$, $b_2 = y\bar{\rho}$

$$\lambda_1 = 1.8401 \, N\bar{\rho}, \quad \lambda_2 = -0.8401 \, N\bar{\rho} \tag{B-21}$$

and thus

$$\rho_n = \{1.8401 - 0.8401[n^{5/3} - (n-1)^{5/3}]/N^{2/3}\}\bar{\rho} \qquad \text{(B-22)}$$

The generalized inverse method is, of course, also applicable for $N \rightarrow \infty$. In this case the sums in Eqs. (4) and (5) are replaced by integrals

$$\int f_i(r)\rho(r)r^2dr = b_i \qquad i = 1,2 \qquad \text{(B-23)}$$

where

$$f_1(r) = 3/R^3, \quad f_2(r) = 5r^2/R^5, \quad b_1 = \bar{\rho}, \quad b_2 = y\bar{\rho} \qquad \text{(B-24)}$$

According to (B-15), $\rho(r)$ can be expressed as a linear combination of $f_1(r)$ and $f_2(r)$, and

$$\underline{A}\,\underline{\tilde{A}} = \frac{3}{R^3}\begin{pmatrix} 1 & 1 \\ 1 & 25/21 \end{pmatrix} \qquad \text{(B-25)}$$

The eigenvalues of this matrix turn out to be

$$\mu_1^2 = 2.09976(3/R^3), \quad \mu^2 = 0.09071(3/R^3) \qquad \text{(B-26)}$$

which, apart from a different scale factor, are close to those obtained for N = 100 (Eq. (B-19)). Similarly

$$\underline{U} = \begin{pmatrix} 0.67275 & 0.73987 \\ 0.73987 & -0.67275 \end{pmatrix} \qquad \text{(B-27)}$$

is close to (B-20), and the generalized inverse solution for the density is

$$\rho(r) = [1.84 - 0.84 \cdot \tfrac{5}{3}(r/R)^2]\bar{\rho} \qquad \text{(B-28)}$$

APPENDIX C

Entropy Expressed in Terms of Expectation Values

The probability distribution for the density values ρ_n is (Eq. (14))

$$p(\underline{\rho}) = \exp(-\textstyle\sum h_n\rho_n)/Z \qquad \text{(C-1)}$$

where

$$h_n = 1/\bar{\rho}_n = (\lambda_1 + I_n\lambda_2) \qquad \text{(C-2)}$$

$$Z = \int \exp(-\textstyle\sum h_n\rho_n)d\rho = \prod_n 1/h_n \qquad \text{(C-3)}$$

and sums and products over n extend from 1 to N.

Equations (C-1) and (C-3) have the forms generally found when the entropy of a probability distribution is maximized under linear constraints. Substitution of (C-1) for $p(\underline{\rho})$ in expression (12) for the entropy ($w(\underline{\rho})$ is constant and can be set to unity) leads to

$$\begin{aligned} H &= \int [\exp(-\sum h_n\rho_n)/Z][\sum h_n\rho_n + \ln Z]d\rho \\ &= \sum h_n\bar{\rho}_n + \ln Z \end{aligned} \tag{C-4}$$

Because of (C-2) and (C-3)

$$\begin{aligned} H &= N - \sum \ln(h_n) \\ &= \sum \ln(e/h_n) \end{aligned} \tag{C-5}$$

For $N \to \infty$, that is when the number of shells goes to infinity, the expression (C-5) for the entropy diverges in general. Replacing the entropy by the entropy rate H/N (the entropy per shell) while retaining the same symbol we get

$$H = \sum \ln(e/h_n)\,\frac{1}{N} \tag{C-6}$$

For large N

$$\begin{aligned} h_n &= \lambda_1 + \lambda_2[n^{5/3} - (n-1)^{5/3}]/N^{2/3} \\ &= \lambda_1 + \lambda_2\,\frac{5}{3}\,(\frac{n}{N})^{2/3} + O((nN^2)^{-1/3}) \end{aligned} \tag{C-7}$$

For $N \to \infty$ and (see Eq. (13))

$$(\frac{n}{N}) \to x = (\frac{r}{R})^3 \tag{C-8}$$

we get

$$H = \int \ln(e/[\lambda_1 + \frac{5}{3}\lambda_2 x^{2/3}]dx = \frac{3}{R^3}\int \ln(e\rho(r))r^2 dr \tag{C-9}$$

Since scale factors and additive constants have no influence on the maximum of H we get the same result independent of whether

$$H = \int \ln[\rho(r)]r^2 dr \tag{C-10}$$

or (C-9) is maximized.

Figure 1. Comparison of the density distribution within the earth obtained with the Backus-Gilbert technique (solid line) with Bullen's density profile (dashed line). The dotted line represents the shape of the resolving kernel for a depth of 2000 km.

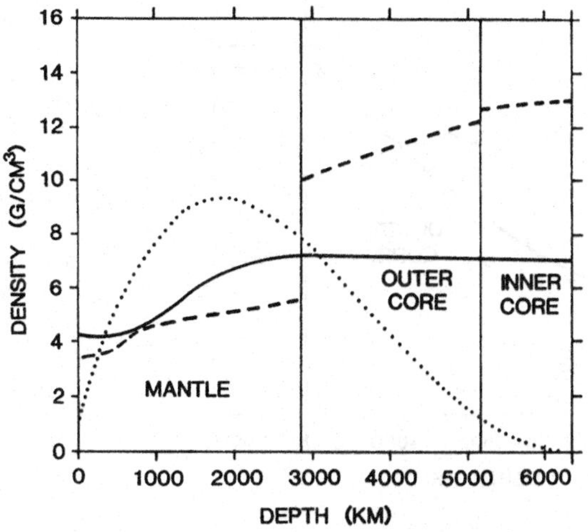

Figure 2. Comparison of the discrete (100 shells) and continuous density distribution within the earth obtained by the generalized inverse method (solid line) with Bullen's density profile (dashed line).

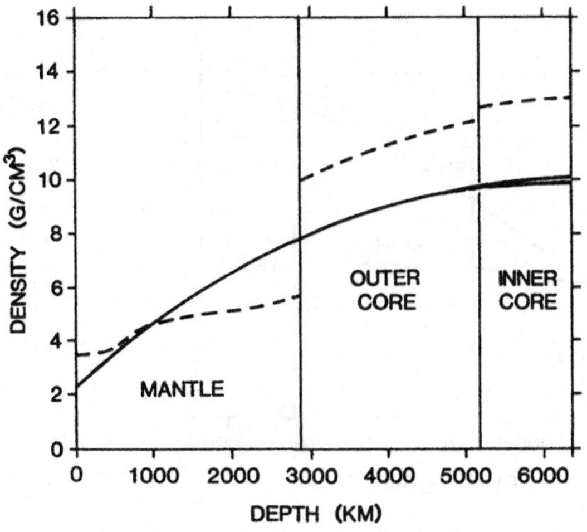

Figure 3. Comparison of the discrete (100 shells) and continuous density distribution within the earth obtained by the maximum entropy method with Bullen's density profile (dashed line).

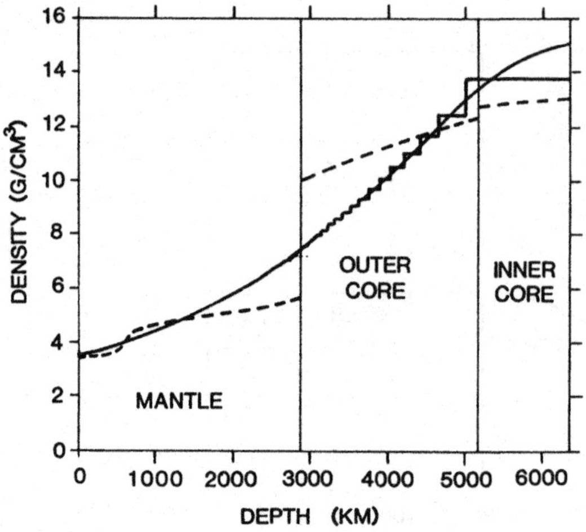

Figure 4. Comparison of the density distribution obtained with Graber's approach (solid line) with Bullen's density profile (dashed line).

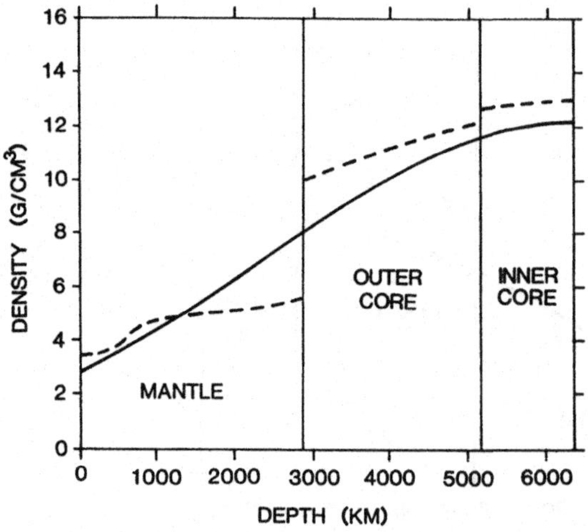

ENTROPY AND SOME INVERSE PROBLEMS IN EXPLORATION SEISMOLOGY

James H. Justice

1 INTRODUCTION

The ultimate problem in exploration seismology is the reconstruction by inference of the structure of portions of the earth's crust to depths of several tens of kilometers from seismic data recorded on the earth's surface. These measurements represent data obtained by recording the arrival of wave fields which are reflected back to the earth's surface due to variations in acoustic impedance within its interior. This data is complicated by multiple travel paths, conversion of wave modes (compressional or shear) at boundaries between homogeneous layers, and is corrupted by additive noise.

Over a long period of development, various procedures have been conceived which, when applied sequentially to seismic data, attempt to accurately reconstruct an earth model which most likely generated the observed data. This process is hampered not only by the complications mentioned above but also by the severely band-limited nature of seismic data (typically 5 to 100 Hz with most energy concentrated around 30 hz, for example) which introduces limitations on resolution, and difficulties in applying certain operators to the data in an attempt to improve resolution.

Most inverse procedures applied to seismic data today are deterministic procedures whose derivation is based on the "convolution model" to be introduced later. Recently, however, some new approaches to inversion of seismic data have been suggested. In contrast to previous methods devised for inverting data, these methods do not rely on operators applied directly to the data but rely instead on directly estimating an earth model which would generate data consistent with the observed data (which may still require some processing).

In recent years, significant results have been obtained by applying this kind of approach to the problem of image reconstruction. One principle which has met with some success in estimating the most probable (true) image which generated a blurred or noisy observed image is the principle of maximum entropy.

Certain inverse problems in exploration seismology can be formulated, based on the convolution model which is analogous to the model used in image reconstruction. In keeping with the idea of estimating a most probable earth model which generated an observed data set (which is dangerous to operate on directly, due to its inherent limitations), the principle of reconstruction based on maximizing certain probabilities subject to constraints imposed by prior knowledge suggests itself as a reasonable avenue for seismic inversion.

In this paper we shall suggest a variety of approaches to formulating certain inverse problems in exploration seismology consistent with the point of view outlined above.

2 PLANE WAVES AND LAYERED MEDIA

In our discussion, we shall restrict our attention to a special case of wave propagation in a heterogeneous medium. In particular, we shall restrict our attention to plane waves normally incident on a stack of plane layers. The velocity and density within each individual layer will be assumed to be constant and we shall consider only compressional waves, that is waves in which particle motion is in the direction of wave propagation (a similar analysis could be carried out for the shear wave field as well).

Assuming that the medium of interest (the earth) is an elastic solid satisfying the conditions of linearity in the stress-strain relationship for small displacements, the elastic wave equation [27]

$$\rho \frac{\partial^2 \tilde{u}}{\partial t^2} = (\lambda+\mu)\ \nabla\theta + \mu\nabla^2 \tilde{u} .$$

may be derived where ρ is the density, $\tilde{u}$ the displacement vector, λ and μ are the Lamé (elastic) constants of the medium and θ = div $\tilde{u}$.

When $\tilde{u}$ can be expressed in the form (Helmholtz Theorem)

$$\tilde{u} = \nabla\phi - \nabla \times \psi$$

with $$\nabla\cdot\psi = 0$$

we find that the scalar wave potential ϕ, satisfies the equation (by substitution above)

$$\frac{\partial^2\phi}{\partial t^2} = c^2\nabla^2\phi$$

where the displacement

$$\tilde{w} = \nabla\phi$$

is in the direction of wave propagation and the velocity c is related to the Lame constants by

$$c = ((\lambda + 2\mu)/\rho)^{\frac{1}{2}}$$

ϕ is called the scalar or compressional wave potential.

We now restrict our attention to a plane wave potential propagating in a vertical direction in our layered earth model

$$\phi(z,t) = \exp(j(\omega t \pm 2\pi kz))$$

where k is the wave number and the $\pm$ sign determines propagation downward or upward through the medium.

Since we are interested in a layered medium, we must consider what happens at a boundary between layers. We require that pressure and the normal component of displacement $\tilde{w} = \nabla\phi$ be continuous across a boundary. Since elastic waves suffer both reflection and transmission at any boundary where acoustic impedance (the product of density and velocity) changes, we must account for the amplitudes of the incident, reflected, and transmitted waves.

Assuming a downward propagating incident wave, (z positive downwards)

$$\phi_I = \exp(j(\omega t - 2\pi kz))$$

the reflected and transmitted waves at the boundary can be denoted by

$$\phi_R = R \exp(j(\omega t + 2\pi kz))$$

$$\phi_T = S \exp(j(\omega t - 2\pi kz))$$

Using our boundary conditions, with density and velocity ρ_1 and V_1 above the boundary and ρ_2 and V_2 below the boundary, we obtain

$$\rho_1 (\phi_I + \phi_R) = \rho_2 \phi_T$$

$$\frac{\partial}{\partial z} (\phi_I + \phi_R) = \frac{\partial}{\partial z} \phi_T$$

Now it can be checked that

$$R = \frac{\text{pressure of reflected wave}}{\text{pressure of incident wave}}$$

and

$$\frac{\rho_2}{\rho_1} S = \frac{\text{pressure of transmitted wave}}{\text{pressure of incident wave}}$$

(where pressure $= -\rho \dfrac{\partial^2 \phi}{\partial t^2}$

If we wish to keep track of pressures in the medium due to passage of the waves then we can easily derive (substituting our wave potentials into the boundary conditions and solving)

$$R = \frac{\rho_2 V_2 - \rho_1 V_1}{\rho_2 V_2 + \rho_1 V_1}$$

and

$$T = \frac{\rho_2}{\rho_1} S \equiv 1 + R = \frac{2\rho_2 V_2}{\rho_2 V_2 + \rho_1 V_1}$$

where T is called the (pressure) transmission coefficient. Note that we could also derive similar coefficients for particle displacement or velocity if these are the quantities of interest. We simply replace R by -R above to obtain the coefficients for these other two cases.

R is called the reflection coefficient for the boundary between the two media and the product $Z = \rho v$ is called the acoustic impedance. In terms of acoustic impedances, we may write

$$R = \frac{Z_2 - Z_1}{Z_2 + Z_1}$$

$$T = \frac{2Z_2}{Z_2 + Z_1}$$

3 RESPONSE OF A LAYERED MEDIUM AND THE INVERSE PROBLEM

Let us now consider a layered medium sandwiched between two half spaces. The medium will be assumed to have constant density and velocity (i.e., acoustic impedance will be constant) within each layer, and reflection and transmission coefficients are computed from our previous equations at each boundary. We shall abstract our results from plane waves and identify each (normally propagating) plane wave by its amplitude and direction of propagation. In accordance with our previous derivations, we shall take acoustic pressure as the measurement of interest. Let us introduce a unit pulse (unit amplitude normally propagating plane wave) into the first layer of the medium and follow its progress through the medium, accounting for our measurement (pressure) along the way. Our medium shall be set up as shown in Figure 1 (ΔT is a fixed travel time between layers). The sequence $\{c_k\}$, $0 \leq k \leq N$ is the sequence of reflection coefficients derived earlier.

If we take $c_0 = -1$ (a free surface) then we may write, (taking ΔT = 1/2) using z-transforms

$$R_N(z) = \frac{z^N G_N(z^{-1})}{A_N(z)} \equiv r_1 z + r_2 z^2 + \ldots. + r_N z^N + \ldots$$

Figure 1.

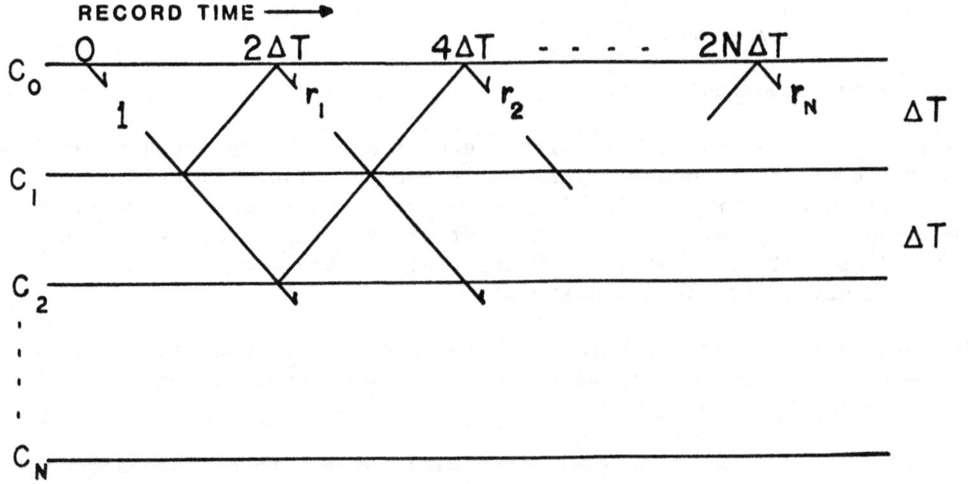

where $R_N(z)$ is the z-transform of the downgoing acoustic pressure response in layer 1 for an N-layer medium [15]. The polynomials G_N and A_N are given recursively by

$$A_k(z) = F_k(z) - z^k G_k(z^{-1})$$

$$F_k(z) = F_{k-1}(z) + c_k z\, G_{k-1}(z)$$

$$G_k(z) = c_k F_{k-1}(z) + z\, G_{k-1}(z)$$

with $F_0(z) = 1$

$G_0(z) = 0$

$A_k(z)$ also satisfies the recursion

$$A_k(z) = A_{k-1}(z) - c_k z^k A_{k-1}(z^{-1})$$

We see then, that given this rather ideal model and the known reflection coefficients, we can quickly generate the response in the first layer which we may take to be a model or synthetic seismogram for an impulsive source in this medium.

It is easy to verify, using the principle of superposition, that if we introduced a time varying disturbance w (a wavelet) into the medium in place of the impulse, our observed seismogram would take the form

$$S(z) = W(z)\, R_N(z) \qquad \text{(frequency domain)}$$

or $$s(t) = w * r \qquad \text{(time domain)}$$

Since a real seismogram always contains additive noise, we may modify our ideal seismogram to the form

$$s(t) = (w*r)(t) + n(t)$$

where n(t) is noise.

This simple model for a seismogram generated by the response of an ideal layered earth to a normally incident plane wave field is called the convolution model for reflection seismology. This model is widely used and though it appears to be rather far removed from reality in many situations, data is processed in exploration seismology so that this model is often closely approximated.

An interesting inverse problem now presents itself; that is, to infer the reflection coefficient sequence $\{c_k\}$, $1 \leq k \leq N$ from an observed response in the first layer.

Though our model may take other (similar) forms with analogous recursive formulae for the response ($|c_0| < 1$, for example), we have chosen this particular representation for discussion here because there is an elegant and simple direct inversion procedure available in the ideal case.

In particular, if the noise-free response to an ideal impulse has been observed, then $R_N(z)$ can be inverted to obtain the reflection coefficients, $\{c_k\}$, $1 \leq k \leq N$ by the procedure [15]:

$$c_1 = r_1, \; A_o(z) = 1$$

$$A_k(z) = A_{k-1}(z) - c_k z^k A_{k-1}(z^{-1})$$

$$\bar{r}_{k+1} = -(a_{k1} r_k + a_{k2} r_{k-1} + \ldots + a_{kk} r_1)$$

$$c_{k+1} = (r_{k+1} - \bar{r}_{k+1}) / \prod_{j=1}^{k} (1-c_j^2)$$

We begin with k=1 and stop when c_N has been calculated.

While this inversion procedure solves the ideal inverse problem in the noise-free case for a plane wave normally incident on a stack of plane layers, the procedure is extremely sensitive to noise and has not found widespread practical application.

We may now pose the inversion problem which we wish to discuss in the remainder of this paper: Given an observed seismogram representing the (assumed) acoustic pressure response of a horizontally layered medium to an input wavelet (pressure wave field) corrupted by additive noise, infer the reflection coefficient sequence, $\{c_k\}$ or, equivalently, the acoustic impedances $\{Z_k\}$ of the individual layers comprising the model.

Acoustic impedances may be obtained from the reflection coefficients using the relation

$$Z_k = Z_{k-1} \frac{(1 + c_{k-1})}{(1 - c_{k-1})}$$

or, equivalently

$$Z_k = Z_1 \prod_{\ell=1}^{k-1} \frac{(1+ c_\ell)}{(1- c_\ell)}$$

where Z_1 is assumed known and $c_k \neq 1$ for any k, $k \geq 1$.

4 DECONVOLUTION AND INVERSION

We shall now direct our attention to the convolution model for a seismic trace (seismogram) and the related inversion problem of estimating the reflection coefficients for this model. The direct inversion procedure referred to earlier, to be successful, would require perfect noise suppression on the data followed by a perfect (non band-limited) deconvolution. These requirements exceed our reasonable expectation in dealing with real data. As a result, we must look further to identify procedures which are robust in the presence of insufficient information or data from which information has been irretrievably lost due to band width constraints.

Let us review the approach which is widely used to deconvolve seismic data, and discuss its effect on subsequent attempts at inversion. Given the convolution model for a seismic trace,

$$s = w * r + n$$

where s is the trace, w is the source wavelet input into the earth, r the impulse response (not to be confused with the reflection coefficient sequence c_k which it may approximate) and n is additive noise.

Depending on the form and coherency of the noise, various procedures may be applied to the trace (or traces) to enhance signal-to-noise ratio. No matter how effective these may be, however, we may always assume that some residual noise remains on the traces and it is this residual noise which appears in the remainder of our discussions.

Standard procedure is then to estimate in some fashion a deconvolution filter which, ideally, both removes the smearing effect of the wavelet and minimizes the error due to noise. There is usually a trade-off between these two criteria, and doing well with one usually implies a poorer performance with the other.

A simple and widely used procedure is to correlate the seismic trace, s, with itself. Under the assumptions that noise is white and uncorrelated with w and r, and further that r is white we obtain

$$S \otimes S = (w \otimes w) * (r \otimes r) + n \otimes n =$$
$$= (w \otimes w) * (\sigma_r^2 \delta) + \sigma_n^2 \delta$$

where σ_r and σ_n are the energy associated with the r and n sequences. The result is a scaled and biased estimate of the wavelet autocorrelation. If we now assume that the wavelet is minimum phase (not a bad assumption for dynamite sources, for example) we can uniquely infer the wavelet, w, provided that estimates of σ_r and σ_n are available.

Many other procedures have been suggested and are used including direct monitoring of the wavelet, which is possible in marine work. In any event, if the wavelet is minimum phase, we may use a variety of procedures to obtain an approximation of any desired length to its inverse, $f \cong w^{-1}$. Applying $f \cong w^{-1}$ to the original trace yields

$$f * s = f * w * r + f * n \cong \delta * r + f * n$$
$$= r + f * n$$

Of course, we would like to estimate r but the factor f*n can be quite a large error. The error due to the term f*n manifests itself especially at the higher frequencies. In terms of z-transforms we have

$$F(z)\ S(z) \cong \frac{W(z)\ R(z)}{W(z)} + \frac{N(z)}{W(z)}$$

If W(z) has low power in the high frequencies (as is usually the case) and N(z) does not, the quotient $\left|\frac{N(z)}{W(z)}\right|$ can be quite large at high frequencies. The deconvolution filter is most often estimated by seeking a least-squares approximation of a given length to the inverse of w (Wiener "spiking" filter). This process tends to be reasonably robust in practice and it may be modified to include a factor to control noise amplification, a procedure which necessarily degrades its performance.

We may also note that deconvolution carried out by convolving the trace with a deconvolution operator cannot increase the bandwidth of the original data and so a perfect impulse response $\{r_k\}$ is simply not achievable in practice using this procedure, no matter how the deconvolution filter, f, may have been arrived at. We may now inquire what effect an imperfect (and band-limited) estimate of the impulse response r_k may have on a subsequent inversion. There are several approaches to direct inversion of an impulse response but all are similar in spirit to the solution given earlier and all exhibit sensitivity to error (either from imperfect deconvolution or from additive noise) which may render the results useless if the deconvolved data is not essentially perfect, a case not likely to be met in practice. A good comparison of a variety of approaches to inversion along the lines discussed here may be found in [22].

An analysis of the error which may occur in a direct inversion of this kind may be found in [1]. In this study it is shown first that multiples (events which have reflected more than once in the medium as opposed to "primaries") in general cannot be ignored in the inversion process (as is often done by attempting to remove all multiple reflections before inversion). Second, it is shown that even when the full impulse response (assumed to be properly corrected to fit the assumptions) is used, then even under the assumption that all previous reflection coefficients have been estimated with an accuracy of .025 and that no errors at all have been made in deconvolution, the uncertainty in estimating the next reflection coefficient may significantly exceed its expected value. This error increases as the index of the reflection coefficient increases (see also [1a]).

Finally, it should be pointed out that, in conversion to acoustic impedance, the low frequency component which is missing in seismic data must be estimated. This is often done using estimated velocities from move-out characteristics of offset seismic data together with an assumed relationship between velocity and density which may take the form

$$\rho = KV^{\frac{1}{4}}$$

for example [13] where K is a constant (K = .31, for example, if V is in m/sec and ρ is in g/cm^3).

We have pointed out a few of the errors associated with deconvolution and with the attempt to apply direct inversion algorithms to less-than-perfectly deconvolved data. It seems reasonable to seek approaches different from those described thus far, but before abandoning the concepts of deconvolution and direct inversion, let us consider some applications of the maximum entropy concept to these problems.

5 APPLICATIONS OF THE BURG ALGORITHM IN DECONVOLUTION AND INVERSION

The first applications of maximum entropy analysis to deconvolution and inversion were developed through use of the Burg algorithm [5].

If we recall our earlier discussion, it was pointed out that one of the most widely used deconvolution procedures in seismic data processing is to obtain an approximate least-squares inverse of a fixed length for the estimated source wavelet. One procedure for doing this was based on estimating the autocorrelation coefficients of the wavelet directly from the seismic trace under the assumption that the noise and reflectivity sequences are white and uncorrelated with each other.

Under these assumptions, let us now estimate a prediction error filter $\{1, -a_1, ---, -a_N\}$ of some fixed length for the trace. We may then write

$$S_t = \sum_{1}^{N} a_k S_{t-k} + e_t$$

where e_t is the prediction error. Let us further assume that the trace satisfies the convolution model

$$S_t = \sum_{o}^{M} r_j w_{t-j} + n_t$$

where r_t is the impulse response of the medium and n_t is noise. We shall now apply the prediction error filter for the trace to the convolution model obtaining, after some computation:

$$\sum_{o}^{M} r_j [w_{t-j} - \sum_{1}^{N} a_k w_{t-k-j}] + n_t - \sum_{1}^{N} a_j n_{t-j} = e_t$$

we recognize the first term on the left as a convolution of the sequence $\{r_t\}$ with the term in brackets, i.e., $r*(w-a*w)$.

Under our previous assumptions (white reflectivity sequence, r; white noise sequence, n; r, n and w uncorrelated), the trace autocorrelation is a scaled version of the wavelet autocorrelation with an additive term at the origin, i.e.,

$$\phi_{ss} = \sigma_r^2 \phi_{ww}(t) + \sigma_n^2 \delta_t$$

But this tells us that the prediction error filter which we have estimated for the trace is actually a scaled version of the prediction error filter of the wavelet plus uncorrelated white noise. This means that there is a prediction error sequence ε_t satisfying

$$(n'_t + w_t) - \sum_{1}^{N} a_k(n'_{t-k} + w_{t-k}) = \varepsilon_t$$

where n' is a scaled version of n.

Substituting this expression into our previous expression for the prediction error of the trace, we obtain

$$\sum_{o}^{M} r_j \{\varepsilon_{t-j} - [n'_{t-j} - \sum_{1}^{N} a_k n'_{t-k-j}]\} + n_t - \sum_{1}^{N} a_j n_{t-j} = e_t$$

The presence of noise obviously obscures the interpretation of the prediction error sequence, e_t , but at least its influence on the output from our deconvolution with the prediction error operator is clear. If we now make the assumption that the noise can be assumed to be zero (the ideal case) the above formulation reduces to

$$\sum_{0}^{M} r_j \varepsilon_{t-j} = e_t$$

where

$$w_t - \sum_{1}^{N} a_k w_{t-k} = \varepsilon_t$$

since the prediction error operator for the trace is now the prediction error operator of length N+1 for the wavelet, we conclude that $\varepsilon_t = 0$, $1 \leq t \leq N$.

This in turn, implies

$$\sum_{0}^{M} r_k \varepsilon_{t-k} \equiv \varepsilon_o r_t \quad , \quad 0 \leq t \leq N$$

In other words, in the noise-free case, our deconvolution with the prediction error filter is a scaled version of the impulse response of the medium for the first N+1 output values. Beyond this point, the output is a weighted average of the true response.

If the wavelet itself were, in fact, an autoregressive process satisfying the prediction equation

$$w_t = \sum_{1}^{N} a_k w_{t-k} + w_o$$

then it is clear that (in the noise-free case)

$$\sum_{0}^{M} r_k \varepsilon_{t-k} \equiv w_o r_t$$

for all t. For this reason, it is obviously advantageous to assume that the wavelet is an autoregressive process and some inversion procedures model it as such.

To obtain the prediction error filter, we may solve the traditional Yule-Walker equations, or use the Burg algorithm to obtain increased accuracy. Having obtained the (noise-free) deconvolution, we may proceed with inversion as described earlier.

A direct approach to inversion using the Burg algorithm results from noting that the transmission response of our N-layer model satisfies [21]

$$T(z) = \frac{\prod_{j=1}^{N} (1-c_j)}{A_n(z)}$$

where T(z) is the z transform of the wave transmitted into the lower half-space. $A_n(z)$ satisfies

$$A_j(z) = A_{j-1}(z) + c_j z^j A_{j-1}(z^{-1})$$
$$A_o(z) = 1$$

where we have assumed $c_o=1$ in this case.

Clearly the transmission response is an autoregressive process. We may either use a Yule-Walker approach or the Burg algorithm to invert the transmission response for the reflection coefficients. If we use the Burg algorithm to estimate the successive approximations to $A_N(z)$ then the reflection coefficients may be obtained from [21]

$$c_j = \frac{-2\sum_{n=0}^{M-j}\left[\left(\sum_{k=0}^{j-1} a_k^{(j-1)} t_{k+n}\right)\left(\sum_{k=0}^{j-1} a_k^{(j-1)} t_{j-k+n}\right)\right]}{\sum_{n=0}^{M-j}\left[\left(\sum_{k=0}^{j-1} a_k^{(j-1)} t_{k+n}\right)^2 + \left(\sum_{k=0}^{j-1} a_k^{(j-1)} t_{j-k+n}\right)^2\right]}$$

where t_j is the impulse response of the seismogram. $A_j(z)$ is obtained from the recursion above.

Another use of the Burg algorithm in inversion for the acoustic impedance has been suggested by Walker and Ulrych [32]. In this approach it is assumed that the data has been corrected to remove all effects of multiples and that transmission effects have been corrected as well, so that the impulse response in the convolution model actually coincides with the reflection coefficient sequence $\{c_k\}$. They further assume that the data has been perfectly deconvolved within the bandwidth of the wavelet. The result, then, is a noisy bandlimited estimate of the reflection coefficients. The problem is to remove the band limited constraint by predicting information in the missing bands, hopefully in a manner not too sensitive to noise.

Under these assumptions, the Fourier transform of the seismogram takes the form

$$Y = CW + N$$

where Y is the transformed seismogram, C is sum of weighted exponentials (transform of reflection coefficients) W is 1 on some portion of the frequency band (determined by the wavelet) and zero elsewhere and N is the noise transform. The method is based on the fact that C (the transform to be extended to the full band by prediction) is an ARMA process which may be approximated by an AR process. To extend C over the low frequency end of the band, Yule-Walker equations or Burg estimation may be used. Walker and Ulrych [32] state a preference for a modified minimum entropy extension for the high frequency band. The acoustic impedance is obtained from our previous equations once the reflection coefficient sequence has been estimated.

6 ALTERNATIVE APPROACHES TO INVERSION

In our previous discussions, we have considered applications of an algorithm derived from the maximum entropy principle but we have not considered the application of that principle directly to the inversion problem. The maximum entropy principle provides us with a way to infer a solution to our problem which, ideally, honors only constraints resulting from prior knowledge. These may be imposed, for example, by the convolution model.

It seems appropriate, then, to formulate direct approaches to inversion which may avoid the problems associated with deconvolution but which honor the type of model such as the convolution model on which such procedures are based.

The band-limited nature of seismic data poses problems in inversion as we have seen and implies a non-uniqueness to the solution. As a result, it seems reasonable to seek a solution which satisfies the usual constraints but which in some sense either makes no unwarranted assumption about the solution or is the most probable by some reasonable criterion.

While we may begin directly with the definition of an entropy measure which we seek to maximize, we might better be guided by probabilistic considerations in formulating our model as is commonly done in image reconstruction, for example.

In image reconstruction, we imagine that every possible picture can be obtained by assigning silver grains to pixels in every possible way. The probability of occurrence of a particular image is then determined by the number of different ways in which it can be formed. Maximizing the probability of occurrence of an image, subject to constraints, leads in a natural way to maximizing the entropy measure.

$$E = - \sum p_k \ell n \; p_k$$

subject to these constraints [11].

We might follow a similar analysis for the reflection coefficient sequence in the geophysical inverse problem but there are some differences between this problem and the image reconstruction problem though the models are identical.

In the geophysics case, it does not seem warranted to assign probabilities as it is done in image reconstruction since many reflection coefficient sequences, while being theoretically possible are highly improbable on physical grounds. This is to say that not every possible assignment of reflection coefficients should be equally probable with any other. To assume this would surely result in many physically unacceptable inversions. In addition, reflection coefficients unlike reflectances, are not constrained to be positive, but they are bounded $|c_k| \leq 1$, and rarely exceed a few tenths in magnitude.

As a result, it seems reasonable to look for a new approach to assigning probabilities to reflection coefficient sequences in which we have more control over what is probable and what is not.

In an effort to find a reasonably simple approach which will agree with the limitations mentioned above, we might assign probabilities to reflection coefficient sequences as follows.

Let us imagine the reflection coefficient sequence as a series of bins to which may be added various small (but equal) quanta of "reflectivity", much as is done in image analysis. We are therefore selecting reflection coefficients from a finite set which simplifies the problem somewhat. We shall fill each bin separately based on a fixed number L of trials of a Bernoulli experiment. The outcome for each bin is assumed to be independent of the other outcomes and the probability of success may vary from bin to bin to reflect prior geological knowledge.

The probability that n_k quanta will be assigned to bin k is, therefore (the kth reflection coefficient will eventually be determined by n_k)

$$p_k(n_k) = \binom{L}{n_k} p_k^{n_k} (1-p_k)^{L-n_k}$$

where P_k is the probability of success assigned to the kth bin, and this single probability is all that needs to be determined if prior information is available. For example, if it is known that

$E(n_k) = n_e$, where E is expected value,

then $p_k = n_e/L$

We may now infer the probability of occurrence of a particular sequence

$$P(n_1,\dots,n_N) = \prod_{k=1}^{N} \binom{L}{n_k} p_k^{n_k} (1-p_k)^{L-n_k} =$$

$$= \frac{(L!)^N}{\prod_{k=1}^{N} n_k!\,(L-n_k)!} \prod_{k=1}^{N} p_k^{n_k} (1-p_k)^{L-n_k}$$

and so

$$\ell n\, P(p_1,\dots,p_N) = N\, \ell n\, L! - \sum \ell n\, n_k! - \sum \ell n\, (L-n_k)! + \sum n_k\, \ell n\, p_k$$
$$+ \sum (L-n_k)\, \ell n\, (1-p_k)$$

We now invoke Stirlings formula (for large L and n_k) $\ln n! \cong (n\, \ell n\, n)-n$. After some algebraic manipulations this reduces to

$$-L \left\{ \sum_1^N \frac{n_k}{L} \ell n \left(\frac{n_k}{L}\right) + \sum_1^N \left(1-\frac{n_k}{L}\right) \ell n \left(1-\frac{n_k}{L}\right) - \sum_1^N \frac{n_k}{L} \ell n \left(\frac{p_k}{1-p_k}\right) - \sum_1^N \ell n\, (1-p_k) \right\}$$

The ratio $\frac{n_k}{L}$ satisfies

$$0 \leq \frac{n_k}{L} \leq 1$$

There are now a variety of ways in which we could relate the ratio n_k/L to the reflection coefficient, c_k, and our choice might better be left open in any particular case.

A simple and reasonable approach, however, might be to assign

$$|c_k| = \frac{n_k}{L}$$

and let the sign of c_k be determined by the constraints. By doing this we allow the possibility that if p_k is small, c_k will be expected to be small but may be either positive or negative with equal probability. With this assignment, then, our probability becomes

$$\ln P(c_1,\ldots,c_n) = -L\left\{\sum_1^N |c_k| \ln |c_k| + \sum_1^N (1-|c_k| \ln (1-|c_k|)) - \sum_1^N |c_k| \ln \left(\frac{p_k}{1-p_k}\right) - \sum_1^N \ln (1-p_k)\right\}$$

Maximizing the probability of occurrence (without regard to constraints), we may ignore the last term above so that the expression to be maximized is

$$F(c_1,\ldots,c_n) = -\sum_1^N |c_k| \ln |c_k| - \sum_1^N (1-|c_k|) \ln (1-|c_k|) + \sum_1^N |c_k| \ln \left(\frac{p_k}{1-p_k}\right)$$

An alternative definition of c_k,

$$c_k = \frac{2n_k}{L} - 1$$

which satisfies $-1 \le c_k \le 1$

results in the symmetric form

$$F(c_1,\ldots,c_n) = -\sum_1^N \frac{1+c_k}{2} \ln \left(\frac{1+c_k}{2}\right) - \sum_1^N \frac{1-c_k}{2} \ln \left(\frac{1-c_k}{2}\right) + \sum_1^N \frac{1+c_k}{2} \ln \left(\frac{p_k}{1-p_k}\right)$$

In this latter case, assigning $p_k = 1/2$ to insure $E(c_k) = 0$ for $1 \le k \le N$ causes the last term to vanish, leaving us with a much simpler result, and one which would not be unreasonable in many geophysical situations. This particular choice of p_k is the only one in this case which will not bias the sign of the reflection coefficient to be either positive or negative.

If we now add a set of constraints (convolution model, etc.)

$$g_j = 0$$

then we may solve for the reflection coefficients which maximize the objective function F, subject to the constraints, using Lagrange multipliers, for example.

In the latter (simplest) case we arrive at:

$$c_j = \frac{1 - \exp\left[2 \sum \lambda_k \dfrac{\partial g_k}{\partial c_j}\right]}{1 + \exp\left[2 \sum \lambda_k \dfrac{\partial g_k}{\partial c_j}\right]}$$

Before looking further into the form of the constraints, let us delve more deeply into the assignment of prior probabilities in our objective function.

We would like in particular, to investigate the implications which result from treating the vertical lithologic sequence as a Markov process. Geologically this seems most reasonable and has been supported by some field studies [31]. Assignment of transitional probabilities for a Markov process is clearly dependent on the depositional environment and may change with both spatial location as well as depth. That is to say that the site of a deltaic sequence on a continental margin at one era of geologic time, may have been an area of carbonate build-up accompanied by anhydrite sequences in another era of time. These scenarios would each determine its own set of transitional probabilities within its section of the vertical sequence.

In cases such as these, assignment of transitional probabilities would be most effectively aided by well control, and the concept of inversion being discussed here would depend on the successful integration of geological and geophysical data. We must re-emphasize that this approach lends strong credence to the point of view that probabilities associated with reflection coefficients should be assigned on an individual basis depending on the position of their occurrence within the stratigraphic section.

With these thoughts in mind, let us suppose that M distinct lithologies $\ell_1, \cdots, \ell_M$ (associated with different values of velocity, density and/or Q for example) have been identified in a particular section of the stratigraphic column. We assign a set of transition probabilities to all possible pairs,

$$q_{kj} \equiv q(\ell_j \mid \ell_k) \equiv \text{probability that lithology } \ell_k \text{ is followed by lithology } \ell_j$$

There are M^2 of these transition probabilities which we may envision as an array

$$\begin{bmatrix} q_{11} & q_{12} & \cdots & q_{1M} \\ q_{21} & q_{22} & \cdots & q_{2M} \\ \vdots & \vdots & & \vdots \\ q_{M1} & q_{M2} & \cdots & q_{MM} \end{bmatrix}$$

with $q_{jj} = 0 \quad 1 \leq j \leq M$ (we shall not allow a lithology to succeed itself) and

$$\sum_{j=1}^{M} q_{kj} = 1 \quad , \quad 1 \leq K \leq M$$

Again, based on prior information (well data, geophysical/ geological data, etc.) an expected or average reflection coefficient is calculated for each transitional state:

c_{kj} = normal incidence pressure reflection coefficient at the boundary when lithology ℓ_k is followed by lithology ℓ_j.

We recall from our previous discussions that if velocity and density of each lithology can be estimated, then

$$c_{kj} = \frac{\rho_j v_j - \rho_k v_k}{\rho_j v_j + \rho_k v_k}$$

(normal incidence pressure reflection coefficient), where ρ = density and v = velocity.

If only velocity is known, then density can be estimated from Gardner's relation [13] given earlier

$$\rho = kv^{\frac{1}{4}}$$

Finally, we must assign a set of probabilities to the M lithologies in the sequence, again, consistent with whatever prior knowledge is available

q_j = probability of occurrence of lithology ℓ_j

with

$$\sum_{j=1}^{M} q_j = 1$$

We are now in a position to estimate the expected value of reflection coefficients in this sequence

$$E(c) = \sum_{j=1}^{M} \sum_{k=1}^{M} c_{kj} \, q_{kj} \, q_j$$

using the fact that the transitional sequences are mutually exclusive and account for all possibilities.

We may now use these expected values to assign the probabilities in our binomial distributions introduced earlier for determining the function to be maximized for the reflection coefficient sequences.

Given the procedure just outlined, the choice

$$c_k = \frac{2n_k}{L} - 1$$

would again be a good choice since assignment of p_k (probability of success for the kth reflection coefficient) allows us to clearly bias the probabilities for c_k to be either positive or negative depending on the expected values obtained above.

We may now go one step further, if desired, and consider expected thicknesses of each of the identified lithologies, ℓ_j.

If lithology ℓ_j has an expected "thickness", T_k, (in terms of two way travel time) then we may also calculate an expected value of thickness

$$E(T) = \sum_{j=1}^{M} T_j q_j$$

This expected value together with its inferred distribution function may be used to determine the probable placement of reflection coefficients (which we have been analyzing) on the two-way time axis.

At this point, let is re-emphasize that we are not building reflection coefficient sequences! We are simply building constraints based on prior knowledge to help guide in the selection and placing of these sequences subject to the ultimate constraint which is that the inferred inversion must be consistent with the observed data (we still have not discussed this constraint).

Before going on, a few remarks are in order. We have confined our attention thus far to a rather particular approach to a formulation of the inversion problem based on some rather general principles. These same principles can be used to construct other implementations which may in some ways be more desirable than the one outlined here.

For example, the binomial distribution was arrived at from simple assumptions assuming very little prior knowledge. It affords a simplistic approach to the problem with a single parameter (p_k) to control. In the absence of any information at all, it was suggested that p_k = 1/2 would be a reasonable choice if

$$c_k = \frac{2n_k}{L} - 1$$

is used.

In areas which have been well studied, other distributions such as the normal distribution involving a larger number of parameters may be preferred [31]. In these cases, we may choose to take any preferred form for the measure to be maximized. Finally, we may combine these other approaches with the Markov chain concepts, for example, and even these may be replaced by other possibly preferable constructs. The point is that prior knowledge may affect not only the assignment of probabilities but the formulation of the inversion problem itself.

In the absence of strongly compelling reasons to do otherwise, however, the approach we have outlined here may be as far as we dare to go in most areas where more reliable information is won at high cost.

Finally, let us consider the ultimate constraint, that the solution arrived at must be in some acceptable way consistent with the observed data. We re-emphasize that all of our discussion to this point has been directed at the possibility of obtaining the highest possible and most probable resolution in the solution of the inversion problem which is, at the same time, consistent with our observations.

The primary constraint which we wish to impose on a solution to our inversion problem, as we have approached it, is that the convolution model shall provide predictions consistent with the data.

If the wavelet and noise are perfectly known, then our constraint would take the form

$$s_t - \sum w_k \hat{r}_{t-k} - n_t = 0$$

where s is the observed seismogram, w the wavelet, $\hat{r}$ the impulse response of the medium derived from the estimated reflection coefficients and n is the noise. In this formulation, $\hat{r}$ could be obtained from the forward modelling approach for plane waves normally incident on a horizontally stratified medium given earlier:

$$R_N(z) = \frac{z^N G_N(z^{-1})}{A_N(z)} \equiv \sum \hat{r}_k z^k$$

for example.

Rarely would we expect to know the wavelet and noise well enough to justify the use of the equality constraint above. Instead, it seems reasonable to require that this constraint be approximately satisfied by requiring, for example

$$|| S - (\hat{w} * \hat{r} + \hat{n}) || < \varepsilon$$

where $\hat{w}$, $\hat{r}$ and $\hat{n}$ may all be estimated, and the choice of norm is to be specified. While procedures do exist in geophysical data analysis for estimating w and n, it would seem that a maximum probability estimate for the noise might result in a "cleaner" inversion. If we therefore use a similar model for the noise as we developed for the reflection coefficients, but uniformly set $p_k = 1/2$ to provide a symmetric probability distribution about zero (assuming that the noise has zero mean) then we would find that we must maximize the quantity

$$- \sum_1^N \frac{1+c_k}{2} \ln \left(\frac{1+c_k}{2}\right) - \sum_1^N \frac{1-c_k}{2} \ln \left(\frac{1-c_k}{2}\right) + \sum_1^N \frac{1+c_k}{2} \ln \left(\frac{p_k}{1-p_k}\right)$$

$$- \alpha \sum_1^N \frac{\nu_k + \nu}{2\nu} \ln \left(\frac{\nu_k + \nu}{2\nu}\right) - \alpha \sum_1^N \frac{\nu - \nu_k}{2\nu} \ln \left(\frac{\nu - \nu_k}{2\nu}\right)$$

subject to the constraint

$$E(S_t - \sum w_k \hat{r}_{t-k} - \nu_t) = 0$$

where ν_t is the estimated noise and

$$-\nu \leq \nu_t \leq \nu$$

The factor α is used to control the emphasis given to noise versus reflection coefficients in the optimization process.

An alternative to the convolution model constraint suggested above would be the constraint used by Gull and Daniell [14] (see also [4]).

$$\sum_t |\hat{S}_t - S_t|^2 / \sigma_t^2 = N$$

where N is the number of data points, $\hat{s}$ is the observed seismogram, S is the predicted seismogram,

$$\hat{S}_t = \Sigma w_k \hat{r}_{t-k} + \nu_t$$

and σ_t = estimated error in the observed seismogram.

Needless to say there are many variations possible for quantities to be represented by entropy-like expressions and for constraints. We shall not dwell further on this topic here, but surely some possibilities will prove to be more suitable than others when applied to inversion problems.

The inversion method developed above is motivated by the geological fact that, at any particular site on the earth, the probability distribution for each reflection coefficient is very much a function of the depositional history of the site. The approaches which we have discussed allow us to directly apply knowledge of the geology of the area, data from well logs, etc., to the determination of the probability distribution of each individual reflection coefficient. The inversion is then effected by determining that reflection coefficient sequence which is compatible with the observed data (the constraints) and which is the "most probable" of all reflection coefficient sequences which are compatible with the data, based on our assignment of probabilities. We have seen that this approach results in the problem of maximizing "entropy-like" measures subject to constraints.

Our purpose here has been to examine the relationship of the maximum probability estimators, which result from assuming certain probability distributions, to the maximum entropy estimator. Our discussion has been motivated, in part, by a very few published studies which have examined the statistical distribution of reflection coefficients in geological settings.

There is, of course, essentially only one function which may properly be referred to as the "entropy" of a probability distribution if we impose the usual conditions required of the entropy measure [24].

An alternative inversion procedure using the entropy measure

$$H = -\sum P(c_k)\ \ell n\ P(c_k)$$

where $\{c_k\}$ is the reflection coefficient sequence (or its equivalent), could be formulated in the classical manner [16], allowing the maximization of the entropy function, subject to constraints, to determine the required probability distribution. That is to say that the (unknown) probability distribution of the reflection coefficients is taken to be the one which maximizes the value of H subject to the constraints. The expected value of each reflection coefficient could then be taken as the solution to the inverse problem, for example. A nice example of this approach to inversion is given in [24] where the procedure is used to estimate the density structure of the earth.

Finally, we should like to point out that the random variables chosen for the inversion analysis need not be restricted to the reflection coefficients but could be taken to be interval velocities, densities, or both, for example, and there may be advantages in choosing the random variables to be one or more of these alternatives [26].

7 OTHER APPROACHES TO DIRECT INVERSION

Having outlined an approach to direct inversion based on maximum probability or entropy considerations, we shall now consider several other approaches which have been suggested for direct inversion.

Bamberger et al [1, 2] consider a similar inversion problem to the one considered here. They assume that the wavelet is known and seek an impedance function from a set of impedances with finite total variation which minimizes the objective function

$$J(\sigma) = \| Y(\sigma) - Y_d \|_2^2$$

where Y is the synthetic seismogram resulting from the impedance estimate, σ, and Y_d is the observed seismogram.

Obviously the problem is posed in a similar way to the maximum entropy problem just outlined with the major constraint (convolution model) built directly into the objective function to be minimized. They state that if the data can be modeled closely enough by the convolution model, then the problem admits a unique solution which depends in a continuous way on the data. The method is claimed to be very robust in the presence of noise. It is noted that the missing low-frequency component of the impedance curve cannot be reconstructed with this procedure alone.

The objective function for this approach is basically the same as some of the constraints mentioned above for the maximum entropy formulation. A comparison of the methods would tend to suggest that with the proper constraint, the maximum entropy approach would tend to approach a minimum of this objective function but with potentially higher resolution in the reflection coefficient sequence.

Maximum probability (or maximum likelihood) estimators have also been suggested as the basis for several other formulations of the inverse problem. In particular, we mention the work of Mendel [19,22a] and Theriault and Baggeroer [28].

In the maximum likelihood approach to deconvolution developed by Kormylo and Mendel [19], they assume the usual model for the observed seismogram

$$S = w * r + n$$

where s is the observed seismogram, w the wavelet, r the impulse response of the layered earth and n is noise. In this very general approach, n and w are not assumed known, but they do assume something about the structure of r. In particular, r is assumed to be of the form

$$r(k) = \alpha(k)\, \tau(k)$$

where α is a zero mean Gaussian white noise process determining the amplitude of the spike and $\tau(k)$ is a Benoulli process taking the values 0 or 1 which determines the spacing of the spikes in the impulse response. It is further assumed that the wavelet, w, can be modeled by an ARMA (n,n) process

$$W(z) = \frac{\sum_{k=0}^{n} b_k z^{-k}}{\sum_{j=0}^{n} a_j z^{-j}}$$

Statistics, including the variance, c, of α, the probability, λ, of success in the Bernoulli process, τ , and the variance of the noise, R, are to be estimated (n is also assumed to be a Gaussian white noise sequence).

Given N observations from the seismogram, s, the problem is to estimate the parameter vector, θ , containing the ARMA model coefficients, for the wavelet, and the parameters R, C, and λ . In addition, we must estimate α and τ .

Using Bayesian principles, they point out that to estimate θ , given s, we might observe

$$P(\theta|s) = p(s|\theta)\ p(\theta)/p(s)$$

and so we might choose to maximize

$$F(\theta|s) = p(s|\theta)\ p(\theta)$$

To estimate θ, α, and τ they suggest integrating out α and estimating θ and τ by maximizing $F_1(\tau,\theta|s) = p(s|\tau,\theta).p(\tau|\theta)$
α is then estimated from

$$F_2(\alpha|s,\hat{\tau},\hat{\theta}) = p(s|\alpha,\hat{\tau},\hat{\theta})\ p(\alpha|\hat{\tau},\hat{\theta})$$

where the estimates, $\hat{\tau}$ and $\hat{\theta}$ are taken as known from the first step. In various publications, Mendel <u>et al</u> have considered many variations on this theme from "everything known" to "nothing known".

Theirault & Baggeroer [28] also consider a maximum likelihood approach to estimate the reflection coefficients. In this approach, it is assumed that the data has not been deconvolved, but that the source waveform is known and the Goupillaud model is assumed to be the generator of the seismogram which contains additive noise. The reflection coefficient sequence, g, is chosen to be that estimate, $\hat{g}$, which maximizes an assumed conditional probability, $p(s|\hat{g})$, where s is the observed seismogram, subject to the constraints

$$|\hat{g}_k| \leq 1$$

They obtain the Cramer-Rao bound on performance of the estimator and give some reason to believe that the estimator is robust in the presence of noise by carrying out Monte Carlo simulations.

Wiggins [33] introduced yet another direct approach to seismic inversion which uses the varimax norm from statistics in place of the entropy measure. Wiggins' goal was to derive a sparse spike train approximation to the impulse response of our layered earth model consistent with the observations. His method is to find a deconvolution operator for the observed data which will yield as its deconvolved output (the estimated impulse response) a sequence which minimizes the varimax norm and which would tend, therefore, to resemble a sparse spike train with relatively large spikes.

In the derivation of the deconvolution operator, Wiggins assumed that several traces were available with the same source wavelet but with the spikes in the impulse response separated by varying time increments. These requirements are met in a standard field gather where the shot is recorded at several receiver locations, each at a different offset, for example.

Let $\{S_{ij}\}$ be the set of signals with $1 \leq i \leq N_s$, $1 \leq j \leq N_t$ where N_s is the number of trace segments or sample signals and N_t is the number of time samples per segment.

If $\{f_k\}$ is the deconvolution operator, then

$$Y_{ij} = \sum_{k=1}^{N_f} f_k \, S_{i,j-k}$$

is the deconvolved output from each segment or trace.

The method for finding the operator $\{f_k\}$ is to select it in such a way that the varimax norm, V, is minimized, where

$$V = \sum V_i$$

and

$$V_i = \sum_j y_{ij}^4 \Big/ \left(\sum_j y_{ij}^2\right)^2$$

Solving the equations obtained by setting derivatives with respect to the coefficients $\{f_k\}$ equal to zero leads to a set of nonlinear equations which can be solved iteratively.

We might raise the objection that the approach outlined above ignores many constraints which might reasonably be imposed on the solution and so we would like to consider a reformulation of the problem which allows for the incorporation of constraints on prior information.

Let is simplify our notation, for the sake of discussion, by working with a single trace segment. In this case, we may drop one index in the formulation above (it can be easily replaced later, if desired). Our problem then is to estimate the impulse response, $\{y_j\}$ using the principle of minimizing the varimax norm while requiring as a constraint that $\{y_j\}$ should, in some sense, generate the observed seismogram. This means that we shall require an estimate of wavelet and noise, if available, but if not, they could be estimated using maximum entropy, for example, subject to suitable band-width, etc., constraints.

We recall that

$$V = \sum_j y_j^4 \;/\; (\sum_j y_j^2)^2$$

(other authors have used different normalizations, but these are unimportant to the problem).

We shall assume that the impulse response $\{y_k\}$ which is to be estimated satisfies the convolution model

$$S_t = \sum_j w_j y_{t-j} + n_t$$

Rather than treat the problem as an unconstrained optimization problem as originally proposed, it seems reasonable to formulate constraints and to incorporate them into the solution as appropriate. In particular, the wavelet, w, may have a known phase structure in bandwidth and these should be incorporated. We may even have a reasonable estimate of w or its statissics. We may also wish to incorporate constraints which acknowledge the noise term and which incorporate its statistics into the solution.

For the sake of illustration, now, we shall consider a particularly simple case in which an estimate of the wavelet is available and the expected value and variance of the noise are known. Making the simplest assumptions (w, y, and n are pairwise uncorrelated) we would then wish to minimize the varimax norm subject to the constraints

$$\text{(A)} \quad E(s_t - \hat{s}_t) = E(n_t)$$

where

$$\text{(B)} \quad \hat{S}_t = \sum_j \hat{w}_j y_{t-j}$$

and $\hat{w}_j$ is the (known) wavelet estimate, and

$$\text{(C)} \quad \mathrm{Var}\,(s_t - \hat{s}_t) = \sigma_n^2$$

where σ_n^2 is the variance of the noise.

Rewriting our constraint equations (A)-(C) in the form $g_k = 0$

and introducing Lagrange multipliers, our problem reduces to solving the following equations for the impulse response, $\{y_k\}$:

$$\frac{4y_k^3}{(\overline{y^2})^2} - \frac{4y_k\,\overline{y^4}}{(\overline{y^2})^3} + \sum \lambda_j \frac{\partial g_j}{\partial y_k} = 0$$

and $g_k = 0$ for all k.

The first equation suggests the additional constraint $\overline{y^2} = 1$ where our notation is defined by

$$\overline{y^4} = \Sigma y_k^4 \; , \; \overline{y^2} = \Sigma y_k^2$$

If the constraint $\overline{y^2} = 1$ is incorporated, then the impulse response must be re-scaled after solving. This could be done by calibrating the subsequent inversion to either known reflection coefficients or acoustic impedance using the formulations in Section 3, for example.

Other types of constraints would lead to corresponding alternative formulations of the problem. These alternative formulations may or may not be easy to solve in practice.

Deeming [8] suggests the general measure $V = \overline{zF(z)}$ (Wiggins' original measure results from this if we take $F(z) = z$), and suggests several choices for $F(z)$. If we take $F(z) = \ell n\, z$, then $V = \overline{z\, \ell n\, z}$ where

$$\overline{z\, \ell n\, z} = \frac{1}{N} \sum z_k \;\; \ell n\, z_k \quad \text{and} \quad z_k = y_k^2 / \overline{y^2}$$

In this case,

$$V = \frac{1}{N\, \overline{y^2}} \sum y_k^2\, \ell n\, (y_k^2) - \ell n\, (\tfrac{1}{N} \Sigma\, y_k^2)$$

If we normalize so that

$$\overline{y^2} \equiv \frac{1}{N} \sum y_k^2 = 1$$

then this becomes suggestive of an entropy measure on the square of the estimated impulse response, $\{y_k\}$

$$V = \frac{1}{N} \sum y_k^2 \;\; \ell n\, (y_k^2)$$

Deeming points out that Ooe and Ulrych's exponential transform method [23] corresponds approximately to the choice

$$F(z) = 1 - \exp\,(-\,\alpha z)$$

and this could be treated by our direct approach as well.

A few comments now seem to be in order. Wiggins' approach, while suffering the disadvantages of requiring a deconvolution of the (noisy) seismic trace with its attendant problems, does appear to be attractive since so little is assumed. However, we might suggest that a solution as unconstrained as this is might conceivably have little relation to reality when the solution is non-unique. Thus, in spite of its apparent attractiveness, we would suggest that opting for the direct solution with appropriate constraints (even if these involve prior knowledge) may be preferable, particularly in cases where prior knowledge may be reliable or reasonably so.

8 SUMMARY

In this paper we have reviewed a particularly simple form of inversion for seismic traces based on the response of a horizontally layered earth to normally incident plane waves, a commonly used model in seismic data processing.

Our purpose here has not been to review the very extensive literature related to this problem, but to present a particular approach to it based on maximizing probabilities or entropy. While much more work remains to be done, this approach seems well founded in view of the problems associated with purely deterministic methods and in view of the successes which have been obtained by using this kind of analysis in image reconstruction.

Several similar approaches to this problem have also been discussed which lead to different formulations of the inverse problem. We have also pointed out that minimum entropy deconvolution, as formulated by Wiggins and others, while attractive in some ways, is a highly unconstrained problem which relies on deconvolving the seismic trace with a suitable operator to obtain an estimate of the impulse response. We have suggested alternative approaches which incorporate constraints and which may avoid deconvolution of the seismic trace.

Finally, many issues, particularly the best choices for constraints, have been treated lightly, in passing. This comprises a very large topic in itself and should form a part of more detailed analyses to follow.

REFERENCES

1 Bamberger, A., G. Chavent, CH. Hemon, and P. Lailly (1982). Inversion of Normal Incidence Seismograms. Geoph., Vol. 47, No. 5, pp. 757-770.

1a Bamberger, A. (1983). Discussion on "Inversion of Normal Incidence Seismograms". Geoph., Vol. 48, No. 10, pp. 1411-1412.

2 Bamberger, A., G. Chavent and P. Lailly (1980). An Optimal Control Solution of the Inverse Problem of Reflection Seismics. Computing Methods in Applied Sciences and Engineering, R. Glowinski, J.L. Lions (eds.), North-Holland Publishing Company INRIA. pp. 529-545.

3 Berteussen, K.A. and B. Ursin (1983). Approximate Computation of the Acoustic Impedance from Seismic Data. Geoph., Vol. 48, No. 10, pp. 1351-1358.

4 Bryan, R.K. and J. Skilling (1980). Deconvolution by Maximum Entropy, As Illustrated by Application to the Jet of M87. Mon. Not. R. Astr. Soc., 191, pp.69-79.

5 Burg, J.P. (1975). Maximum Entropy Spectral Analysis, Ph.D. Thesis, Stanford University.

6 Chi, Chong-Yung and Jerry Mendel (1984). Improved Maximum-Likelihood Detection and Estimation of Bernoulli-Gaussian Processes. IEEE Transactions on Information Theory, Vol. IT-30, No. 2, pp. 429-435.

7 Daniell, G.J. and S.F. Gull (1980). Maximum Entropy Algorithm Applied to Image Enhancement. IEEE Proc., Vol. 127, Pt. E., No. 5, pp. 170-172.

8 Deeming, T.J. (1981). Deconvolution and Reflection Coefficient Estimation Using a Generalized Minimum Entropy Principle. 1981 Annual International Meeting and Exposition, Society of Exploration Geophysicists, Los Angeles. Technical Papers, Vol. 6 - Processing II, S21.7 pp. 3871-3903.

9 Frieden, B. Roy (1980). Statistical Models for the Image Restoration Problem. Computer Graphics and Image Processing 12, pp. 40-59.

10 Frieden, B. Roy (1979). Image Enhancement and Restoration, Picture Processing and Digital Filtering, T.S. Huang, ed., Springer Verlag, N.Y. pp. 177-248.

11 Frieden, B. Roy (1972). Restoring with Maximum Likelihood and Maximum Entropy. Journal of the Optical Society of America, Vol. 62, No. 4, pp. 511-518.

12 Frieden, B. Roy (1973). Statistical Estimates of Bounded Optical Scenes by the Method of "Prior Probabilities". IEEE Transactions on Information Theory, Vol. IT-19, pp. 118-119.

13 Gardner, G.H.F., L.W. Gardner and A.R. Gregory (1974). "Formation Velocity and Density - the Diagnostic Basis for Stratigraphic Traps". Geophysics, Vol. 39, pp. 770-780.

14 Gull, S.F. and G.J. Daniell (1978). Image Reconstruction from Incomplete and Noisy Data. Nature, Vol. 272, 20, pp. 686-690.

15 Hubral, P. (1978). On Getting Reflection Coefficients from Waves. Geophysical Prospecting, 26, pp. 617-630.

16 Jaynes, Edwin T. (1968). Prior Probabilities. IEEE Transactions on Systems Science and Cybernetics, Vol. SSC-4, No. 3, pp. 227-241.

17 Jaynes, Edwin T. (1982). On the Rationale of Maximum-Entropy Methods. Proceedings of the IEEE, Vol. 70, No. 9, pp. 939-952.

18 Jaynes, Edwin T. (1984). Prior Information and Ambiguity in Inverse Problems. SIAM-AMS Proceedings, Vol. 14, pp. 151-166.

19 Kormylo, John J. and Jerry M. Mendel (1983). Maximum-Likelihood Seismic Deconvolution. IEEE Transactions on Geoscience and Remote Sensing, Vol. GE-21, No. 1, pp. 77-82.

20 Lindseth, R.O. (1979). Synthetic Sonic Logs - A Process for Stratigraphic Interpretation. Geoph., Vol. 44, No. 1, pp. 3-26.

21 Loewenthal, D., P.R. Gutowski and S. Treitel (1978). Direct Inversion of Transmission Synthetic Seismograms. Geoph., Vol. 43, No. 5, pp. 886-898.

22 Mendel, Jerry M. and Farrokh Habibi-Ashrafi (1980). A Survey of Approaches to Solving Inverse Problems for Lossless Layered Media Systems. IEEE Transactions on Geoscience and Remote Sensing, Vol. GE-18, No. 4, pp. 320-330.

22a Mendel, J.M. (1983). Optimal Seismic Deconvolution. Academic Press, New York.

23 Ooe, M. and T.J. Ulrych (1979). Minimum Entropy Deconvolution with An Exponential Transformation. Geophysical Prospecting, 27, pp. 458-473.

24 Rietsch, E. (1977). The Maximum Entropy Approach to Inverse Problems: Spectral Analysis of Short Data Records and Density Structure of the Earth. J. Geophys., 42, pp. 489-506.

25 Rietsch, E. (1978). Extreme Models from the Maximum Entropy Formulation of Inverse Problems. J. of Geophys., 43, pp. 272-275.

26 Rietsch, E. (1984). Personal communication.

27 Sommerfeld, A. (1950). Mechanics of Deformable Bodies. Lectures on Theoretical Physics, Vol. II, Academic Press, New York, New York.

28 Theriault, Kenneth B. and Arthur B. Baggeroer (1977). Structure Estimation from Acoustic Reflection Measurements. In Aspects of Signal Processing, Part I, by D. Reidel Publishing Company, Dordrecht-Holland. pp. 279-295.

29 Tikochinsky, Y., N.Z. Tishby and R.D. Levine (1984). Consistent Inference of Probabilities for Reproducible Experiments. Physical Review Letters, Vol. 52, No. 16, pp. 1357-1360.

30 Ulrych, Tad J. and Colin Walker (1982). Analytic Minimum Entropy Deconvolution. Geophy., Vol. 47, No. 9, pp. 1295-1302.

31 Velzeboer, C.J. (1981). The Theoretical Seismic Reflection Response of Sedimentary Sequences. Geophy., Vol. 46, No. 6, pp. 843-853.

32 Walker, Colin and Tad. J. Ulrych (1983). Autoregressive Recovery of the Acoustic Impedance. Geophy., Vol. 48, No. 10, pp. 1338-1350.

33 Wiggins, Ralph A. (1978). Minimum Entropy Deconvolution. Geoexploration, 16, pp. 21-35.

PRINCIPLE OF MAXIMUM ENTROPY AND INVERSE SCATTERING PROBLEMS

Ramarao Inguva
Physics Department
University of Wyoming
Laramie, Wyoming 82071
and
Department of Physics
University of Albuqurque
Albuqurque, New Mexico

James Baker-Jarvis*
Laramie Projects Office
Laramie, Wyoming 82070

*AWU Postdoctoral Fellow through DOE Laramie Projects Office

ABSTRACT

Using the principle of maximum entropy, a procedure is outlined to study some aspects of the inverse scattering problem. As an application we study a) the quantum mechanical inverse scattering problem using the Born Approximation, b) the electromagnetic inverse problem, and c) the solutions to the Marchenko equation of inverse scattering.

1 INTRODUCTION

Following the pioneering work by Jaynes [1], concerning the information theoretic approach to statistical mechanics, there have been several novel applications of the principle of maximum entropy in a number of areas such as image processing, and geophysical data analysis [2]. Of particular importance to the present work is a paper by Jaynes on time series analysis using the principle of maximum entropy in which Jaynes [3] derived Burg's [4] spectral method. The main goal of this paper is to demonstrate the applicability of the maximum entropy method for analyzing the generalized inverse problem. In Section 2 we develop the formulation for tackling generalized inverse problems suitable for the case when the available information is either incomplete or noisy. The efficacy of the formulation of Section 2 is demonstrated in Section 3 by studying its application to the inverse problem in quantum mechanical scattering theory. We present two more examples in Sections 4 and 5 where we present a solution to the inverse problem associated with the Marchenko integral equation of scattering theory and the electromagnetic inverse problem. Finally, in Section 6 we make some concluding remarks.

2 PRINCIPLE OF MAXIMUM ENTROPY AND THE LINEAR INVERSE PROBLEM

We consider the inverse problem consisting of the solution of the following set of integral equations

$$\int_0^\infty \alpha_k(r)F(V(r))dr = A_k \qquad (2.1)$$

$$k=1,2,\ldots M$$

where V(r) is the unknown function of r for which a solution is desired, F(r) is some functional of the unknown V(r) and $\alpha_k(r)$ is a known function of r. An example of F(V(r)) is the linear inverse problem in which

$$F(V(r)) = V(r) \tag{2.2}$$

The quantities $(A_1, \ldots A_M)$ are assumed to be given. We are particularly interested in the case where the available information on A_i is noisy. In this later case it is well known that in general there is no unique solution to the inverse problem of Eq. (2.1).

We consider the following discretized version of Eq. (2.1)

$$\vec{r} \qquad (r_1, r_2, \ldots\ldots, r_N) \tag{2.3}$$

$$\alpha_k(\vec{r}) \qquad (\alpha_{k1}, \alpha_{k2}, \alpha_{k3} \ldots \alpha_{kN}) \tag{2.4}$$

$$\bar{V} \qquad (V_1, V_2, V_3, \ldots, V_N) \tag{2.5}$$

$$F(V(r)) \qquad [F_1(\bar{V}), F_2(\bar{V}), \ldots\ldots F_N(\bar{V})] \tag{2.6}$$

$$\sum_{j=1}^{N} \alpha_{kj} F_j(\bar{V}) = A_k \quad ; \ k=1,2,..,M \tag{2.7}$$

The available information on A_k apart from being noisy may also be insufficient to carry out the inversion indicated in Eq. (2.7), i.e., $M \ll N$. We now outline a procedure for inverting Eq. (2.7) to obtain an estimate for the vector V using the principle of maximum entropy adapting a method given earlier by Jaynes for time series analysis. We treat the vector V as a random vector allowing V_i to take any values lying in the interval

$$-B \le V_i < \infty \qquad i=1,2,\ldots\ldots,N \tag{2.8}$$

where B is a finite positive semi-definite constant. We next define $P(\bar{V}) = P(V_1, V_2, \ldots V_N)$ by

$$P(\bar{V}) = \text{The probability for the occurrence of the sequence } (V_1, V_2, \ldots V_N) \tag{2.9}$$

The information entropy of this probability distribution is defined by

$$S = - \sum_{\bar{V}} P(\bar{V}) \ln P(\bar{V}) \tag{2.10}$$

Following Jaynes we maximize the information entropy of Eq. (2.10) subject to the following constraints, to obtain P(V)

$$\sum_{\bar{V}} P(\bar{V}) = 1 \quad , \tag{2.11}$$

$$\sum_{j=1}^{N} \alpha_{kj} \langle F_j(\bar{V}) \rangle = A_k \quad , \tag{2.12}$$

$$\langle F_j(\bar{V}) \rangle = \sum_{\bar{V}} P(\bar{V}) F_j(\bar{V}) \quad . \tag{2.13}$$

The maximization of S is conveniently carried out using Lagrange's method of undetermined multipliers. The resulting equation for $P(\bar{V})$ is given by

$$P(\bar{V}) = Z^{-1} \exp\{-\sum_{k=1}^{M} \lambda_k \sum_{j=1}^{N} \alpha_{kj} F_j(\bar{V})\} \tag{2.14}$$

where

$$Z = \sum_{\bar{V}} \exp\{-\sum_{k=1}^{M} \lambda_k \sum_{j=1}^{N} \alpha_{kj} F_j(\bar{V})\} \tag{2.15}$$

$$= Z(\lambda_1, \lambda_2, \ldots \lambda_M)$$

In Eqs. (2.14) and (2.15) $\lambda_1 \ldots \lambda_M$ are Lagrange multipliers to be determined by the constraint conditions of Eq. (2.12). We rewrite Eqs. (2.14) and (2.15) as

$$P(\bar{V}) = Z^{-1} \exp\{-\sum_{j=1}^{N} \Gamma_j F_j(\bar{V})\} \tag{2.16}$$

$$Z(\lambda_1 \ldots \lambda_M) = \sum_{\bar{V}} \exp\{-\sum_j \Gamma_j F_j\} \tag{2.17}$$

where

$$\Gamma_j = \sum_{k=1}^{M} \lambda_k \alpha_{kj} \qquad ; j=1,2,..,N \tag{2.18}$$

The prediction for any quantity G(V) is given by the expectation value

$$\langle G(\bar{V}) \rangle = \sum_{V} G(\bar{V}) P(\bar{V}) \tag{2.19}$$

In particular the solution of the inverse problem posed by Eq. (2.7) is now given by

$$\langle V_i \rangle = \sum_{\bar{V}} V_i P(\bar{V}) \tag{2.20}$$

Our procedure for solving the inverse problem defined by Eq. (2.7) can now be summarized as follows. The M pieces of data $A_1 \ldots A_M$ (which may or may not be noisy) provide M constraints given in Eq. (2.12) to determine the probability distribution. The coupled system of non-linear equations

$$Z^{-1} \sum_{j=1}^{N} \alpha_{kj} F_j(\bar{V}) \exp\{-\sum_{i=1}^{N} \Gamma_i F_i(\bar{V})\} = A_k \qquad k=1,2,..,M \tag{2.21}$$

is solved to obtain $\lambda_1 \ldots \lambda_M$.

Once the set $(\lambda_1 \ldots \lambda_M)$ is obtained one can then obtain V(r), the solution to the inverse problem is obtained from Eq. (2.20). The special case of the linear inverse problem is of great interest and our results are summarized below.

$$F_j(\bar{V}) = V_j \tag{2.22}$$

$$P(\bar{V}) = Z^{-1} \exp\{-\sum_{i=1}^{N} \Gamma_i V_i\} \tag{2.23}$$

$$Z = \sum_{V} \exp\{-\sum_{i=1}^{N} \Gamma_i V_i\} \tag{2.24}$$

Replacing the summations over $V_1 \ldots V_M$ by integrations we obtain

$$P(\bar{V}) = \prod_{i=1}^{N} \Gamma_i \exp\{-\Gamma_i (V_i - B)\} \tag{2.25}$$

Using Eq. (2.25) we obtain the following estimate for $\langle V_i \rangle$ from Eq. (2.20)

$$\langle V_i \rangle = \Gamma_i^{-1} - B \tag{2.26}$$

Thus the solution to the linear inverse problem is given by

$$\langle V(\vec{r}) \rangle = \Gamma(\vec{r})^{-1} - B$$
$$\Gamma(\vec{r}) = \Sigma \alpha_k(\vec{r}) \lambda_k \tag{2.28}$$

The Lagrange multipliers $\lambda_1 \ldots \lambda_M$ are now determined from solutions of the M nonlinear equations given by

$$\sum_{i=1}^{N} \alpha_{ki}\Gamma_i^{-1} = A_k \qquad ; \; k=1,2..,M \tag{2.29}$$

From Eqs. (2.25)-(2.29) we note that the linear inverse problem is particularly simple to implement numerically. In the next section we present an application to scattering theory in the Born Approximation. In a recent paper the formulation outlined in this section has been successfully applied to the problem of interpolation and extrapolation of noisy data.

3 INVERSE PROBLEM IN APPLICATION TO QUANTUM SCATTERING

In this section we apply the formulation given in Section 2 to study the problem of extraction of information on the potential from scattering measurements. We consider for simplicity only potential scattering [3].

We begin with a summary of the relevant results from potential scattering theory. We focus our attention on the quantum mechanical problem of collision of two spinless particles described by the time independent Schroedinger equation. The interaction between the particles V(r) is assumed to be energy independent and is considered to be a function only of the magnitude relative distance of the particles (spherically symmetric central potentials). The differential scattering cross section (the ratio of the outgoing flux to the incoming flux in scattering measurement) is given by

$$\frac{d\sigma}{d\Omega} = |f(\vec{k},\Theta)|^2 \tag{3.1}$$

where $d\Omega$ is the solid angle. In Eq. (3.1) k and Θ are the center of mass momentum and the scattering angle respectively. The quantity $f(k,\Theta)$ in Eq. (3.1) is the scattering amplitude given by

$$f(\vec{k},\Theta) = F(\vec{k},\vec{k}') = -\frac{1}{4\Pi}\int d^3r e^{i\vec{k}.\vec{r}} V(\vec{r})\omega_k(\vec{r}) \tag{3.2}$$

where $\vec{k}$, $\vec{k}'$ are the center of mass momenta before and after the collision. The wave function $\omega_k(r)$ is the solution of the Schroedinger equation in the center of mass system

$$\frac{\hbar^2}{2\mu}(\nabla^2 + k^2)\omega_k(\vec{r}) = V(\vec{r})\omega_r(\vec{r}) \tag{3.3}$$

where μ is the reduced mass.

For the case of spherically symmetric potentials it is convenient to make the partial wave decomposition defined by

$$\omega(\vec{r}) = \frac{1}{r} \sum_{L=0}^{\infty} (2L+1) R_L(r) P_L(\cos\Theta) \tag{3.4}$$

The scattering amplitude $f(k,\Theta)$ is identified from the asymptotic behaviour of $\omega(r)$:

$$\omega(r) \xrightarrow[r\to\infty]{} e^{ikr} + e^{ikr} \frac{f(k,\Theta)}{r} \tag{3.5}$$

Then the partial wave decomposition of $f(k,\Theta)$ is given by

$$f(k,\Theta) = \sum_L (2L+1) P_L(\cos\Theta) e^{i\delta_L} \mathrm{Sin}\,\delta_L \tag{3.6}$$

where the quantities δ_L are the phase shifts and are given by

$$A_L(k) = e^{i\delta_L} \mathrm{Sin}\,\delta_L = -\frac{2\mu k}{\hbar 2} \int_0^{\infty} V(r) j_L(kr) R_L(r) r^2 dr \tag{3.7}$$

where $j_L(kr)$ are the spherical Bessel functions in order L. The differential scattering cross sections $\frac{d\sigma}{d\Omega}$ is related to the phase shifts via (3.1) and (3.7). From scattering cross sections one can extract information on the phase shifts corresponding to the lowest few partial waves (L=0,1,2...). From Eqs. (3.1) and (3.7) it is clear that the extraction of phase shifts from scattering cross sections is a nontrivial problem and we do not dwell on this aspect any further however. We assume hereafter that $\delta_L(k)$ are available from measurements.

Since data is available for $\delta_L(k)$ only for a limited range of energies and a few partial waves, the inversion of Eq. (3.7) to obtain V(r) is highly nonunique, a point discussed earlier, also by Chadan and Sebatier [6]. Moreover the right side is a complicated nonlinear functional of V(r) which makes the inversion even more difficult. In the high energy limit under the so called Born Approximation one makes the approximation

$$R_L(r) = j_L(kr) \tag{3.8}$$

to obtain

$$A_L = e^{i\delta_L} \mathrm{Sin}\,\delta_L = -\frac{2\mu}{\hbar 2} \int_0^{\infty} V(r) r^2 j_L^2(kr) dr \tag{3.9}$$

This is a linear inverse problem to obtain V(r) and can be handled by techniques given in Section 2. Identifying the quantity $\alpha_j(r)$ of Eq. (2) by

$$\alpha_L(r) = -\frac{2\mu k}{\hbar^2} r^2 j_L^2(kr) \tag{3.10}$$

we obtain the following system of equations to obtain an estimate for V(r)

$$\langle V_i \rangle = \Gamma_i^{-1} - B \tag{3.11}$$

$$\Gamma_i = \sum_{L=1}^{M} \alpha_{Li} \lambda_L \tag{3.12}$$

$$\alpha_{Li} = -2\mu k\hbar^{-2} j_L^2(kr_i) \tag{3.13}$$

$$A_j = \sum_{i=1}^{N} \alpha_{ji} [\Gamma_i^{-1} - B] \tag{3.14}$$

The reconstructed potential V(r) is given by

$$V(r)=[\Gamma(r)]^{-1}-B \tag{3.15}$$

To test this procedure we have first calculated phase shifts at various values of L for standard potentials such as exponential and Yukawa potentials. We then used the resultant phase shifts to carry out the inversion to obtain ⟨V(r)⟩. In Figures 1-2 we give the results for exponential and Yukawa potentials. As we note from these figures the reconstruction is satisfactory. Next we took the calculated values of the phase shifts at various L and added gaussian noise up to 50% to each data point. We then carried out the above procedure for this noisy data. Again the reconstructions as seen from Figures 3-6 are quite good. This shows the efficacy of our method. Work on applying this formulation to extract information on nucleon-nucleon potentials using the nucleon-nucleon scattering data is in progress.

We conclude this section by making a few comments on the problem of obtaining the inverse from Eq. (3.7) without making the Born approximation. From the right hand side of Eq. (3.7) we note that one needs a procedure for handling the radial wave function R(r). One method would be to carry out a perturbation expansion keeping the next order term (second Born approximation) and develop an iterative procedure to extract ⟨V(r)⟩ by generalizing the above procedure. A more interesting approach would be to start with an expression for R(r) from variational methods or other considerations. Once a form for R(r) as a guess is available one can again use an iterative procedure to extract V(r). Thus one would first solve a linearized inverse problem using the known R(r). In the second stage one would obtain a better $R_L(r)$ by solving the radial Schroedinger equation for $R_L(r)$, i.e.,

$$\left\{\frac{d^2}{dr^2} + \left[k^2 - \frac{2\mu V(r)}{\hbar^2} - \frac{L(L+1)}{r^2}\right]\right\}R_L(r) = 0 \qquad (3.16)$$

using the V(r) obtained in the first approximation. The resulting $R_L(r)$ could then be used to repeat the whole procedure until a convergent solution is obtained for <V(r)>. These and other similar techniques are currently under investigation to improve the Born inversion.

Figure 1. The method developed in Sec. 3 was applied to the Born Approximation (Eqn.(3.7)) when the phase shifts were first calculated exactly. In this case the potential is $V(r) = r^2\exp(-r)$, the exact solution is plotted in the solid line, the reconstructed solution in the dashed line. (Note: the factor of r^2 is absorbed into the potential and not α).

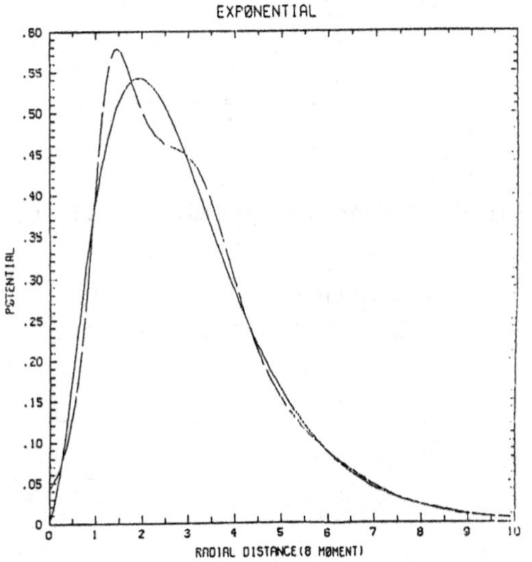

4 MARCHENKO EQUATION OF INVERSE SCATTERING

In this section we study the inverse problem in scattering theory via the Marchenko integral equation and the principle of maximum entropy [6].

The inverse problem in scattering theory for one-dimensional potential problems can be recast into an integral equation for the scattering function. An excellent review of this formulation is available in reference [7]. In this formulation one relates the reflection coefficient for waves incident from the right or left ($r_R(x)$ or $r_L(x)$) to the scattering functions from the right or left ($A_R(x,ct)$ or $A_L(x,ct)$) as a function of the position variable and time t. The Marchenko integral equations are given by

Figure 2. Same as Fig. 1, but for the Yukawa potential V(r)=r exp(-4).

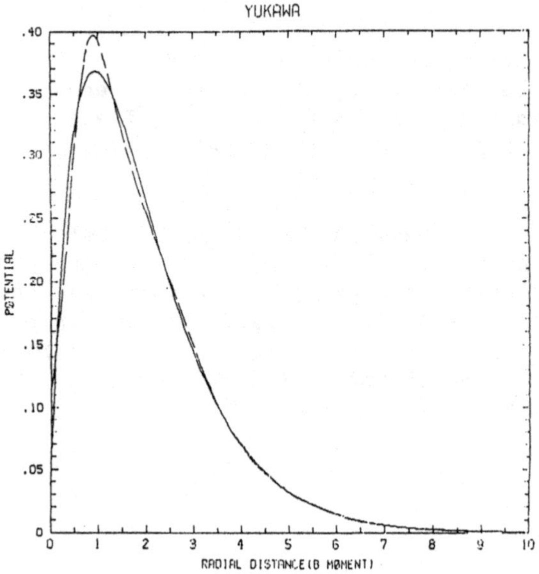

Figure 3. A simulated exponential potential with Gaussian noise added.

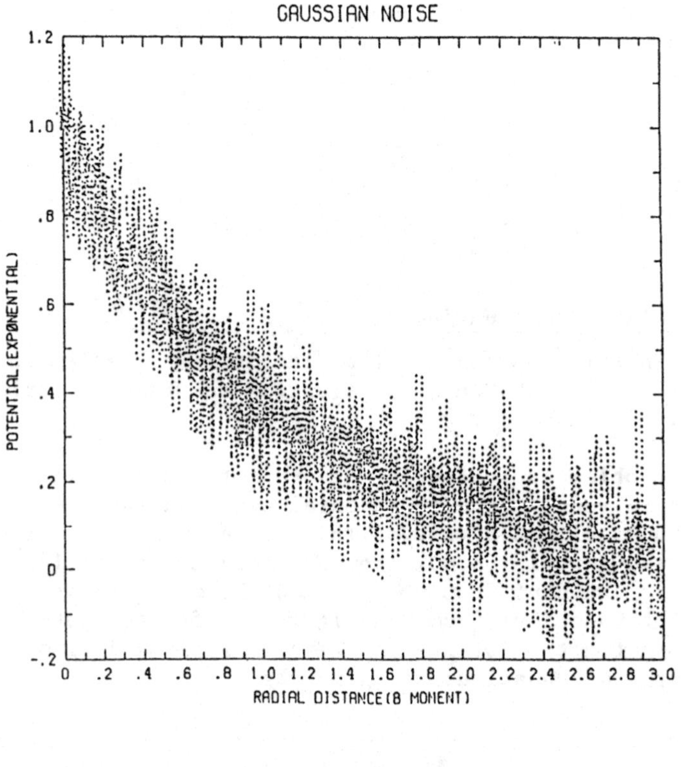

Figure 4. The reconstructed potential of Fig. 3 using the Born Approximation.

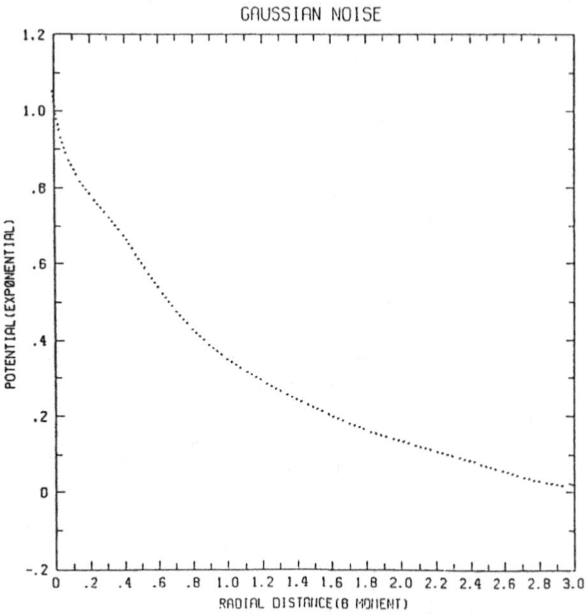

Figure 5. A simulated Yukawa potential with Gaussian noise added.

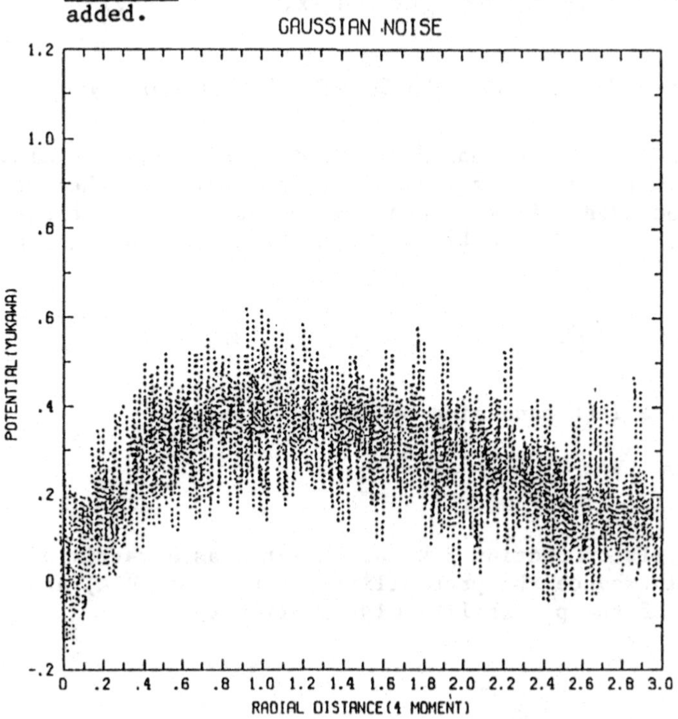

Figure 6. The reconstructed potential of Fig. 5 using the Born Approximation.

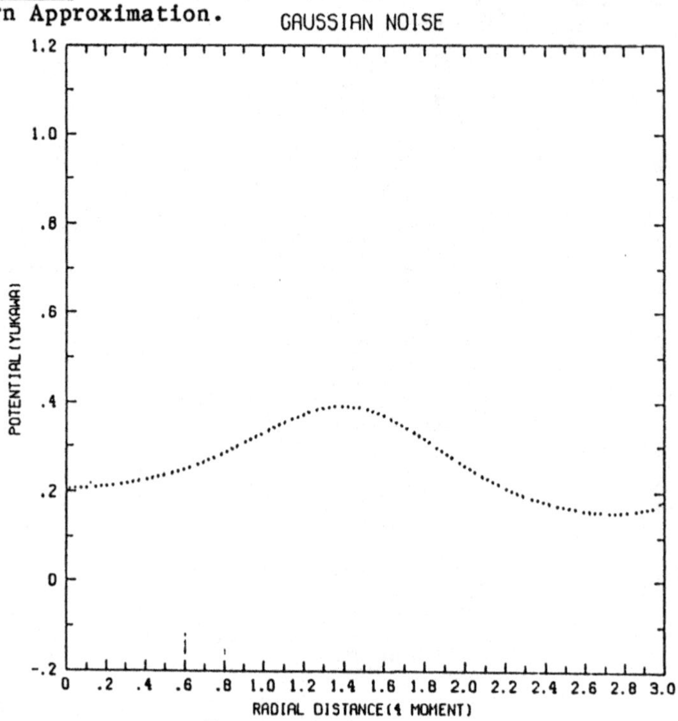

$$r_R(x+y)+A_R(x,y)+\int_x^\infty dx' r_R(x'y)A_R(x,x') = 0 \quad (x<y)$$

$$r_L(x+y)+A_L(x,y) + \int_{-\infty}^{x} dx' r_L(x'+y)A_L(x,x') = 0 \quad (x>y) \tag{4.1}$$

where x' = ct', y=-ct (c is the speed of light). Eq. (4.1) is also valid when there are bound states present except that r(x) then becomes a function which includes information on the bound states (see Eqs. (3.2.7) and (3.2.8) of reference [7]). Once the A(x,ct) are known then the potential V(x) is given by

$$V(x) = -2 \frac{dA(x,x')}{dx} \tag{4.2}$$

We now discretize the left hand side of Eq. (4.1) as

$$r_{ij} + A_{ij} + \sum_k A_{ik} r_{kj} = 0 \tag{4.3}$$

Following the analysis of Section 2 we now treat A as a matrix of random variables A_{ij} and introduce the probability distribution P(A) and define the entropy of the probability distribution by

$$S(\bar{A}) = - \int P(\bar{A}) \ln P(\bar{A}) \prod_{jk} dA_{jk} \tag{4.4}$$

we maximize S of Eq. (4.4) subject to the following constraints

$$\int P(\bar{A}) \prod dA_{jk} = 1 \tag{4.5}$$

$$\langle A_{ij} \rangle + \sum_k \langle A_{ik} \rangle r_{kj} = -r_{ij} \tag{4.6}$$

$$\langle A_{ik} \rangle = \int A_{ik} P(\bar{A}) \prod_{jl} dA_{jl} \tag{4.7}$$

The resulting solution to $P(\bar{A})$ is given by

$$P(\bar{A}) = Z^{-1} \exp\{ - \sum_{lk} A_{lk} \Gamma_{lk} \} \tag{4.8}$$

$$\Gamma_{lk} = \sum_m [r_{km} + \delta_{km}] \lambda_{lm} \tag{4.9}$$

$$Z = \exp\{- \sum_{lm} A_{lm} \Gamma_{lm} \} \prod_{ij} dA_{ij} \tag{4.10}$$

Performing the integrations in (4.10) we obtain

$$P(\bar{A}) = \prod_{ij} \Gamma_{ij} \exp\{- \sum_{lm} \Gamma_{lm} (A_{lm} - B) \tag{4.11}$$

$$\langle A_{lm} \rangle = \Gamma_{lm}^{-1} - B \tag{4.12}$$

The matrix of Lagrange multipliers can be determined by solving the coupled system of nonlinear equations

$$\sum_k r_{ik} [\Gamma_{kj}^{-1} - B] = - r_{ij} \tag{4.13}$$

Once the Lagrange multipliers λ_{ij} are determined $\langle A_{ij} \rangle$ can be obtained and hence the reconstruction of V(r) is then straightforward. The discussion at the end of Section 2 would equally well apply here. Further work on the analysis of the inverse problem using the Marchenko equation is in progress.

5 ELECTROMAGNETIC AND ACOUSTIC INVERSE PROBLEMS

Optical properties of heterogeneous media in which the scale of the inhomogeneties is small relative to the wavelength reveals anomalous properties that do not exist in homogeneous systems [8-10]. Some systems of interest are aggregated materials such as metallic films, suspensions of fine metal particles in colloidal media, and layered composite materials. The study of dielectric properties of heterogeneous media has become important in recent years [8]. The study of wave propagation in random media is also receiving a great deal of attention. In this section we set up the general study of inverse problems associated with this area.

We consider a two dimensional network with nodes labelled by a pair of integers (n_1,n_2). A resistive element of admittance $\varepsilon_{nn'}$ connects the nodes (n,n') and $\varepsilon_{nn'} = \varepsilon_{n'n}$. At every site (node) there can be a current I_n. If $I_n > 0$ then the node is a source of current I_n. If I_n is negative then the node acts as a sink for current I_n. The potential at the nth node is denoted by V_n. It is easily shown that Ohm's Law for this lattice takes the form

$$\sum_{n'} \varepsilon_{nn'}(V_n - V_{n'}) = I_n \tag{5.1}$$

The central problem of interest in the study of heterogeneous media is the calculation of the effective dielectric constant of the medium. Thus we are interested in finding the effective resistance R_n between any two nodes say between $(0,0)$ and (n_1,n_2). The effective resistance R_n is given by

$$R_n = (V_o - V_n)/I \tag{5.2}$$

for the case where there is a source current I at the node o and a sink current -I at the node n and all other currents are zero. The calculation therefore involves the inverse problem of obtaining V_n. Defining a vector F by

$$F_n = \sum_{n'} \varepsilon_{nn'} \tag{5.3}$$

we can rewrite (5.1) as

$$[B - \varepsilon]\, \bar{V} = \bar{I} \tag{5.4}$$

where

$$B_{nn'} = F_n\, \delta_{nn'} \tag{5.5}$$

Unless the matrix $[B - \varepsilon]$ is square it is not possible to obtain V from Eq. (5.4) by matrix inversion. The example of Eq. (5.5) is a typical inverse problem of the type discussed in Section 2 and the principle of maximum entropy can be used to invert Eq. (5.4). A similar treatment can be used to solve the inverse problem posed by Poisson's equation

$$\nabla\cdot[\varepsilon(\vec{r})\nabla\phi(\vec{r})] = -\,4\Pi\,\rho(\vec{r}) \tag{5.6}$$

Discretization converts Eq. (5.6) to a form similar to Eqs. (5.1-5.6) and the same techniques can be used to obtain the effective medium properties.

An inverse problem of great interest is the following. Given the potential distribution $\phi(r)$ one requires information on the source $\rho(r)$ in Eq. (5.6). We study here the case when ε is a constant. The solution to Poisson's equation is given by

$$\Phi = \int_V \frac{\rho(\vec{r}\,')}{|\vec{r}-\vec{r}\,'|} d^3r' \tag{5.7}$$

with appropriate boundary conditions. In practice the information available on Φ is limited and noisy making the inverse problem for $\rho(r)$ nonunique. We can again apply the techniques of Section 2 treating $\rho(r)$ as a random variable to obtain information regarding $\rho(r)$ (Note that the function $\alpha(r)$ of Eq. (2.7 Section 2) now contains the factor $1/|\vec{r}-\vec{r}\,'|$.).

The same techniques can be used to handle the inverse problem posed by the time dependent wave equation [11] (or any other equation which can be put in the form of Eq. (5.9))

$$\nabla^2 f(\vec{r},t) - \frac{1}{v^2}\frac{\partial^2 f(r,t)}{\partial t^2} = \rho(\vec{r},t) \tag{5.8}$$

As in the preceding paragraph one has limited information on f(r,t) and one wants to invert for the sources $\rho(r,t)$. Such a situation exists for example in sound propagation in heterogeneous media and electromagnetic waves generated by sources.

The solution of Eq. (5.8) (apart from surface contributions) is given by the standard form

$$f(\vec{r},t) = \int d^3r' \int dt' G(r,t\,;\,\vec{r}\,',t')\,\rho(\vec{r}\,',t') \tag{5.9}$$

where $G(\vec{r},t,\vec{r}\,',t')$ is the Green function satisfying the differential equations

$$[\nabla^2 - \frac{1}{v^2}\frac{\partial^2}{\partial t^2}]G(\vec{r},t,\vec{r}\,',t') = \delta(t-t')\delta(\vec{r}-\vec{r}\,') \tag{5.10}$$

The form of Eq. (5.9) for f(r,t) is similar to the linear inverse problem discussed in Section 2 except that the weighting function $\alpha(r)$ of Section 2 now depends on time (i.e., $= \alpha(r,t)$). These problems are currently under investigation.

6 CONCLUSIONS

A general method to study inverse problems for the case when the available data is incomplete and/or noisy is proposed (see Section 2) by adapting a method proposed earlier by Jaynes. The case of linearized inverse is studied in detail with applications to quantum scattering problems and the electromagnetic inverse problem. A preliminary analysis of the inverse problem associated with the Marchenko integral equation in scattering theory has been presented. The results of this work show, in our opinion, the versatility of the method outlined in Section 2. We believe that this method can be applied in several problems involving the computation of an inverse in integral or integrodifferential equations, for example in other quantum mechanical problems, geophysical inverse problems.

ACKNOWLEDGEMENTS

The authors express their deep gratitude to Professor E.T. Jaynes for several useful discussions. Part of this work was completed while RI was visiting the Institute of Modern Optics, Dept. of Physics, University of New Mexico, Albuqurque, New Mexico. RI is grateful to Professor Marlan Scully for his hospitality and financial support through a AFOSR grant.

REFERENCES

1 Jaynes, E.T. (1957). Phys. Rev. 106, 620.

2 For a good bibliography in this context see, 'Maximum Entropy Inverses in Physics', by C.R. Smith, R. Inguva, and R.L. Morgan, SIAM-AMS Proceedings 14, 127 (1984).

3 Jaynes, E.T. (1982). Proc.-IEEE 70, 939, and references therein.

4 Burg, J.P. (1975). Maximum Entropy Spectral Analysis. Ph.D. Dissertation, Stanford University, Stanford, California.

5 For a good review see for example L.I. Schiff, 'Quantum Mechanics', Third Edition, McGraw-Hill, U.S.A. (1975).

6 Chadan and P.C. Sebatier (1977). Inverse Problems in Quantum Scattering Theory. Springer-Verlag, New York.

7 Lamb, G.L. (1980). Elements of Soliton Theory. Wiley, New York.

8 See for example, I. Webner, J. Jortner, and M.H. Cohen, Phys. Rev. B15, 5712 (1977).

9 Chow, P.L., W.E. Kohler, and G.C. Papanicolaou (eds.) (1981). Multiple Scattering and Waves in Random Media. North Holland.

10 Baltes, H.P. (1980). Inverse Scattering in Optics. Vol. 20. Springer-Verlag.

11 Bevensee, R.M. (1981). Solution of Underdetermined Electromagnetic and Seismic Problems by the Method of Maximum Entropy. IEEE Trans. on Antennas and Propagation. Vol. Ap-29, No. 2, March.

INDEX